THE HIDDEN LINK BETWEEN EARTH'S MAGNETIC FIELD AND CLIMATE

THE HIDDEN LINK BETWEEN EARTH'S MAGNETIC FIELD AND CLIMATE

NATALYA A. KILIFARSKA
Geophysics, National Institute of Geophysics, Geodesy and Geography, Bulgarian Academy of Sciences, Sofia, Bulgaria

VOLODYMYR G. BAKHMUTOV
Institute of Geophysics, National Academy of Sciences of Ukraine, Kyiv, Ukraine

GALYNA V. MELNYK
Institute of Geophysics, National Academy of Sciences of Ukraine, Kyiv, Ukraine

ELSEVIER

Elsevier
Radarweg 29, PO Box 211, 1000 AE Amsterdam, Netherlands
The Boulevard, Langford Lane, Kidlington, Oxford OX5 1GB, United Kingdom
50 Hampshire Street, 5th Floor, Cambridge, MA 02139, United States

© 2020 Elsevier Inc. All rights reserved.

No part of this publication may be reproduced or transmitted in any form or by any means, electronic or mechanical, including photocopying, recording, or any information storage and retrieval system, without permission in writing from the publisher. Details on how to seek permission, further information about the Publisher's permissions policies and our arrangements with organizations such as the Copyright Clearance Center and the Copyright Licensing Agency, can be found at our website: www.elsevier.com/permissions.

This book and the individual contributions contained in it are protected under copyright by the Publisher (other than as may be noted herein).

Notices
Knowledge and best practice in this field are constantly changing. As new research and experience broaden our understanding, changes in research methods, professional practices, or medical treatment may become necessary.

Practitioners and researchers must always rely on their own experience and knowledge in evaluating and using any information, methods, compounds, or experiments described herein. In using such information or methods they should be mindful of their own safety and the safety of others, including parties for whom they have a professional responsibility.

To the fullest extent of the law, neither the Publisher nor the authors, contributors, or editors, assume any liability for any injury and/or damage to persons or property as a matter of products liability, negligence or otherwise, or from any use or operation of any methods, products, instructions, or ideas contained in the material herein.

Library of Congress Cataloging-in-Publication Data
A catalog record for this book is available from the Library of Congress

British Library Cataloguing-in-Publication Data
A catalogue record for this book is available from the British Library

ISBN: 978-0-12-819346-4

For information on all Elsevier publications
visit our website at https://www.elsevier.com/books-and-journals

Publisher: Candice Janco
Acquisitions Editor: Peter J. Llewellyn
Editorial Project Manager: Amy Moone
Production Project Manager: Anitha Sivaraj
Cover Designer: Matthew Limbert

Typeset by SPi Global, India

Working together
to grow libraries in
developing countries

www.elsevier.com • www.bookaid.org

Contents

7. Mechanisms of geomagnetic influence on climate

8. Geomagnetic field and internal climate modes

List of abbreviations

aa a simplified index describing global geomagnetic activity. It is derived from the *K* indices from two approximately antipodal observatories and has units of 1 nT (nanotesla). Current observatories used are Hartland in the United Kingdom, operated by BGS, and Canberra in Australia, operated by Geoscience Australia. The observatories used have changed over the time span of the series.

AD Anno Domini, meaning 'in the year of the Lord' (Medieval Latin). In the Julian calendar it counts the number years from the birth of Jesus Christ. Traditionally the 'AD' abbreviation is placed before the year number, e.g. AD 2019.

AE an auroral electrojet (refer to the Specific terminology) index obtained from a number (usually greater than 10) of stations distributed in local time in the latitude region that is typical of the Northern Hemisphere auroral zone.

Ap a measure of the general level of geomagnetic activity over the globe for a given day. It is derived from measurements of the variation of the geomagnetic field due to currents flowing in the earth's ionosphere (and to a lesser extent in Earth's magnetosphere), made at a number of stations worldwide.

Aura-MLS microwave limb sounder (MLS) experiments, aboard the Aura satellite, measured (naturally occurring) microwave thermal emission from the limb (edge) of Earth's upper atmosphere launched in 2004. The data are used to create vertical profiles of atmospheric gases, temperature, pressure, and cloud ice.

BC Before Christ, traditionally BC is placed after the year number (e.g. 1200 BC).

CCM chemistry climate models couple the stratospheric chemical models with climate models in one model, representing both stratospheric chemistry and atmospheric climate. Coupling both processes in a single model allows the investigation of the feedback processes between these two components (e.g. addressing the question of how global climate change, associated with the production of anthropogenic greenhouse gases, will interfere with the anticipated ozone recovery in the 21st century). However, coupling these two processes into a single model complicates the interpretation of results, compared to models that treat the processes separately (e.g. chemistry transport models and general circulation models).

CMB core–mantle boundary of the Earth; this lies between the planet's silicate mantle and its liquid iron-nickel outer core. It is observed via the discontinuity in seismic wave velocities at that depth. The boundary is thought to harbour topography, much like Earth's surface, that is supported by solid-state convection within the overlying mantle.

CRs cosmic rays are highly energetic atomic nuclei or other particles travelling through space at a speed approaching that of light. When originating from the galaxy or beyond it, they are known as galactic cosmic rays (GCRs).

CRM chemical remanent magnetization is a magnetization formed by phase change, physicochemical changes (such as oxidation or reduction), dehydration, recrystallization, or precipitation of natural elements at low temperatures.

ERA 40, ERA Interim, and ERA 20C products of the European Centre for Medium-Range Weather Forecasts (ECMWF) providing gridded 4D data for a great variety of atmospheric parameters, based on the reanalysis of multidecadal series of past observations, aimed for use in studies of climate variability.

eV an electronvolt is the amount of kinetic energy gained (or lost) by a single electron, which is accelerated from rest, through an electric potential difference of one volt in vacuum.

GCRs galactic cosmic rays.

GeV a gigaelectronvolt is equal to 10^9 electron volts.

HO_x the family of hydrogen oxide radicals ($HO_x \equiv H + OH + HO_2$), influencing the ozone density in mesosphere.

IGRF the International Geomagnetic Reference Field is a standard mathematical description of the large-scale structure of Earth's main magnetic field, and its secular variation. It was created by fitting parameters of a mathematical model to measured magnetic field data from surveys, observatories, and satellites across the globe.

IMF the interplanetary magnetic field, or heliospheric magnetic field, is the component of the solar magnetic field that is dragged out from the solar corona by the solar wind flow to fill the solar system.

INTERMAGNET International Real-time Magnetic Observatory Network.

IPCC the Intergovernmental Panel on Climate Change, established in 1988 by the WMO and the United Nations Environment Programme (UNEP). Its reports cover the scientific, technical, and socio-economic information relevant to understanding the scientific basis of risk of human-induced climate change, its potential impacts and options for adaptation and mitigation.

IPDP Intervals of Pulsations with Diminishing Period are continuous structured narrow-band geomagnetic pulsations, whose frequency resemble Pc1 pulsation and in addition increases with time (typically from 0.2 to $1 \div 2$ Hz). These pulsations occur in the evening sector, in association with the expansion phase of a substorm (refer to the Specific terminology).

K index quantifying disturbances in the horizontal component of Earth's magnetic field, measured in a given magnetic observatory. Its range of variability is between 0 and 9—where 1 denotes calm, and 5 or more indicates a geomagnetic storm. It is a quasilogarithmic local index of the geomagnetic activity, derived from the maximum fluctuations of **horizontal** components observed on a magnetometer during a 3 h interval.

ka a kiloannus, abbreviated *ka*, is a period of 1000 Julian years, equal to 365,250 days. It is derived from the prefix kilo (in SI system) and the Latin for year, *annus*; *ka* is equivalent to *ky*.

keV a kiloelectronvolt is equal to 1000 electron volts (10^3 eV).

Kp the index of the global geomagnetic activity, based on 3-h measurements from ground-based magnetometers around the world. Each station is calibrated according to its latitude and reports about a certain K-index, which depends on the geomagnetic activity measured at the location of the magnetometer.

LS the lower stratosphere is the lowest part of stratospheric layer bordering the tropopause. It is thinner in equatorial regions and comparatively thicker at high latitudes. Dominant characteristics of the lower stratosphere are the presence of ozone and its extremely low temperature, within the range −40 to −80°C.

Ma an abbreviation from the Latin *mega-annum*, i.e. million years.

MeV one megaelectronvolt is equal to 1 million electron volts (10^6 eV).

NAO North Atlantic Oscillation is a weather phenomenon in the North Atlantic Ocean, manifesting itself as fluctuations in atmospheric sea level pressure between the Icelandic Low and the Azores High. It controls the strength and direction of westerly winds and location of storm tracks across the North Atlantic.

NH Northern Hemisphere.

NM neutron monitor.

NMDB the neutron monitor database, a real-time database for high-resolution neutron monitor measurements.

NOAA the National Oceanic and Atmospheric Administration is an American scientific agency within the United States Department of Commerce that focuses on the conditions of the oceans, major waterways, and the atmosphere.

NO_x the family of atmospheric nitrogen oxides ($NO_x \equiv NO + NO_2$) influencing the mesospheric, stratospheric and tropospheric ozone.

NRM natural remnant magnetization is the permanent magnetism of a rock or sediment. It preserves a record of the Earth's magnetic field at the time when the mineral was laid down as sediment, or crystallized in magma, and also the tectonic movement of the rock over millions of years from its original position.

nT a nanotesla is a unit of measurement of a magnetic field (in SI units system), equal to one billionth of a tesla (T), i.e. 10^{-9} T.

SAGE II Stratospheric Aerosol and Gas Experiments (SAGE) are satellite-based solar occultation instruments spanning over 26 years that have been a cornerstone in studies of stratospheric change.

SH Southern Hemisphere.

SHA spherical harmonic analysis is the procedure of representing a potential function by a sum of spherical harmonic functions.

SpH atmospheric specific humidity, defined as the mass of water vapour per unit mass of the moist air, usually in kg/kg.

T2m air temperature at 2 m above the surface from ERA reanalyses (products of the ECMWF—the European Centre for Medium-Range Weather Forecasts).

TIM/SOURCE abbreviation of the Total Irradiance Monitor (TIM), launched in January 2003 on the NASA Earth observing system—Solar Radiation and Climate Experiment (SORCE). The TIM measures the total solar irradiance, the spatially and spectrally integrated solar radiation incident at the top of the Earth's atmosphere.

TOZ the total ozone at any location on the globe is defined as the sum of all the ozone in the atmosphere directly above that location. Most ozone resides in the stratospheric ozone layer and a small percentage (about 10%) is distributed throughout the troposphere.

TRM thermoremanent magnetization is the magnetization that an igneous rock acquires, usually from the magnetic field in which it is located, when the temperature of the magma or lava from which it forms falls below the Curie point during the cooling and solidification process.

TSI total solar irradiance is a measure of the electromagnetic radiation emitted from the Sun and incident on the Earth's upper atmosphere—per unit area, averaged over all wavelengths.

UARS-HALOE the Halogen Occultation Experiment (HALOE) has been collecting profiles of middle atmosphere composition and temperature on board the Upper Atmosphere Research Satellite (UARS), within the period 1991–2005.

UTLS the upper troposphere and lower stratosphere layer is broadly defined as the region ±5 km around the tropopause, which is the traditional boundary between the troposphere and the stratosphere. The dynamical, chemical, and radiative properties of the UTLS are in many ways distinct from both the lower troposphere and the middle stratosphere.

UV ultraviolet is electromagnetic radiation with wavelengths from 10 to 400 nm, shorter than those of visible light but longer than X-rays. UV radiation is present in sunlight, and constitutes about 10% of the total electromagnetic radiation output from the Sun.

VADM the virtual axial dipole moment describes the intensity of an imaginary axial (along the Earth's rotation axis) centric (located in the centre of the Earth) dipole that would produce the estimated archaeo-/palaeointensity at the sampling site.

VGP virtual geomagnetic pole is the point on the Earth's surface at which a magnetic pole would be located if the observed direction of remanence at a particular location was due to a magnetic dipole at the centre of the Earth.

VIRGO/SOHO VIRGO (Variability of solar IRradiance and Gravity Oscillations) is an experiment on the ESA/NASA Solar and Heliospheric Observatory (SOHO) mission, investigating the irradiance (particularly the TSI) and gravity oscillations of the Sun.

VRM viscous magnetization is remanence that is acquired by ferromagnetic materials by sitting in a magnetic field for some time. The natural remanent magnetization of an igneous rock can be altered by this process.

WMO World Meteorological Organization.

Γ_w wet adiabatic lapse rate of atmospheric temperature, known also as moist or saturated lapse rate.

Specific terminology

adiabatic invariant property of a physical system that stays approximately constant when changes occur slowly. This means that if a system is varied between two end points and the time for the variation between them is close to infinity, the variation of an adiabatic invariant between the two end points goes to zero.

Antarctic convergence zone known also as the Antarctic Polar Front, a curve continuously encircling Antarctica (varying in latitude seasonally), where the cold equatorward-flowing Antarctic waters meet the relatively warmer waters of the sub-Antarctic region. Antarctic waters predominantly sink beneath the warmer sub-Antarctic waters.

atmospheric static stability (also called hydrostatic stability or vertical stability) measures the gravitational resistance of atmosphere to vertical displacements. It is determined by the vertical stratification of density or potential temperature.

atmospheric window wavelengths of the electromagnetic spectrum that can be transmitted through the Earth's atmosphere. Atmospheric windows occur in the visible, infrared, and radio regions of the spectrum.

auroral electrojet the large horizontal currents (Hall currents) that flow in the D and E regions (i.e. 100–150 km) of the **auroral** ionosphere, flowing from noon towards nights.

autocatalytic cycle a set of chemical reactions producing catalysts, stimulating the entire set of chemical reactions and ensuring its self-sustainability, given an input of energy and food molecules.

α-particles alpha particles, also called alpha ray or alpha radiation, which consist of two protons and two neutrons bound together into a particle identical to a helium-4 nucleus.

Brewer–Dobson circulation mean meridional overturning circulation in the stratosphere, characterized by two-cell structure in the lower stratosphere, i.e. ascendance of air in the tropics, its poleward propagation, and its descendance in the middle and high latitudes in both hemispheres. The two-cell structure was first proposed by Dobson and Brewer to explain the observations of ozone and water vapour in the stratosphere. At higher altitudes a single-cell circulation exists with air ascending in the summer hemisphere, crossing the equator, and descending in the winter hemisphere.

cation ionic species with a positive charge. A cation has more protons than electrons, giving it a net positive charge.

chronozone (or chron) time interval in chronostratigraphy, defined by events such as geomagnetic reversals (magnetozones), or based on the presence of specific fossils (biozone or biochronozone).

climatic mode repeating patterns of time-space variability of the climate system.

climatology (in meteorology) long-term mean of a given climate variable.

cosmic ray spallation (in nuclear physics) the process in which a heavy nucleus emits numerous nucleons as a result of being hit by a high-energy particle, thus greatly reducing its atomic weight.

cosmogenic isotopes (i.e. cosmogenic nuclides) rare isotopes created when a high-energy cosmic ray interacts with atomic nucleus, causing nucleons (protons and neutrons) to be expelled from the atom. These isotopes are produced within Earth materials such as rocks or soil, in Earth's atmosphere, and in extra-terrestrial items such as meteorites. There are both radioactive and stable cosmogenic isotopes.

cross section measure of probability that a specific process will take place in a collision of two particles. If the particles interact through some action-at-a-distance force, such as electromagnetism or gravity, their scattering cross section is generally larger than their geometric size.

drift motion movement of charged particles (confined by a magnetic field) in a direction perpendicular to both the magnetic field line and the applied force (electric field, magnetic gradient, or curvature of magnetic field lines) due to the action of Lorentz force.

effective temperature the temperature of an object calculated from the radiation it emits, assuming black-body behaviour.

flood basalt the result of a giant volcanic eruption or series of eruptions that covers large stretches of land (or the ocean floor) with basalt lava.

geomagnetic excursion like a geomagnetic reversal, it manifests with a significant change in Earth's magnetic field. Unlike reversals, however, an excursion does not permanently change the large-scale orientation of the field, but rather represents a dramatic, typically short-lived change in field intensity, with a variation in pole orientation of up to 45 degrees from the previous position. These events, which typically last a few thousand to a few tens of thousands of years, often involve declines in field strength between 0% and 20% of normal.

glaciation formation, existence, or movement of glaciers over the surface of the earth.

guiding centre the centre of the vast circular motion of charged particle around magnetic field line, which is drifting slowly in a direction perpendicular to the field lines in the case of spatially heterogeneous or temporary varying magnetic fields.

interstadial relatively warm period during a glacial epoch, when glaciers temporarily stop or retreat.

knock-on electron secondary electron (ejected by high speed particles through its interaction with matter) having enough energy to escape a significant distance away from the primary radiation beam and produce further ionization.

lithosphere solid, outer part of Earth, including the brittle upper portion of the mantle and the crust.

loss cone solid angle defining the minimum angle between velocity vector of arriving charged particle and magnetic field line, ensuring particle reflection by the magnetic mirror, i.e. its confinement by the magnetic field. Particles approaching the magnetic field at lower angles are lost in the surrounding environment.

magnetic lensing focusing or deflection of moving charged particles, such as electrons or ions, due to the action of the magnetic Lorentz force.

magnetic polarity the orientation of magnetic field poles in space.

magnetic rigidity measure of the momentum of charged particle in magnetic field. It refers to the fact that a higher momentum particle will have a higher resistance to a deflection by magnetic field. It is defined as $R = p/q$, where p is the particle momentum and q is its charge.

Matuyama–Brunes border is Earth's latest magnetic field reversal event. It is an important calibration point on the geological timescale, connecting sediments and volcanic rocks, and has therefore been the focus of a number of palaeomagnetic studies.

Maunder minimum period around 1645–1715 during which sunspots became exceedingly rare.

mean free path the average distance travelled by a moving particle (i.e. atom, molecule, photon, etc.) between collisions with other particles—modifying its direction, energy, or other particle properties.

meson hadronic (i.e. large, massive) subatomic particles composed of one quark and one antiquark, bound together by strong (i.e. nuclear) interactions.

Milankovitch cycles describe the collective effects of changes in the Earth's rotation around its axis, and revolution around the Sun (due to the gravitational interactions with other bodies in the solar system) on its climate over thousands of years.

mode of variability climate pattern with identifiable characteristics, specific regional effects, and often oscillatory behaviour.

muon an elementary particle similar to the electron, with an electric charge of $-1\,e$ and a spin of 1/2, but with a much greater mass (105.66 MeV/c^2, which is about 207 times that of the electron).

obliquity (or axial tilt) the angle between Earth's rotational axis and its orbital axis, or, equivalently, the angle between its equatorial plane and orbital plane, varying between ~22.1 and 24.5 degrees.

Older Dryas stadial (cold) period between the Bølling and Allerød interstadials (warmer phases), about 14,000 years BP, towards the end of the Pleistocene.

planetary albedo percentage of solar irradiance that is reflected immediately back into space by clouds, aerosols, the Earth's surface, etc. The Earth's planetary albedo is approximately 30%.

plate tectonics scientific theory describing the structure of Earth's crust and many associated phenomena as resulting from the interaction of rigid lithospheric plates, moving slowly over the underlying mantle.

potential function mathematical function whose values are a physical potential (scalar or vector potential). In geomagnetism, the magnetic vector field B is presented as a gradient of a scalar field P, i.e. $B = -\nabla P = -\left(\frac{\partial P}{\partial x}, \frac{\partial P}{\partial y}, \frac{\partial P}{\partial z}\right)$, called magnetic potential, which is well described by spherical harmonic functions. Thus a least-squares fit to the magnetic field measurements gives the Earth's field as the sum of spherical harmonics, each multiplied by the best-fitting Gauss coefficient g_m^{ℓ} or h_m^{ℓ}.

primary cosmic rays stable charged particles that have been accelerated to enormous energies by astrophysical sources somewhere in our universe (the Milky Way or distant galaxies, the solar atmosphere and heliomagnetic field, and even Earth's radiation belts). Upon impact with the Earth's atmosphere, cosmic rays can produce showers of secondary particles that sometimes reach the surface.

Quaternary current and most recent (starting about 2.5 million years ago) of the three periods of the Cenozoic Era in the geologic timescale (according to the International Commission on Stratigraphy). The Quaternary Period is typically defined by the cyclic growth and decay of continental ice sheets associated with Milankovitch cycles, and the associated climate and environmental changes that occurred. In this period, modern humans appeared.

Regener–Pfotzer maximum the maximum of the lower atmospheric ionization layer, consisting of so-called 'secondary' cosmic radiation. It is produced by the multiple interactions (nuclear-electromagnetic-muonic-pionic, etc.) of primary cosmic rays with atmospheric atoms and molecules.

secondary electrons electrons generated as ionization products. They are called 'secondary' because they are generated by the primary radiation (i.e. ions, electrons, or photons with energy exceeding the ionization potential of the target atom/molecule).

secular variation geomagnetic secular variation refers to changes in Earth's magnetic field with periods of a year or more, reflecting changes in the Earth's core.

stadials and interstadials phases dividing the Quaternary Period, or the last 2.6 million years. Stadials are periods of colder climate while interstadials are periods of warmer climate.

subduction a geological process that takes place at convergent boundaries of tectonic plates, where the heavier plate (usually oceanic) is sinking gravitationally under the lighter one (usually continental) into the mantle. The subduction rates are typically measured in centimetres per year, with the average rate of convergence being approximately 2–8 cm per year along most plate boundaries. Regions where this process occurs are known as subduction zones.

substorm (also sometimes magnetospheric substorm or auroral substorm) a brief disturbance in the Earth's magnetosphere that causes energy to be released from the 'tail' of the magnetosphere and injected into the high latitude ionosphere. Visually, a substorm is seen as a sudden brightening and increased movement of auroral arcs.

superchron time interval (chron) lasting more than 10 million years between events, especially reversals of the polarity of the Earth's magnetic field.

teleconnection (in atmospheric sciences) causal connection or correlation between meteorological or other environmental phenomena which occur a long distance apart.

Van Allen radiation belt layer of charged and energetic particles which is held by the planetary magnetic field around the planet. The Van Allen belt specifically refers to the radiation belts around the Earth.

virtual geomagnetic pole point on the earth surface at which a magnetic pole would be located if the observed direction of remanence—at a particular location—was due to a magnetic dipole at the centre of the Earth.

Younger Dryas period around 12,900–11,700 years BP, characterized by a return to glacial conditions after the Late Glacial Interstadial. It reversed temporarily the gradual climatic warming after the Last Glacial Maximum (LGM) started around 20,000 BP. The Younger Dryas was the most recent and longest of several interruptions to the gradual warming of the Earth's climate since the severe LGM, about 27,000–24,000 years BP.

CHAPTER

1

Geomagnetic field—Origin, spatial-temporal structure, and variability

OUTLINE

Earth's magnetic field plays an important role in many aspects of Earth sciences. It is one of the key components of the complex integrated system of our planet, because it interacts with all Earth's shells—the atmosphere, the biosphere, Earth's crust, mantle and core—shielding the life on the planet from the harmful effects of cosmic radiation. Therefore, the magnetic field 'contains' information about both the state of near-Earth outer space and the internal structure of the deep Earth's interior. This chapter considers the structure, properties, nature, and methods for investigation of Earth's magnetic field.

1.1 Geomagnetic field structure—Dipole and nondipole components; temporal variability

The observed geomagnetic field on Earth's surface is a vector sum of the magnetic fields of several sources (Fig. 1.1) located in different areas inside the planet and in near-planetary space (Parkinson, 1983; Yanovsky, 1978; etc.):

The Hidden Link Between Earth's Magnetic Field and Climate
https://doi.org/10.1016/B978-0-12-819346-4.00001-2

© 2020 Elsevier Inc. All rights reserved.

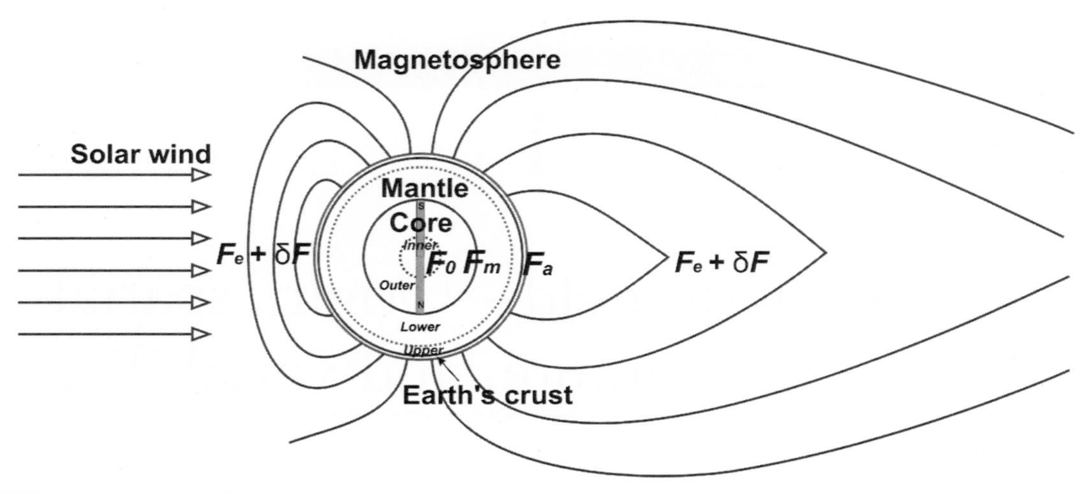

FIG. 1.1 Sources of Earth's magnetic field components.

$$F_T = F_0 + F_m + F_a + F_e + \delta F,$$

where $\boldsymbol{F_0}$ is the dipolar component of geomagnetic field, $\boldsymbol{F_m}$ is the field of world anomalies associated with the heterogeneity of the deep Earth's interior (nondipole field), $\boldsymbol{F_a}$ is the remanent magnetization of the rocks within the upper part of Earth's crust (anomalous field), $\boldsymbol{F_e}$ is the field of external sources, and $\boldsymbol{\delta F}$ is the field of variation, also associated with external causes (Fig. 1.1). The sum of the dipole and nondipole fields is sometimes called the *main* magnetic field of the Earth, i.e. $\boldsymbol{F} = \boldsymbol{F_0} + \boldsymbol{F_n}$.

The geomagnetic field can be described on Earth's surface by its three orthogonal components: $\boldsymbol{X}$ (pointing to the geographic north direction), $\boldsymbol{Y}$ (pointing eastward), and $\boldsymbol{Z}$ (pointing downward in the Northern Hemisphere). The two horizontal components $\boldsymbol{X}$ and $\boldsymbol{Y}$ can be combined, yielding the horizontal component $\boldsymbol{H}$, which is aligned in the direction of the compass needle:

$$H = \sqrt{X^2 + Y^2} \tag{1.1}$$

The sum of all three components defines the total field intensity, directed towards the centre of the planet:

$$F = \sqrt{X^2 + Y^2 + Z^2} \tag{1.2}$$

The declination $\boldsymbol{D}$ is defined as the angle between $\boldsymbol{H}$ and geographic north, while the inclination $\boldsymbol{I}$ is the angle between the horizontal plane and the vector of total field intensity $\boldsymbol{F}$. In the international SI system, the measurable units of geomagnetic field strength are Tesla (**T**) and its subunits: $\boldsymbol{\mu T} = 10^{-6}$ **T** and $\mathbf{nT} = 10^{-9}$ **T**.

The spatial-temporal variability of Earth's magnetic field is one of its most characteristic features. The available information has a different physical basis, accuracy, and resolution, and covers different time ranges. To obtain the most complete information about Earth's magnetic field, data from all sources are used (refer to Table 1.1).

TABLE 1.1 Geomagnetic field variations from various type measurements.

№	Geomagnetic variations	Period ($1 < n < 10$)	Amplitude ($1 < n < 10$)	Measurement accuracy	Data acquisition methods[a]
1	Steady and irregular pulsations	Minutes	$\sim n \times 10^{-1}$ (nT)	0.1–1.0 (nT)	O, S
2	Disturbed and undisturbed variations	Hours	$\sim n \times 10$ (nT)	1.0–5.0 (nT)	O, S
3	Magnetic storms	Hours-days	$n \times 10$–$n \times 10^2$ (nT)	10 (nT)	O, S
4	Secular variations	$n \times 10$–$n \times 10^3$ years	More than ($n \times 10^2 n \times 10^3$) (nT)	1–3 degrees	O, H, A, P
5	Episodes and excursions	$n \times 10^2$–$n \times 10^4$ years	>50 degrees (excursions)	~10 degrees	P
6	Reversals	$n \times 10^3$–$n \times 10^4$ years	–	~10 degrees	P
7	Intervals between reversals	$n \times 10^5$–$n \times 10^6$ years	–	~10–20 degrees	P

[a] A, *archaeomagnetic;* H, *historical;* O, *ground-based observations;* P, *palaeomagnetic methods;* S, *satellite measurements.*
From Bakhmutov, V.G., 2006. Paleosecular Variations of Geomagnetic Field, Kiev. Naukova Dumka, p. 9 (in Russian).

Temporal variations of geomagnetic field cover a broad range of timescales (Table 1.1). Short-term changes (e.g. variations with Nos. 1–3) are caused by the external sources—i.e. the electromagnetic currents in the magnetosphere and ionosphere, which are studied by the use of direct (instrumental) observations. Long-period changes (Nos. 4–6 in Table 1.1) are caused by the internal sources in Earth's core and are studied by using both—direct observations and the results of indirect (i.e. archaeomagnetic and palaeomagnetic) methods. Geomagnetic pulsations (No. 1 in Table 1.1) are very short-lasting oscillations of geomagnetic field. The origin of these fluctuations is ultra-low-frequency hydromagnetic waves, which are excited in solar wind and Earth's magnetosphere. They are divided into two classes: irregular pulsations Pi (individual bursts lasting several minutes), and more stable continuous pulsations Pc (lasting several hours with a quasisinusoidal shape). Among all Pcs, Pc1 pulsations are distinguished with a period of 0.2–5 s (also called 'pearls') and duration of the series from half an hour to several hours. The maximum of their occurrence is observed in the early morning local time hours. In the Pi1 range, several types of pulsations are observed, in particular Intervals of Pulsations with Diminishing Period (IPDPs) associated with the development of the magnetospheric substorm. IPDPs are most often observed in the afternoon and evening sectors in the form of a series of separate wave packets, similar to Pc1 oscillations, but with a gradually decreasing period, i.e. increasing frequency.

The perturbed and unperturbed geomagnetic variations (Nos. 2–3 in Table 1.1) are changes in the Earth's magnetic field over time under the influence of various factors. *Unperturbed* are the small amplitude (~tens of nT) *annual* variations of the monthly average values of Earth's magnetic field, and *diurnal* variations, which are associated with changes in solar activity and the moon phase. They have a maximum during the daytime hours, and when the moon is in opposition. These are smooth periodic variations with intensities reaching 200 nT, increasing from the equator to the poles.

The *perturbed* variations are *magnetic storms* and *substorms*, associated with active processes on the Sun, and irregular processes in the solar wind, which affect Earth's ionosphere and magnetosphere. The duration of geomagnetic storms ranges from several hours to several days. They are initiated by the disturbed solar wind, when arriving at Earth's magnetopause. The intensification of the equatorial ring current (constantly existing in the region of Earth's radiation belts) induces a magnetic field opposing the direction of the main geomagnetic field. As a result, the ground-based observatories detect a sudden drop in the horizontal geomagnetic field component (e.g. Rastogi, 2005; Maksimenko et al., 2008).

Magnetospheric *substorms* are disturbances detected in polar regions, associated with the interaction of a surging solar wind with Earth's magnetosphere. Their amplitude can reach 1000 nT, gradually decreasing towards the equator. The duration of the substorms is up to 1 h. Disturbances are developed in the magnetosphere, ionosphere, and atmosphere, and manifest themselves in perturbations of currents and magnetic field, acceleration of energetic particles, and aurora. In contrast to magnetic storms, which are mainly associated with changes in the ring current near the geomagnetic equator, and lead to almost global (except for regions near the polar regions) geomagnetic disturbances, substorms are local in nature and cover mainly the night side of polar regions. All these phenomena characterize the externally forced geomagnetic activity, for the assessment of which various indices are used (see, e.g. Parkinson, 1983).

This book is focused on the variations associated with changes in Earth's deep interior (i.e. variations Nos. 4–7 in Table 1.1). *Secular* variations of geomagnetic field (i.e. No. 4 in Table 1.1) cover long periods of tens, hundreds, and even thousands of years. They lead to significant changes in the annual mean values of terrestrial magnetic field. The investigations of the secular variations are based on the ground-based observations, as well as on historical, archaeomagnetic, and palaeomagnetic data. According to recent knowledge, they are associated with processes at the core–mantle boundary.

Variations Nos. 5–7 in Table 1.1 are distinguished by palaeomagnetic data. They are discussed in Section 1.3.4.

1.2 Direct and indirect observations of geomagnetic field

Direct observations have contributed to our knowledge about geomagnetic field variability over the past few centuries. Information about the longer-scale variations, however, is accessible only from archaeomagnetic and palaeomagnetic data. Direct instrumental measurements of geomagnetic field, and its variations on Earth's surface, began about 400 years ago. During this time, a great deal of data has been accumulated about the field declination, but much less about its inclination. In the medieval era, most geomagnetic measurements were carried out on board ships, as part of their navigation.

During the 20th century, many measurements of geomagnetic field components were collected during magnetic surveys: ground-based, oceanic, and aeromagnetic. Until the 1960s, the lack of observatory data over the oceans, and some land areas, was compensated for by repeated measurements on specially equipped ships, or at preliminary selected points for determination of geomagnetic secular variations.

The first magnetic observatories were established in the 1820s, and by the mid-1980s, the global geomagnetic network had expanded to 180 observatories. The accumulated global data allowed scientists to investigate not only the short-term disturbances of magnetospheric-ionospheric origin, but also the geomagnetic secular variations. Many digital magnetic observatories are integrated into the global network International Real-time Magnetic Observatory Network (INTERMAGNET). Data collected by INTERMAGNET meet uniform accuracy requirements of 0.1 nT for the magnetometer resolution (INTERMAGNET, 2012). Moreover, the preliminary minute data must be transmitted to a geomagnetic information node within 72 h of acquisition. Access to the final data is open both through the INTERMAGNET network and through global databases. In 2019, the INTERMAGNET portal provided geomagnetic data from 150 geomagnetic observatories from 42 countries (Fig. 1.2).

With the onset of the satellite era in the 1970s, satellite measurements have been involved in global studies of the geomagnetic field and its variations. To date, however, the observatory data have been and remain the most accurate and reliable source of information.

Fundamental knowledge about the spatial-temporal structure of Earth's palaeomagnetism can be obtained using indirect data, derived by archaeomagnetic and palaeomagnetic methods. The physical foundation of archaeomagnetism and palaeomagnetism are similar, so the archaeomagnetic method can be considered a kind of palaeomagnetic research. Palaeomagnetic studies are based on the following fundamental assumptions:

1. A geomagnetic field averaged over a relatively small interval (on a geological timescale) is the field of the central axial magnetic dipole, the axis of which coincides with Earth's rotational axis.
2. Rocks can be magnetized in the direction of an applied external magnetic field, and obtained magnetization, called natural remanent magnetization (NRM), can persist up to the present day (Tauxe, 2010).

The iron minerals with ferromagnetic properties (i.e. magnetite and its varieties, maghaemite, haematite, haemo-ilmenite, iron hydroxides, and others) are of paramount importance for palaeomagnetic studies. Ferromagnetic substances are characterized by magnetic hysteresis, denoting the irreversibility of the curve of normal magnetization during the process of rock demagnetization. Therefore, after the termination of the applied permanent magnetic field, the rocks will have a remanent magnetization, which is not equal to zero.

Rocks magnetization is a complex phenomenon depending on a variety of factors such as conditions of their formation, tectonic evolution, exposure to ambient influences etc. Part of the rocks magnetization has the memory about the strength and direction of the past magnetic field at the time of their formation or chemical alteration. Magnetic component representing this memory is known as a *natural remanent magnetization* (NRM). It is the vector sum of several magnetizations generated over the geological history of the rocks. They depends not only on the rocks properties and the magnitude of the applied constant magnetic field, but also on many factors, such as time, temperature, mechanical stress, chemical transformations, and others. Consequently, the rocks NRM could bear several components 'imprinted' on it by the temporary varying geomagnetic field in different epochs. The magnetization acquired at the time of rock formation—i.e. cooling below the blocking temperatures (i.e. the Curie point)

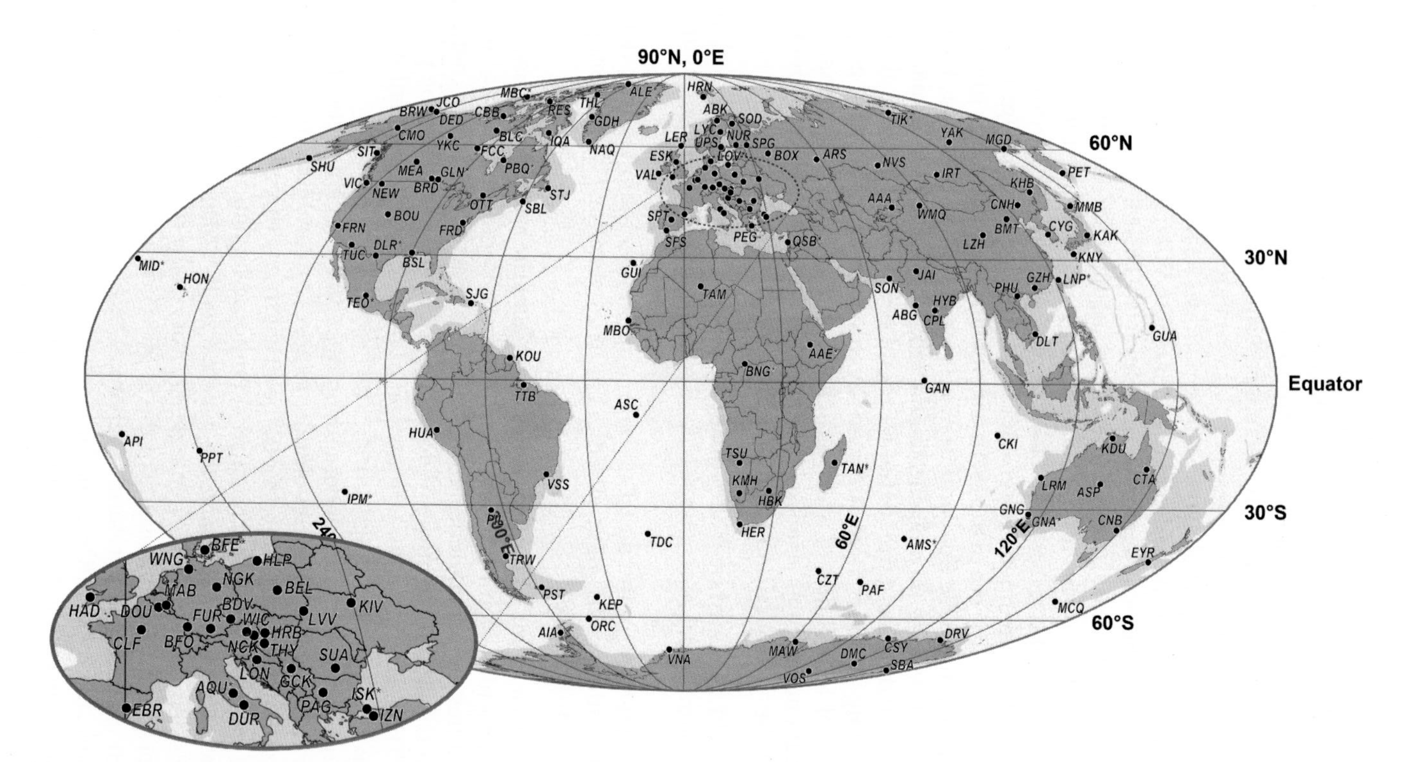

FIG. 1.2 Global network of geomagnetic observatories. *Black dots*, accompanied by the IAGA (International Association of Geomagnetism and Aeronomy) code, denote the position of each observatory. All observatories are members of the INTERMAGNET network. *Based on data from http://www.intermagnet.org/imos/imotblobs-eng.php.*

of volcanic or intrusive rocks, deposition of sediments or metamorphic events—is referred as *primary* magnetization. All other components acquired at later times represent the *secondary* components. Consequently, the main question standing in front of paleomagnetic studies is to distinguish between primary and secondary components in NRM. This is the foundation of the accuracy and reliability of paleomagnetic information.

Rocks are characterized by various types of remanent magnetization. Volcanogenic and intrusive rocks have thermoremanent magnetization (TRM), which is formed when the ferromagnetic material is cooled below the Curie point in Earth's magnetic field (the Curie point for magnetite is 580°C, pyrrhotite 300°C, haematite 675°C). Archaeological objects (furnaces, bricks, ceramics, etc.) are also characterized by the same magnetization, fixing the magnitude and direction of geomagnetic field in the time of their firing.

Time plays an important role in magnetization of ferromagnetic minerals in a weak magnetic field. Therefore, due to the magnetic viscosity (i.e. time-lag in magnetization of magnetizable material under the influence of geomagnetic field), the process of 'viscous' magnetization of rocks continues with time. Viscous magnetization (VRM) is formed under favourable conditions; the rocks are exposed to the influence of Earth's magnetic field for millions and hundreds of millions of years.

Chemical remanent magnetization (CRM) is formed as a result of chemical or other changes in the grains of magnetic minerals placed in a magnetic field, at temperatures below the Curie temperature. Its properties depend both on the magnetic characteristics of initial minerals, and on the newly formed chemical products, such as during the chemical transition of haematite to magnetite, magnetite to maghaemite, etc. Chemical magnetization is widespread in sedimentary rocks, manifested as depositional/postdepositional remanent magnetization.

During the sedimentation of particles on the bottom of the water pool, on the land surface, or in the layer of unconsolidated sediments (in the presence of a certain amount of water in it), the clay fractions (like small magnets) are oriented in the direction of geomagnetic fields (or close to it). However, the primary magnetization of the rock eventually decays, due to the influence of various factors like occurrence of secondary magnetization, not being identical in magnitude and direction to the primary one. The resistance of the primary magnetization to the subsequently imposed influence of other magnetic fields is characterized by its palaeomagnetic *stability* (describing its ability to preserve the primary rock magnetization). It has been experimentally proved that each type of rock magnetization is characterized by a certain resistance to the effects of an alternating magnetic field, temperature, and other factors, leading to the destruction of magnetization. The difference in stability is the basis of the methods of 'magnetic cleaning' of rock samples. The essence of these methods is that a sample placed in a nonmagnetic space is heated or exposed to alternating magnetic fields that grow sequentially. At certain temperatures or amplitudes of an alternating magnetic field, the secondary, less stable forms of magnetization are destroyed. A good indication that an analysed sample is cleaned from the secondary magnetization is the stabilization of the directions of the residual magnetization. The remaining magnetization is considered characteristic, but it is still necessary to prove that it is primary. To achieve this goal, special techniques have been developed, in particular various geological field tests.

In resume, to date there is a full set of methods allowing collection of geomagnetic field data in all ranges of its variations. However, as a rule, these data are unevenly distributed in space and time. To improve the spatial and temporal mapping of geomagnetic components, different types of models of the contemporary and ancient Earth's magnetic field have been developed (see Section 1.5).

1.3 Characteristics of Earth's main magnetic field

1.3.1 Description of the present magnetic field

Analysis of Earth's magnetic field variations over the past century (from instrumental observations), and over the last several hundreds and thousands of years (from historical and archaeomagnetic data), shows that even on the scales of several years to several decades, the field components can vary significantly. For most of the 20th century, the drift velocity of geomagnetic poles was about ~10 km/year. In the 1990s, however, the speed of the pole in the Northern Hemisphere increased sharply, reaching speeds of 40–60 km/year (Olsen and Mandea, 2007). Moreover, since 1840 the magnitude of the geomagnetic dipole moment has decreased by about 5%–7% per century (Gubbins et al., 2006; Mandea and Purucker, 2005). The strongest changes in the intensity of the Earth's magnetic field over the past few centuries have occurred in South America and the South Atlantic (Finlay et al., 2010).

Two regions with a stronger geomagnetic field could be distinguished in the Northern Hemisphere. One of them is known as the Canadian anomaly (containing the recent position of the geomagnetic South Pole), while the other is known as the Siberian one. In the Southern Hemisphere, however, there is a single maximum in geomagnetic field intensity, placed in the Southern Ocean, between Australia and Antarctica (Fig. 1.3, upper panels). Over the 20th century, the intensity of geomagnetic field in the Canadian anomaly has significantly decreased, and its centre has been shifted in a north-west direction. In contrast, the Siberian anomaly has strengthened and its centre has been shifted south-west. Changes in the Canadian anomaly could affect the motion of the geomagnetic South Pole, whose velocity has actually increased since 1990 (Olsen and Mandea, 2007). Temporal variations of geomagnetic field intensity also differ in both hemispheres. Thus, the geomagnetic field strength weakens considerably more slowly in the Northern Hemisphere than in the Southern Hemisphere (Fig. 1.3, lower panels). More specifically, for the past 120 years, the hemispherical mean intensity has decreased by ~1140 nT in the Northern Hemisphere and by ~5530 nT in the Southern Hemisphere.

1.3.2 Secular variations

Secular variations, as indicated above, are long-term changes in the elements of Earth's magnetism over time. Although the periodicities of these variations are still under investigation, some of them are more or less established. For example, variations with periods 60–80, 500, 2000, 5000 years or more have been reported. The geomagnetic secular variations are calculated as a difference between the field intensity in the final and initial moments of an examined period, divided by the number of years in that period, as follows:

$$F_{sv} = \frac{F_{i+n} - F_i}{n} \tag{1.3}$$

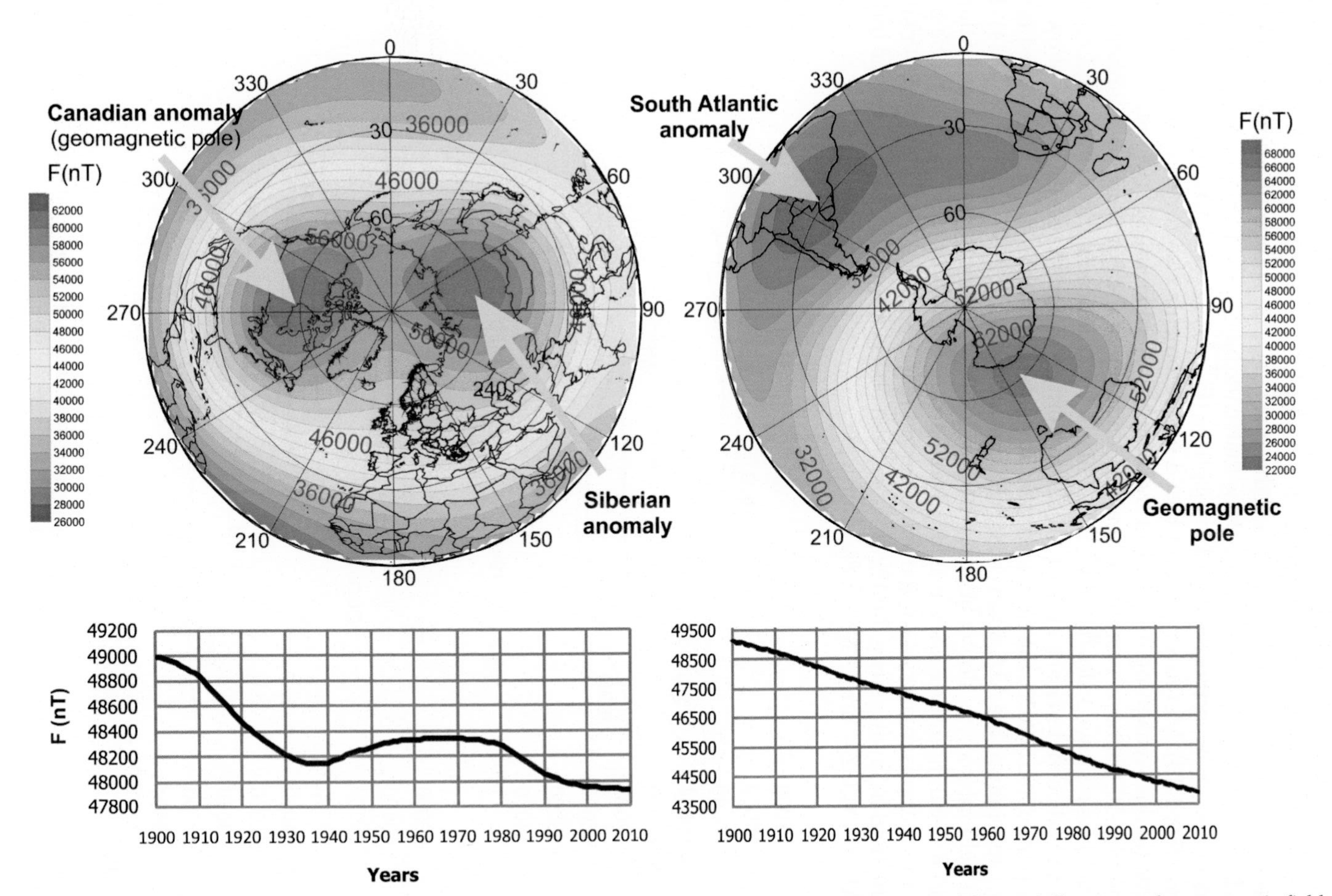

FIG. 1.3 Spatial structure of the centennially averaged geomagnetic field intensity *(top)* and dynamics of the spatially averaged geomagnetic field strength *(bottom)* in the Northern *(left)* and Southern *(right)* Hemispheres. *Data source: International Geomagnetic Reference Field model (https://www.ngdc.noaa.gov/geomag/geomag.shtml).*

where F_{sv} is the secular variation, F_{i+n} is the final year of the period, F_i is the initial year of the period, and n is the number of years within the period.

Data from geomagnetic observatories are the most reliable source of information about changes in Earth's magnetic field, and they show that in the Northern Hemisphere, geomagnetic field intensity decreases in the Western Hemisphere (Fig. 1.4A) and increases in the Eastern Hemisphere (Fig. 1.4B). Moreover, ground-based measurements also reveal that the Southern Hemispheric field is weakening much faster than the Northern Hemispheric one.

Since the mid-1990s, however, the decreasing tendency of the Southern Hemisphere geomagnetic field has 'slowed down' (Fig. 1.5). This is very noticeable in central and eastern Antarctica and adjacent territories (Fig. 1.5B). Data from geomagnetic observatories show that during the period 1957–2010, the field intensity at western Antarctic shores decreased faster (Fig. 1.5A), especially at the AIA observatory (Bakhmutov et al., 2006; Melnyk et al., 2014). The decrease of field intensity is slower at observatories located in the centre of the South Atlantic anomaly (in Russian literature known as the Brazilian anomaly) and at the geomagnetic pole, compared to other observatories for the same period (e.g. VSS and DRV in Fig. 1.5). Positive trends of the field intensity, since the 1990s, have been observed at observatories in the Indian Ocean (PAF, TAN, CZT, and AMS in Fig. 1.5B).

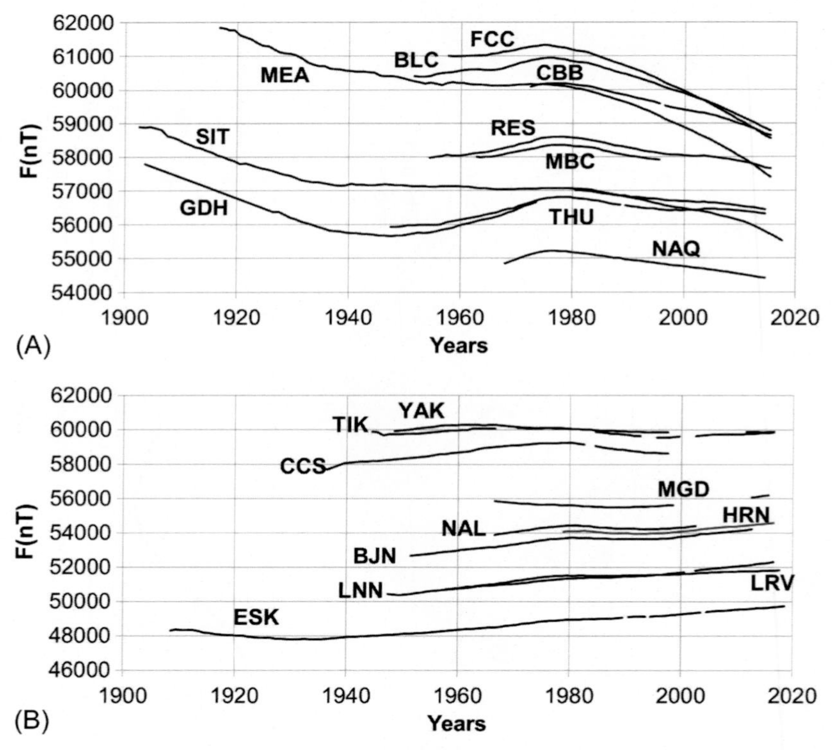

FIG. 1.4 Changes in the Northern Hemisphere geomagnetic field intensity (*F*) registered in various geomagnetic observatories during the 20th and 21st centuries: (A) in North America and (B) in Europe and Asia. *Data source: British Geological Survey (http://www.geomag.bgs.ac.uk/data_service /data/annual_means.shtml).*

that. On the other hand, the equatorial motions of the main field features are directed westward for most of the past 3000 years (Dumberry and Finlay, 2007).

From the foregoing, the ambiguity in estimation of the nondipole field drift velocity becomes obvious, as does the direction of drift motions in different regions over the world. Depending on the analysed time intervals, the amount of data from different regions, methods of data processing, and subjectivity of the interpretation, the estimations of the westward drift velocity vary from 0.08 eastward per year, up to 0.733 westward (for a summary, see Langel, 1987). Today, the concept of westward drift has been largely replaced by more sophisticated models of secular variations, and maps of the evolution of the field at the core–mantle boundary (see below). However, the concept is still used for interpretation of archaeological and palaeomagnetic data.

1.3.4 Geomagnetic reversal

Theoretical estimates show that Earth's magnetic field has existed for more than 3 billion years. Palaeomagnetic studies show evidence for multiple reversal of the field direction during this period, in which the Northern and the Southern geomagnetic poles interchange their places. The recent understanding of geomagnetism and contemporary numerical modes are able to reproduce the reversals of geomagnetic polarity. However, many fundamental questions remain to be clarified. For this reason, the problem of geomagnetic field reversals is still one of the main tasks for geophysicists, particularly for those involved in the problems of palaeomagnetism.

Information about the reversals is 'recorded' in rocks and ores containing ferromagnetic minerals (e.g. magnetite, haematite, titanomagnetite). They are able to preserve the remanent magnetization, storing it in such a way that information about the state of Earth's magnetic field at the time of the rocks' formation is retained. The study of the remanent magnetization in rocks of different ages is the basis for compiling a timescale of geomagnetic field reversals.

During the geomagnetic polarity reversal, the magnetic field strength drops dramatically, leading to a severe weakening of the planetary magnetic shielding, which protects living organisms from the harmful cosmic radiation. Several authors have demonstrated that in such periods, when the magnetic field is dominated by its quadrupole, octupole, etc. moments, the highly energetic cosmic rays (with hundreds MeV and GeV energies) are able to reach the low-latitude troposphere along the open magnetic field lines. This is the motivation for some authors to assume that geomagnetic field reversals could be considered as one of the probable causes for the episodes of mass extinctions that occurred during the evolution of terrestrial biota (Starchenko and Shcherbakov, 1991; Vogt et al., 2007; Channell and Vigliotti, 2019).

The time intervals during which the geomagnetic field maintains constant polarity (normal or reversed) are called *chrons*. Each chron may last hundreds of thousands to millions of years. Palaeomagnetic data show that geomagnetic reversals occur at random separations, so that a sequence of reversals defines polarity chrons of quite variable lengths. The first scale of geomagnetic polarity alteration was built in the 1960s by Cox et al. (1963). The most complete and detailed record of geomagnetic polarity reversals is stored in sequences of lineated marine magnetic anomalies, found in ocean basins (Vine and Matthews, 1963). The sequence of polarity reversal intervals has been carefully studied and used to construct magnetostratigraphic timescales linking biostratigraphy, isotope stratigraphy, and geochronological age determinations (see review by Opdyke and Channell, 1996).

For the last million years, the geomagnetic field has changed its polarity four times. This last happened about 780,000 years ago. In the entire history of the planet, at least several hundred reversals of the magnetic field have occurred. Nevertheless, no systematic pattern in reversals occurrence has been found, and they are considered to be a stochastic process.

The frequency of polarity changes over the past 160 million years is not constant in time. On average, there is one reversal per million, with a maximum value of six reversals per million years. There are periods when the field did not change its polarity for tens of millions of years. For example, the Cretaceous normal superchron occurred between 120 and 83 million years (Ma); Kayama reversed polarity superchron (from late Carboniferous to middle Permian (i.e. between 310 and 260 Ma); and Moyero reversed polarity superchron in the first half of the Ordovician (Pavlov and Gallet, 2005).

Information about the geometry and dynamics of the geomagnetic field during polarity reversal can be obtained from high-resolution palaeomagnetic records of both sedimentary and igneous rocks. The key issue in analysis of these data is the length of the transitional interval of changing from one polarity to another, in order to determine the beginning and end of any chron. The analysis is complicated by the appearance of short-term but strong oscillations, preceding the polarity reversal (Hartl and Tauxe, 1996; Dormy et al., 2000). Such oscillations could be interpreted as increased secular variations in periods of dipolar field weakening. On the other hand, they could be considered as consecutive attempts of the geomagnetic field to change polarity. The determination of intermediate field direction, during the polarity transformation, could also be a problem obstructing the formation of high-quality palaeomagnetic time series.

Despite the difficulties in obtaining reliable palaeomagnetic records, some general conclusions can now be drawn. Regarding the field intensity, for several thousand years before the polarity change, the intensity starts decreasing. It remains low until the polarity reversal is completed. With finalization of the reversal, the geomagnetic field intensity increases to its normal values. In general, the process takes place for several thousand years and varies significantly from one reversal to another. According to Valet et al. (2005), the decay of axial dipole prior to the polarity reversal lasts for 60,000–80,000 years. The restoration of its intensity after the reversal, however, occurs much faster—at most for several thousand years.

The palaeomagnetic records show that there are short periods within the chrons (with a duration of several thousand years), when the field has departed from its near-axial configuration. Such short-term events are called geomagnetic *excursions*. Investigation of these events increases our knowledge about the geomagnetic field oscillations. The latter is very important for understanding the nature of the geomagnetic field and the processes occurring in the liquid upper core of Earth. They are widely used as chronological and stratigraphic benchmarks, in various fields of earth sciences: in stratigraphy and geochronology, in sedimentology and tectonics, in palaeontology and climatology, etc.

There are different formulations of the term 'excursion'. It is usually defined as a deviation of the virtual geomagnetic pole (VGP) equatorward of 45-degree latitude (for comparison, the secular variations of the VGP certainly do not reach 45-degree latitude).

According to other definitions, geomagnetic 'excursion' is a short-term change in the direction of a geomagnetic field, whose amplitude is at least three times greater than the secular variations for a given period of time. If in such a period the field polarity is reversed, it should be unstable. Excursions are short-term impulse fluctuations, which are mostly replaced by the smoother secular variations in geomagnetic field intensity.

Although the excursions are relatively useful as chronological and stratigraphic benchmarks, their number (even for the youngest Brunhes Chron) is still undetermined (Opdyke and Channell, 1996; Lund et al., 1998; Merrill and McFadden, 1994; Singer et al., 1999; Thouveny et al., 2004; Fuller, 2006).

Another highly debated question is whether the excursions are regional or worldwide events. The answer to this question is difficult, since the duration of the excursions is short, data available are insufficient, and the exact dating of the events is of great importance.

The nature of geomagnetic polarity reversals and excursions is not fully understood. Their general characteristics suggest that they could be considered as manifestations of various processes within the Earth's liquid core. Some authors believe that excursions should be considered as interrupted reversals, while others view them as a significant secular variation in the presence of a weak dipole field.

The existing instrumental records indicate a gradual decrease of geomagnetic field intensity during the 20th century, accompanied by a raise of fluctuations at the core–mantle boundary. These facts have led some authors to suggest that recent geomagnetic field is in the period of transition of its polarity (Hulot et al., 2002). Are we on the verge of changing the polarity of the magnetic field, and what is the probability that this will happen? This issue is hotly debated in the scientific literature, spreading even in mass media. Its correct answer depends on our knowledge about the evolution of the geomagnetic field throughout the geological history of our planet.

1.4 Origin of the geomagnetic field

The dipolar structure of Earth's magnetic field is currently explained by the theory of geodynamo. The 'father' of the recent geodynamo theory is accepted to be Walter M. Elsasser (Elsasser, 1956), who suggested that the geomagnetic field is induced by electric currents in Earth's melted iron outer core. The modern theory of a hydromagnetic dynamo explains many of the geomagnetic field features, such as western drift, polarity reversal, etc.

Seismology has shown that Earth's deep environment consists of a solid inner core, with an approximate radius of 1100 km, surrounded by a fluid outer core of radius approximately 3500 km, confined in turn by the mantle (Gubbins and Herrero-Bervera, 2007; *Core properties, physical*). The electrical conductivity of the mantle is very low compared to that of the outer core, so the liquid metal outer core is the most natural seat of the geodynamo.

According to recent knowledge, gravitational differentiation between the inner and outer core had been completed 4.56 milliard years ago, during the solar system accretion. The growth of the proto-planetary mass is accompanied by a rise in temperature, which at some critical point leads to the melting of metallic components (e.g. iron (Fe), nickel (Ni), etc.). The melting permits the denser components to migrate towards the centre of the planet, thus achieving the process of differentiation. Note that recent compositional models of Earth's core estimate that it contains approximately 85% Fe, 5% Ni, and 10% of minor lighter components (e.g., H, C, O, Si, P, S). Partly because of the energy released during differentiation, the core is believed to be initially hot and convecting. Prior to the formation of the inner core, the terrestrial dynamo is supposedly driven by cooling of the liquid core, with possible assistance from radioactive decay. The crystallization of the inner core is suggested to have begun 2.5–1 milliard years ago (Labrosse et al., 2001), comprising recently ~5% of the core's mass. The cooling of the core depends on the rate at which the overlying colder mantle can remove its heat. The cooling forces the further solidification of the inner core, followed by a release of the lighter elements and the onset of *compositional convection* (i.e. upward motion of the lighter elements, due to their higher buoyancy) in the outer liquid core. Since the net effect of this process is to move heavier material towards the planetary centre, gravitational energy is liberated, supporting the thermal gradient between the core and mantle, and consequently *thermal convection* in the liquid outer core.

The movements of the electrically conductive iron in the magnetic field will produce electric currents. The high electric conductivity of the core makes the magnetic field lines 'frozen' (meaning that they are moving together with the fluid motion). Putting all this in the frame of the vastly rotating Earth, we will have the following: (i) the differential rotation within the fluid core (resulting from the different radius of moving fluid to the rotational axis) will shear the existing poloidal magnetic field lines (directed north-south and radial) into a toroidal (east-west) magnetic field; and (ii) the three-dimensional helical trajectories of the outer core fluid will twist the toroidal field lines into a poloidal field, which in turn replaces the continuously decaying initial magnetic field. The field is self-sustaining if, on average, the field generation is balanced by its decay (Gubbins and Herrero-Bervera, 2007; *Geodynamo*).

The quantitative conformation of the geodynamo hypothesis has been provided by numerical simulations of the complex system of magnetohydrodynamical equations, describing the geodynamo model. The first successful modelling of the three-dimensional problem of geomagnetic field generation was elaborated by Glatzmaier and Roberts (1995). Later on, Glatzmaier et al. (1999) succeeded in modelling the geomagnetic reversal, thus confirming the validity of the geodynamo mechanism for geomagnetic field generation. Additional support for plausibility of the geodynamo model comes from the laboratory experiment based on a turbulent flow of liquid sodium, simulating the geomagnetic field reversal (Berhanu et al., 2007).

New studies have shown that subduction flux (on planetary timescales) modulates the geomagnetic polarity reversal rate (Hounslow et al., 2018). These studies became possible thanks to the new data provided by the full plate tectonic models, allowing an estimation of the global subduction rate (see, for example, Vérard et al., 2015, known as the V15 model; Matthews et al., 2016, known as the M16 model; Domeier and Torsvik, 2019). These models take into account the boundaries of all lithospheric plates (continental and oceanic), and make it possible to reconstruct the thickness and age of the oceanic lithosphere before the Jurassic period, as well as the speed of its subduction in the mantle.

More recent numerical modelling of geodynamo shows that the frequency of geomagnetic reversal depends on the heat exchange between the outer core and mantle, as well as on the heterogeneity of the thermal flux at the core–mantle boundary (Olson and Amit, 2014). Cold lithospheric plates, subducting along active continental margins, are the most important convection driver in the mantle. Consequently, the geodynamical processes in the core, determining the polarity of Earth's magnetic field, could be estimated by a single geodynamic model taking into account dynamical relationships between the processes occurring on Earth's surface, at the core–mantle boundary, and in the upper core (Hounslow et al., 2018).

Pétrélis et al. (2009) have proposed a simpler statistical approach for determination of the polarity reversal, suggesting that it is driven by a nonlinear interaction between axisymmetric stationary modes of geomagnetic field (e.g. dipolar and quadrupole) and turbulent fluctuations in the core. According to the authors, the higher modes in Gaussian field decomposition are born from turbulent fluctuations in the motion of electrically conductive iron in the upper core. The geomagnetic field is approximated by the sum of the dipole and quadrupole modes, where the latter acts as white noise on the main dipole component. Due to this noise, the geomagnetic poles deviate from their stable equilibrium positions and pass into the so-called unstable equilibrium points. Two further options are possible: the poles can slowly return to their former stable position (a situation equivalent to the observed excursions), or they will be attracted by the pole with the opposite sign—a situation equivalent to the observed polarity reversals. Supporting their hypothesis with some quantitative mathematical calculations, Pétrélis et al. (2009) tried to explain the nonperiodic fluctuations in the rate of geomagnetic reversals, as well as the presence of long periods without polarity changes; these are known as superchrons. The authors also claimed that their approach was an alternative to the numerical solutions of the system of magneto-thermo-dynamical equations, which itself requires a lot of computational time and huge computational resources.

Thus, the theory of the hydromagnetic dynamo is by far the most popular one on the origin of the Earth's magnetic field. Despite the lack of data about some important characteristics of Earth's core, today it faces fewer difficulties than other alternative theories.

1.5 Empirical models of the geomagnetic field

Since the observatories are located unevenly around the globe, as well as the points of palaeomagnetic and archaeomagnetic studies, empirical model calculations are used to compile maps of the geomagnetic field. One of the most representative models, using ground-based instrumental observations, is the International Geomagnetic Reference Field, or IGRF (Thébault et al., 2015). The model describes the internal geomagnetic field and its secular variations in a spherical coordinate system $(\boldsymbol{r}, \boldsymbol{\theta}, \boldsymbol{\phi}, \boldsymbol{t})$. Introduced by Gauss in 1839, this method of geomagnetic field descriptions uses spherical harmonic functions (Gauss, 1839).

On and above the Earth's surface, the magnetic field vector is defined in terms of the magnetic scalar potential $-\nabla.V$. In a spherical coordinate system, $\boldsymbol{V}$ is approximated by the sum:

$$V(r,\theta,\varphi,t) = a\sum_{n=1}^{N}\sum_{m=1}^{n}\left(\frac{a}{r}\right)^{n+1}.\{g_n^m(t)\cos(m\varphi) + h_n^m(t)\sin(m\varphi)P_n^m(\cos\theta)\} \tag{1.4}$$

where **r** is the radial distance from Earth's centre, $\boldsymbol{a}$ = **6371.2** km is the average Earth's radius, $\boldsymbol{\theta}$ is latitude, $\boldsymbol{\phi}$ is east longitude, and $P_n^m(\cos\theta)$ are the Schmidt quasinormalized associated Legendre functions of degree $\boldsymbol{n}$ and order $\boldsymbol{m}$. The coefficients $\boldsymbol{g}$ and $\boldsymbol{h}$ are now called Gauss coefficients. They are functions of time and are conditionally given in nT. The Gauss coefficients $\boldsymbol{g}$ and $\boldsymbol{h}$ are calculated from all available geomagnetic field measurements for each time interval. The 12th and earlier generations of the IGRF model allow us to calculate the values of the total vector of geomagnetic field ***F***, its components ***H*** (horizontal), ***X*** (east), ***Y*** (north), ***Z*** (vertical), ***D*** (declination), ***I*** (inclination), and their secular variations for the period from 1900 to 2015. Furthermore, for simplicity, we will consider only the maps of geomagnetic field strength ***F*** and its secular variations, the data for which are synthesized using IGRF coefficients every 5 years from the beginning of the 20th century (http://wdc.kugi.kyoto-u.ac.jp/igrf/index.html) and calculated for each year (http://www.ngdc.noaa.gov/geomag-web/#igrfwmm) in grid nodes with a step of 10 degrees in latitude and longitude on the Earth's surface.

The sampling data points of the ancient Earth's magnetic field (derived by archaeomagnetic and palaeomagnetic methods) are located even more unevenly around the globe. Moreover, temporal data binding is also a very difficult task. However, to date, the number of relatively accurately dated samples, from various areas around the globe, is already sufficient to determine the coefficients of spherical harmonic analysis and to build global models of the geomagnetic field on a 1000-year timescale. They represent the evolution of the geomagnetic field and the processes at the core–mantle boundary better than separate time series.

The latest versions of models, as well as their updates, are collected in the GEOMAGIA50 database (http://geomagia.gfz-potsdam.de/index.php). This also contains data on the basis of which the models are created. Today the most complete and relevant modes are as follows:

- The CALS3k.4 model is built from all available data over the past 3000 years (Korte et al., 2011) and associated with the gufm1 model (Jackson et al., 2000) for the last 400 years. Using the gufm1 and IGRF models, calculations of the components of the geomagnetic field from 1590 to the present day are available with temporal discretization of 1 year (https://www.ngdc.noaa.gov/geomag-web/#igrfwmm). However, it should be borne in mind that the accuracy of determination of these components is different; for example, the accuracy is lower for earlier periods and increases for times closer to the present. The gufm1 model was created on the basis of a compilation of historical observations of the magnetic field, mainly marine shipping data, from 1590 onwards (Jackson et al., 2000).
- The models CALS10k.1 and CALS10k.1b cover the last 10,000 years. They are based on the largest number of lakes and marine, lava, and archaeological data available to date. The CALS10k.1b model takes better account of the uncertainties in palaeomagnetic and chronological data than CALS10k.1 and represents averaged values that are strongly smoothed compared to CALS3k.3 and CALS3k.4, which provide reconstructions of earlier eras. CALS10k.1 has been closely associated with gufm1 (Jackson et al., 2000) over the past four centuries. It can be used to calculate the field across the globe. The model is described by Korte et al. (2011).
- ARCH3k.1 was built from the available archaeomagnetic data until 2009 and covers the last 3000 years. The model is valid for the Northern Hemisphere and is not applicable for global research or field reconstruction in the Southern Hemisphere.
- SED3k.1 is based on sedimentary rock data available prior to 2009. It covers the last 3000 years. The data have a better global distribution than ARCH3k.1, and can be used for estimates in the Southern Hemisphere.
- The pfm9k.1a model (Nilsson et al., 2014) covers the last 9000 years. The model is built from the same data as the CALS10k.1b model (Korte et al., 2011), but a new approach in data analysis was performed, especially regarding the sedimentary rocks. It includes redistribution of the weight of different data types and data sources, iterative recalibration

of the relative declination data, and adjustment of the time intervals of sedimentary records using a preliminary model. These data processing, in particular timescale adjustments, reduce inconsistencies in the database, allowing pfm9k.1a to capture large amplitude variations of the paleosecular variations. The model is valid until AD 1900, and, unlike CALS10k.1b, is not limited to gufm1.

- The ARCH-UK.1 model is applicable only to the United Kingdom and is valid only in the range of 49°–61°N and 11°W–2°E. It is part of the global field model using the same data set as ARCH10k.1 (Constable et al., 2016), but data for the United Kingdom have four times greater weight than the global data. The model covers the 10,000 years before 1990 (Batt et al., 2017).
- CALS10k.2 covers the last 10,000 years and contains an updated set of lava and archaeological data, the same as ARCH10k.1, in addition to almost the same sedimentary records as CALS10k.1b. CALS10k.2 has a higher temporal and spatial resolution than CALS10k.1b, due to the improved estimates of data uncertainty of sedimentation rate. The model is closely associated with gufm1 (Jackson et al., 2000) over the past four centuries (Constable et al., 2016).
- ARCH10k.1 is built using all available archaeomagnetic and lava flow data. Due to the very uneven distribution of data, it cannot be used for calculation of global characteristics, and even regional forecasts should be considered with caution for periods earlier than the past 3000 years (Constable et al., 2016).
- HFM.OL1.AL1 covers the same period of time and includes the same data as CALS10k.2, but differs in modelling strategies. HFM.OL1.A1 has a higher temporal but slightly lower spatial resolution than CALS10k.2. It is not limited to any other model (Constable et al., 2016; Panovska et al., 2015).

In resume, the present palaeomagnetic models are able to evaluate the spatial-temporal variations of geomagnetic field intensity, with some uncertainties. Moreover, they can be used to estimate the expected future evolution of the geomagnetic field. For example, the palaeomagnetic modelling performed by Brown et al. (2018) suggests that hypothesized consecutive polar reversal of the contemporary geomagnetic field is unlikely to happen in the near future. This conclusion is based on the comparison of the spatial distribution of the geomagnetic field strength, obtained from the palaeomagnetic modelling for the past 30,000–50,000 years (built from volcanic and sedimentary rocks). The model data, corresponding to the period of instrumental measurements, are in good agreement with the IGRF model (see Fig. 1 in Brown et al., 2018). The comparison has shown that none of the observed palaeogeomagnetic excursions has a structure similar to the spatial distribution of the modern field. Thorough analysis of palaeomagnetic data reveals that a weakening of the geomagnetic field intensity, similar to that observed nowadays, was found about 49,000 and 46,000 years ago, but it did not lead to an excursion or a polar reversal. This suggests that the modern field will be restored without a change of its polarity.

References

Bakhmutov, V., Yaremenko, L., Melnyk, G., Shenderovska, O., 2006. Magnetic field of Antarctic according to modelling calculations IGRF and supervision of observatories. Ukr. Antarct. J. 4–5, 64–71 (in Russian).

Bakhmutov, V.G., Martazinova, V.F., Kilifarska, N.A., Melnyk, G.V., Ivanova, E.K., 2014. Geomagnetic field and climate variability. 1. Spatial—temporal distribution of geomagnetic field and climatic parameters during XX century. Geophys. J. 36 (1), 81–104 (in Russian).

Batt, C.M., Brown, M.C., Clelland, S.-J., Korte, M., Linford, P., Outram, Z., 2017. Advances in archaeomagnetic dating in Britain: new data, new approaches and a new calibration curve. J. Archaeol. Sci. 85, 66–82. https://doi.org/10.1016/j.jas.2017.07.002.

Benkova, N.P., Kolomiytseva, G.I., Cherevko, T.N., 1974. Analytical model of geomagnetic field and its secular variations for the last 400 years (1550-1950). Geomagn. Aeron. 14 (5), 881–887 (in Russian).

Berhanu, M., Monchaux, R., Fauve, S., Mordant, N., Pétrélis, F., Chiffaudel, A., Daviaud, F., Dubrulle, B., Marié, L., Ravelet, F., Bourgoin, M., Odier, P., Pinton, J.-F., Volk, R., 2007. Magnetic field reversals in an experimental turbulent dynamo. Europhys. Lett. 77, 59001. https://doi.org/10.1209/0295-5075/77/59001.

Braginsky, S.I., 1967. Magnetic waves in the Earth's core. Geomagn. Aeron. 7, 1050–1060 (English translation, 851–859).

Brown, M., Korte, M., Holme, R., Wardinski, I., Gunnarson, S., 2018. Earth's magnetic field is probably not reversing. PNAS 115, 5111–5116. https://doi.org/10.1073/pnas.1722110115.

Bullard, E.C., Freedman, C., Gellman, H., Nixon, J., 1950. The westward drift of the Earth's magnetic field. Philos. Trans. R. Soc. Lond. Ser. A Math. Phys. Sci. 243, 67–92. https://doi.org/10.1098/rsta.1950.0014.

Channell, J.E.T., Vigliotti, L., 2019. The role of geomagnetic field intensity in late quaternary evolution of humans and large mammals. Rev. Geophys. https://doi.org/10.1029/2018RG000629.

Constable, C., Korte, M., Panovska, S., 2016. Persistent high paleosecular variation activity in southern hemisphere for at least 10000 years. Earth Planet. Sci. Lett. 453, 78–86. https://doi.org/10.1016/j.epsl.2016.08.015.

Cox, A., Doell, R.R., Dalrymple, G.B., 1963. Geomagnetic polarity epochs and pleistocene geochronometry. Nature 198, 1049–1051. https://doi.org/10.1038/1981049a0.

Domeier, M., Torsvik, T.H., 2019. Full-plate modelling in pre-Jurassic time. Geol. Mag. 156, 261–280. https://doi.org/10.1017/S0016756817001005.

Dormy, E., Valet, J.-P., Courtillot, V., 2000. Numerical models of the geodynamo and observational constraints. Geochem. Geophys. Geosyst. 1, https://doi.org/10.1029/2000GC000062.

Dumberry, M., Finlay, C.C., 2007. Eastward and westward drift of the Earth's magnetic field for the last three millennia. Earth Planet. Sci. Lett. 254, 146–157. https://doi.org/10.1016/j.epsl.2006.11.026.

Elsasser, W.M., 1956. Hydromagnetic dynamo theory. Rev. Mod. Phys. 28, 135–163. https://doi.org/10.1103/RevModPhys.28.135.

Finlay, C.C., Maus, S., Beggan, C.D., Bondar, T.N., Chambodut, A., Chernova, T.A., Chulliat, A., Golovkov, V.P., Hamilton, B., Hamoudi, M., Holme, R., Hulot, G., Kuang, W., Langlais, B., Lesur, V., Lowes, F.J., Lühr, H., Macmillan, S., Mandea, M., McLean, S., Manoj, C., Menvielle, M., Michaelis, I., Olsen, N., Rauberg, J., Rother, M., Sabaka, T.J., Tangborn, A., Tøffner-Clausen, L., Thébault, E., Thomson, A.W.P., Wardinski, I., Wei, Z., Zvereva, T.I., 2010. International geomagnetic reference field: the eleventh generation. Geophys. J. Int. 183, 1216–1230. https://doi.org/10.1111/j.1365-246X.2010.04804.x.

Fuller, M., 2006. Geomagnetic field intensity, excursions, reversals and the 41,000-yr obliquity signal. Earth Planet. Sci. Lett. 245, 605–615. https://doi.org/10.1016/j.epsl.2006.03.022.

Gauss, C., 1839. Allgemeine Theorie des Erdmagnetismus. In: Resultate aus den Beobachtungen des Magnetischen Verein im Jahre 1838. Göttinger Magnetischer Verein, Leipzig, pp. 1–52.

Glatzmaier, G.A., Coe, R.S., Hongre, L., Roberts, P.H., 1999. The role of the Earth's mantle in controlling the frequency of geomagnetic reversals. Nature 401, 885–890. https://doi.org/10.1038/44776.

Glatzmaier, G.A., Roberts, P.H., 1995. A three-dimensional self-consistent computer simulation of a geomagnetic field reversal. Nature 377, 203–209. https://doi.org/10.1038/377203a0.

Gubbins, D., Herrero-Bervera, E. (Eds.), 2007. Encyclopedia of Geomagnetism and Paleomagnetism. Encyclopedia of Earth Sciences SeriesSpringer Netherlands.

Gubbins, D., Jones, A.L., Finlay, C.C., 2006. Fall in Earth's magnetic field is erratic. Science 312, 900–902. https://doi.org/10.1126/science.1124855.

Halley, E., 1692. An account of the cause of the change of the variation of the magnetical needle. With an hypothesis of the structure of the internal parts of the earth: as it was proposed to the Royal Society in one of their late meetings. Philos. Trans. R. Soc. Lond. 17, 563–578. https://doi.org/10.1098/rstl.1686.0107.

Hartl, P., Tauxe, L., 1996. A precursor to the Matuyama/Brunhes transition-field instability as recorded in pelagic sediments. Earth Planet. Sci. Lett. 138, 121–135. https://doi.org/10.1016/0012-821X(95)00231-Z.

Hide, R., 1966. Free hydromagnetic oscillations of the Earth's core and the theory of the geomagnetic secular variation. Philos. Trans. R. Soc. Lond. Ser. A Math. Phys. Sci. 259, 615–647. https://doi.org/10.1098/rsta.1966.0026.

Hounslow, M.W., Domeier, M., Biggin, A.J., 2018. Subduction flux modulates the geomagnetic polarity reversal rate. Tectonophysics 742–743, 34–49. https://doi.org/10.1016/j.tecto.2018.05.018.

Hulot, G., Eymin, C., Langlais, B., Mandea, M., Olsen, N., 2002. Small-scale structure of the geodynamo inferred from Oersted and Magsat satellite data. Nature 416, 620–623. https://doi.org/10.1038/416620a.

INTERMAGNET, 2012. In: St-Louis, B. (Ed.), INTERMAGNET Technical Reference Manual. Version 4.6. 92 pp. http://www.intermagnet.org/publications/intermag_4-6.pdf.

Jackson, A., Jonkers, A.R.T., Walker, M.R., 2000. Four centuries of geomagnetic secular variation from historical records. Phil. Trans. R. Soc. A 358, 957–990.

Korte, M., Constable, C., Donadini, F., Holme, R., 2011. Reconstructing the Holocene geomagnetic field. Earth Planet. Sci. Lett. 312, 497–505. https://doi.org/10.1016/j.epsl.2011.10.031.

Labrosse, S., Poirier, J.-P., Le Mouël, J.-L., 2001. The age of the inner core. Earth Planet. Sci. Lett. 190, 111–123. https://doi.org/10.1016/S0012-821X(01)00387-9.

Langel, R.A., 1987. The main field. In: Jacobs, J.A. (Ed.), Geomagnetism. In: vol. 1. Academic Press, New York (Chapter 4).

Lund, S.P., Acton, G., Clement, B., Hastedt, M., Okada, M., Williams, T., 1998. Geomagnetic field excursions occurred often during the last million years. EOS Trans. Am. Geophys. Union 79, 178–179. https://doi.org/10.1029/98EO00134.

Maksimenko, O.I., Melnik, G.V., Shenderovska, O.J., 2008. Spatial distribution of magnetic storm fields. In: Proceedings of the 7th International Conference "Problems of Geocosmos", St. Petersburg, Russia, 26–30 May 2008, pp. 158–163.

Mandea, M., Purucker, M., 2005. Observing, modeling, and interpreting magnetic fields of the solid earth. Surv. Geophys. 26, 415–459. https://doi.org/10.1007/s10712-005-3857-x.

Matthews, K.J., Maloney, K.T., Zahirovic, S., Williams, S.E., Seton, M., Müller, R.D., 2016. Global plate boundary evolution and kinematics since the late Paleozoic. Glob. Planet. Chang. 146, 226–250. https://doi.org/10.1016/j.gloplacha.2016.10.002.

Melnyk, G., Bakhmutov, V., Shenderovska, O., 2014. Antarctic geomagnetic field changes in the last century. Ukr. Antarct. J. 13, 75–80 (in Russian).

Merrill, R.T., McFadden, P.L., 1994. Geomagnetic field stability: reversal events and excursions. Earth Planet. Sci. Lett. 121, 57–69. https://doi.org/10.1016/0012-821X(94)90031-0.

Nilsson, A., Holme, R., Korte, M., Suttie, N., Hill, M., 2014. Reconstructing Holocene geomagnetic field variation: new methods, models and implications. Geophys. J. Int. 198, 229–248. https://doi.org/10.1093/gji/ggu120.

Olsen, N., Mandea, M., 2007. Will the magnetic North Pole move to Siberia? EOS Trans. Am. Geophys. Union 88, 293–300. https://doi.org/10.1029/2007EO290001.

Olson, P., Amit, H., 2014. Magnetic reversal frequency scaling in dynamos with thermochemical convection. Phys. Earth Planet. Inter. 229, 122–133. https://doi.org/10.1016/j.pepi.2014.01.009.

Opdyke, N.D., Channell, J.E.T., 1996. Magnetic Stratigraphy. Academic Press, San Diego, CA 341 pp.

Panovska, S., Korte, M., Finlay, C.C., Constable, C.G., 2015. Limitations in paleomagnetic data and modelling techniques and their impact on Holocene geomagnetic field models. Geophys. J. Int. 202, 402–418. https://doi.org/10.1093/gji/ggv137.

Parkinson, W.D., 1983. Introduction to Geomagnetism. Scottish Academic Press, Edinburgh.

Pavlov, V., Gallet, Y., 2005. A third superchron during the Early Paleozoic. Episodes 28, 78–84.

Pétrélis, F., Fauve, S., Dormy, E., Valet, J.-P., 2009. Simple mechanism for reversals of Earth's magnetic field. Phys. Rev. Lett. 102, 144503. https://doi.org/10.1103/PhysRevLett.102.144503.

Rastogi, R.G., 2005. Magnetic storm effects in H and D components of the geomagnetic field at low and middle latitudes. J. Atmos. Sol. Terr. Phys. 67, 665–675. https://doi.org/10.1016/j.jastp.2004.11.002.

Singer, B.S., Hoffman, K.A., Chauvin, A., Coe, R.S., Pringle, M.S., 1999. Dating transitionally magnetized lavas of the late Matuyama Chron: toward a new 40Ar/39Ar timescale of reversals and events. J. Geophys. Res. Solid Earth 104, 679–693. https://doi.org/10.1029/1998JB900016.

Skiles, D.D., 1970. A method of inferring the direction of drift of the geomagnetic field from paleomagnetic data. J. Geomagn. Geoelectr. 22, 441–462. https://doi.org/10.5636/jgg.22.441.

Starchenko, S.V., Shcherbakov, V.P., 1991. Magnitosphere during reomagnetic reversal. Trans. USSR Acad. Sci. 321 (1), 69–74 (in Russian).

Tauxe, L., 2010. Essentials of Paleomagnetism. xvi + 489 pp University of California Press, Berkeley.

Thébault, E., Finlay, C.C., Beggan, C.D., Alken, P., Aubert, J., Barrois, O., Bertrand, F., Bondar, T., Boness, A., Brocco, L., Canet, E., Chambodut, A., Chulliat, A., Coïsson, P., Civet, F., Du, A., Fournier, A., Fratter, I., Gillet, N., Hamilton, B., Hamoudi, M., Hulot, G., Jager, T., Korte, M., Kuang, W., Lalanne, X., Langlais, B., Léger, J.-M., Lesur, V., Lowes, F.J., Macmillan, S., Mandea, M., Manoj, C., Maus, S., Olsen, N., Petrov, V., Ridley, V., Rother, M., Sabaka, T.J., Saturnino, D., Schachtschneider, R., Sirol, O., Tangborn, A., Thomson, A., Tøffner-Clausen, L., Vigneron, P., Wardinski, I., Zvereva, T., 2015. International geomagnetic reference field: the 12th generation. Earth Planets Space 67, 79. https://doi.org/10.1186/s40623-015-0228-9.

Thompson, R., 1984. Geomagnetic evolution: 400 years of change on planet Earth. Phys. Earth Planet. Inter. Special Issue Origin of Main Fields and Secular Changes of the Earth and Planets. 36, 61–77. https://doi.org/10.1016/0031-9201(84)90099-2.

Thouveny, N., Carcaillet, J., Moreno, E.M.O., Leduc, G., Nerini, D., 2004. Geomagnetic moment variation and paleomagnetic excursions since 400 kyr BP : a stacked record from sedimentary sequences of the Portuguese margin. Earth Planet. Sci. Lett. https://doi.org/10.1016/s0012-821x(03)00701-5.

Tretyak, A., Yaremenko, L., Bakhmutov, V., 2002. Geomagnetic secular variation in Antarctica. Ukr. Antarct. Cent. Bull. 4, 83–89 (in Russian).

Valet, J.-P., Meynadier, L., Guyodo, Y., 2005. Geomagnetic dipole strength and reversal rate over the past two million years. Nature 435, 802–805.

Vérard, C., Hochard, C., Baumgartner, P.O., Stampfli, G.M., Liu, M., 2015. Geodynamic evolution of the Earth over the Phanerozoic: plate tectonic activity and palaeoclimatic indicators. J. Palaeogeogr. 4, 167–188. https://doi.org/10.3724/SP.J.1261.2015.00072.

Vestine, E.H., Kahle, A.B., 1968. The westward drift and geomagnetic secular change. Geophys. J. R. Astron. Soc. 15, 29–37. https://doi.org/10.1111/j.1365-246X.1968.tb05743.x.

Vestine, E.H., et al., 1947. Description of the Earth's Main Magnetic Field and Its Secular Change 1905-1945, 578 ed. Carnegie Institution of Washington Publication.

Vine, F.J., Matthews, D.H., 1963. Magnetic anomalies over oceanic ridges. Nature 199, 947–949. https://doi.org/10.1038/199947a0.

Vogt, J., Zieger, B., Glassmeier, K.-H., Stadelmann, A., Kallenrode, M.-B., Sinnhuber, M., Winkler, H., 2007. Energetic particles in the paleomagnetosphere: reduced dipole configurations and quadrupolar contributions. J. Geophys. Res. Space Physics. 112https://doi.org/10.1029/2006JA012224.

Xu, W., Wei, Z., Ma, S., 2000. Dramatic variations in the Earth's main magnetic field during the 20th century. Chin. Sci. Bull. 45, 2013–2016. https://doi.org/10.1007/BF02909699.

Yanovsky, B.M., 1978. Earth's Magnetism. Leningrad State University. 591 pp. (in Russian).

Yukutake, T., 1979. Review of the geomagnetic secular variations on the historical time scale. Phys. Earth Planet. Inter. 20, 83–95. https://doi.org/10.1016/0031-9201(79)90030-X.

Yukutake, T., Tachinaka, H., 1969. Separation of the earth's magnetic field into drifting and standing parts. Bull. Earthquake Res. Inst. 47, 65–97.

CHAPTER

2

Variations and covariation in palaeoclimate and palaeomagnetic field

OUTLINE

A possible connection between Earth's magnetic field and climate has been discussed since the 1970s (Wollin et al., 1971a,b; Bucha and Zikmunda, 1976; Gallet et al., 2005; Courtillot et al., 2007; Knudsen and Riisager, 2009; Kitaba et al., 2013; Rossi et al., 2014). Such a relation could reflect the climate influence on the transport and rate of deposition of the radionuclides (used as proxies of palaeomagnetic temporal variations) created in Earth's atmosphere by cosmic rays (Kent, 1982; Guyodo et al., 2000; Roberts et al., 2003). On the other hand, a lot of arguments and evidence have been presented about the reversed relation—i.e. about the geomagnetic influence on climate (e.g. Doake, 1978; Courtillot et al., 1982, 2007; Bucha and Bucha, 1998; Le Mouël et al., 2005; Gallet et al., 2005; Bakhmutov, 2006; Vieira and da Silva, 2006; Bakhmutov et al., 2014). Analysing the current state of knowledge, Courtillot et al. (2007) have drawn attention to the fact that possible relations between geomagnetic field and climate are likely to be underestimated, suggesting that they could be manifest at various timescales, from decadal and centennial to thousands and millions of years.

The Hidden Link Between Earth's Magnetic Field and Climate
https://doi.org/10.1016/B978-0-12-819346-4.00002-4

© 2020 Elsevier Inc. All rights reserved.

The decadal and multidecadal covariations between geomagnetic field and climate could be easily detected from the direct instrumental measurements of both variables. Possible synchronization at a longer timescale, however, could be found only through analysis of palaeomagnetic and palaeoclimatic data records. The palaeo-data reconstructions are based on different indirect methods, aimed to detect variations on centennial, millennial, and longer timescales. We must emphasize that the further back in time the paleomagnetic and paleoclimate reconstruction return, the more inaccurate they becomes due to the poor spatial coverage of the available data. Consequently, the possibility for detection of geomagnetic-climate relations is restricted up to the Late Pleistocene–Holocene. Any relations found in earlier periods become highly speculative.

2.1 Introduction to the problem of geomagnetic-climate coupling

It is worth remembering that geomagnetic field variability is determined by different sources: *internal* and *external* (see Chapter 1). The variations ranging from seconds to a few decades are caused by *external* sources, and are related to the heterogeneity of solar wind and electric currents induced in Earth's magnetosphere-ionosphere system. These geomagnetic variations are described by geomagnetic aa, AE, Kp, Ap, etc. indices. In the scientific literature, there is a lot of information about the existing correlations between surface temperature and geomagnetic indices, solar irradiance, Wolf numbers, cosmic rays, etc., on daily, seasonal, interannual, and interdecadal timescales (e.g. Svensmark and Friis-Christensen, 1997; Cliver et al., 1998; Svensmark, 1998, 2000; Carslaw et al., 2002; Shea and Smart, 2004; El-Borie and Al-Thoyaib, 2006; Valev, 2006; Gray et al., 2010; Mufti and Shah, 2011; Stauning, 2011; Seppälä et al., 2013; etc.).

Unlike the mechanisms of energy transfer from the Sun to magnetosphere and ionosphere, which are generally well understood, the solar and geomagnetic 'signals', found in the surface temperature variability are still unexplained. The huge amount of interpretations could be classified in three main groups: (1) long-term changes in total solar irradiance, which impacts the energy balance of Earth's atmosphere; (2) quasidecadal variations of solar ultraviolet (UV) radiation, affecting stratosphere and upper tropospheric composition and dynamics; and (3) quasidecadal variability of galactic cosmic rays (GCRs) intensity, and their possible effect on cloud cover or atmospheric transparency.

Among the three forcing factors, the total solar irradiance is the only one that is able to affect the surface temperature directly. However, its variability on decadal, centennial, or millennial timescales is proved to be very low at the current stage of solar evolution (see Chapter 3). The solar variability in the UV band is well detected in the middle atmosphere, but the mechanism(s) of solar 'signal' transfer down to the surface are still unclear. The less studied (and therefore more promising) is the possible influence of GCRs on the lower atmosphere. Consisting mainly of protons, GCRs intensity is modulated by the solar and geomagnetic fields. Consequently, the temporal variability of the latter should be imprinted on the charged particles flux entering Earth's magnetosphere and atmosphere.

In addition, the cosmic rays (of solar or galactic origin) are the main source of ionization in the lower atmosphere, particularly near the tropopause. Already in the early works, three main mechanisms for the lower energetic (i.e. with $E \leq 10^7$ eV) cosmic rays influence on the lower atmosphere has been proposed: (i) *chemical* mechanism, describing the effect of charged particles on the atmospheric ozone density (Jackman et al., 2016; Rusch et al., 1981; Krivolutsky et al., 2006; Hauchecorne et al., 2007; Seppälä et al., 2008; etc.); (ii) *electrical* mechanism, focused on changes in atmospheric electric conductivity, which supposedly influence cloud microphysics (Tinslev, 1996; Rycroft et al., 2000; Harrison, 2004; Lucas et al., 2015); and (iii) *condensational* mechanism, related to the formation of condensational nuclei in the upper troposphere, potentially affecting the processes of cloud formation or atmospheric transparency (Dickinson, 1975; Vitinskii et al., 1976; Pudovkin and Babushkina, 1991; Pudovkin and Veretenenko, 1992; Pudovkin and Raspopov, 1992; Veretenenko and Ogurtsov, 2012; etc.). According to the latter authors, despite the small amount of energy deposited by particles in the atmosphere, it is enough to create an atmospheric optical screen, affecting the thermodynamics of the lower atmosphere. In addition to this indirect effect of cosmic rays on the atmospheric properties, Vitinskii et al. (1976) proposed that highly energetic protons and α-particles (with $E \geq 10^8$ eV) could directly affect the tropospheric circulation, transferring their momentum to the atmospheric motions.

Currently, the suggestive relations between geomagnetic field and climate are based mainly on statistical analyses, and in this context, the correlations based on the direct instrumental measurements are considered to be the most reliable. The possible relations between palaeomagnetic field and palaeoclimate, however, are much more poorly investigated. Indirect information about the state of Earth's magnetic field, cosmic ray fluxes, climate, and their changes in the distant past can be obtained from data recorded in various natural 'archives'. They are restored by analyses of cosmogenic isotopes ^{14}C (e.g. in tree rings over the past several hundred years, up to few thousand years), ^{10}Be, and ^{18}O in ice cores and in marine sediments. The variations of cosmogenic radionuclides (^{14}C, ^{10}Be, ^{18}O, etc.) are interpreted as a temporary 'record' of parameters affecting the rate of their formation, transformation, transportation, etc. in the atmosphere. Consequently, radioisotope records obviously contain information about the environmental conditions affecting the processes of transportation and sedimentation of radionuclides.

On the other hand, cosmic rays (consisting mainly of charged particles) are modulated by heliomagnetic and geomagnetic fields. This inverse relationship between the geomagnetic dipole moment and the rate of ^{10}Be formation in the atmosphere was demonstrated for the first time by Elsasser et al. (1956), and over the subsequent decades this relationship was repeatedly confirmed and quantified. The possible mechanisms standing behind such a relationship have been continuously improved (e.g. Beer et al., 1990; Masarik and Beer, 1999, 2009; Raspopov et al., 2000; Kovaltsov and Usoskin, 2010; Vasiliev et al., 2012 and references therein).

However, bearing in mind that the concentration of cosmogenic radionuclides depends on climatic conditions, on the latitude and altitude of their formation in the atmosphere, and on the processes of their transportation and deposition, it becomes clear that interpretation of the record of cosmogenic isotopes is an ambiguous task. This complicated problem has been solved through comparison of two or more records of different isotopes, or with a geomagnetic field recording, obtained by other methods. This approach allows the separation of variations associated with the nonmagnetic factors.

The long-term periodicities (found in ^{10}Be from ice cores and cores of marine deposits) reflect the large amplitude fluctuations of a geomagnetic dipole, on a millennial timescale. The combined studies of marine and ice records, together with modelling estimations, provide convincing arguments that ^{10}Be in the atmosphere is rapidly mixed (for several years) and removed from the water column over a period of about 100 years (Ménabréaz et al., 2014; Heikkilä et al., 2013). The significantly faster deposition of ^{10}Be data allows their use for detection of the variations in Earth's geomagnetic dipole moment.

The most detailed records from marine sediments and ice cores show evidence for a large episode of ^{10}Be superproduction, which corresponds to the collapse of geomagnetic dipole moment at the Matuyama–Brunhes geomagnetic reversal boundary (Carcaillet et al., 2003, 2004; Raisbeck et al., 2006; Suganuma et al., 2010; Ménabréaz et al., 2014; Valet et al., 2014). Recent papers (Simon et al., 2016, 2017, 2018 and references therein) further support the evidence that superproduction of ^{10}Be should be associated with geomagnetic dipole minima during the polarity change, or excursions, over the Quaternary.

The variations in strength and direction of ancient geomagnetic field could be detected by the use of: (1) continuous records of magnetic parameters derived from lake and marine sediments; (2) magnetic anomalies in deep-sea basalts; (3) lava flows; and (4) archaeomagnetic data. These studies give both continuous and discrete records, which are combined and averaged to build a global curve of variations of geomagnetic palaeointensity. In this case, the relative data can be calibrated using the absolute values of the palaeointensity from the lava flows, converted into records of the virtual axial dipole moment. However, it should be taken into account that the distortions in depositional (DRM) and postdepositional (PDRM) remnant magnetization in sediments can lead to smoothing and delays in data recording. Such distortions complicate the accuracy of geomagnetic interpretations (see Tauxe et al., 2006; Roberts et al., 2013). Comparison of the long-lived cosmogenic radionuclides with palaeomagnetic records is the most promising tool for searching for coherent variances in geomagnetic field and palaeotemperatures (e.g. for thousands of years or more). In addition, cores of marine deposits contain information about the variations of geomagnetic dipole, and cosmogenic isotopes. Therefore, the problem of mutual age correlation does not arise.

In one of the first studies of deep-sea sediments in the Northern Pacific and North Atlantic, a repeated synchronization has been found between higher geomagnetic intensity and colder climate, and vice versa (Wollin et al., 1971a,b). The authors have suggested that Earth's magnetic field is able to modulate the corpuscular radiation and somehow climate. The idea has been further investigated and developed by Bucha and Zikmunda (1976), Nurgaliev (1991), Butchvarova and Kovacheva (1993), and Gallet et al. (2005).

On the other hand, another group of authors have reported other evidence (mutually excluding the previous study) according to which the periods of *geomagnetic field weakening* and/or magnetic poles reversals correspond to a *cooler* climate (Valet and Meynadier, 1993; Worm, 1997; Knudsen and Riisager, 2009; Kitaba et al., 2013). Such an apparent controversy has led some authors to the conclusion that geomagnetic-climate relation is a consequence of the inadequate normalization and correction of geomagnetic data (Kent, 1982; Worm, 1997; Frank, 2000).

An alternative explanation of the observed coherent variability between geomagnetic and climatic variables has been offered by Kent and Opdyke (1977). They have suggested that the long-term periodicity of palaeomagnetic field intensity could be attributed to the inserted precessional forcing on Earth's core geodynamo, with a period of about 43,000 years (Malkus, 1968). Other authors (e.g. Yamazaki and Oda, 2002) have reported the presence of 100,000-year periodicity in field inclination and intensity, relating it to Earth's orbit eccentricity. Kent and Carlut (2001) have come to the opposite conclusion, that "no discernible tendency is found for astronomically-dated geomagnetic reversals in the Plio–Pleistocene (0 to 5.3 million years, Ma), or excursions in the Brunhes (0 to 0.78 Ma), to occur at a consistent amplitude or phase of obliquity, nor of orbital eccentricity." Nevertheless, the relationship between Earth's magnetic field and Milankovitch's orbital cycles has been hotly debated during the past 20 years (e.g. Channell et al., 1998; Guyodo and Valet, 1999; Channell, 1999; Carcaillet et al., 2003; Thouveny et al., 2004, 2008; Fuller, 2006; Courtillot et al., 2007).

Some studies have discussed the weakening of geomagnetic dipole moment (due to geomagnetic excursions) shortly before the onset of cold climatic intervals (e.g. Thouveny et al., 2008; Kitaba et al., 2013). A different point of view, on the geomagnetic-climate relations over the past 12,000 years, has been proposed by Petrova and Raspopov (1998). The authors have suggested that relation between field intensity and surface temperature depends on the geomagnetic dipolar moment (M). Thus for $M \leq (0.5–0.6)M_0$—geomagnetic strength and surface temperature are almost linearly connected. With a further increase of geomagnetic moment, however, the relation is weakened, and for $M > M_0$, the temperature does not respond to further strengthening of geomagnetic field intensity. Furthermore, with the nonstationarity of geomagnetic climate relation, Petrova et al. (1992) have explained the fact that a well-pronounced relationship between geomagnetic moment and cosmogenic radionuclide ^{18}O (a proxy for the climatic variations) had been found only during the last 90 ky, as well as between 342 and 198 ky.

In conclusion, due to the serious obstacles in dating of palaeomagnetic and palaeoclimate data, the most reliable data for studying the relationship between geomagnetic field and climate can be obtained for the last several thousand years. For this time interval, there is a sufficient set of archaeomagnetic and palaeomagnetic data, with exact age references that can be compared with palaeoclimate records, both globally and regionally.

In addition to the surface temperature, the imprint of palaeomagnetic dipole variability is found in galactic cosmic ray fluxes, reaching the surface (Usoskin et al., 2004; Kovaltsov and Usoskin, 2007). These results suggest that geomagnetic field imprint on the intensity and spatial distribution of energetic particles, precipitating in Earth's atmosphere, could be a mediator of geomagnetic field influence on climate.

2.2 Geomagnetic jerks and climate changes

The concept of *geomagnetic jerk* was introduced in 1969–70 by Courtillot et al. (1978) regarding the sharp decrease of the secular variation of geomagnetic-Y component at European observatories. For the period of instrumental measurements (i.e. since the late 19th century) other geomagnetic jerks have been observed around 1925, 1978, 1991, and 1999, and possibly also around 1901 and 1913 (e.g. Alexandrescu et al., 1996; Macmillan, 1996; Mandea et al., 2000). The general meaning of geomagnetic jerk is an abrupt and sharp change in the second-time derivative (with respect to time) of the geomagnetic field. In other words this is a relatively short-term impulse in geomagnetic secular variation, possibly related to processes near the boundary of Earth's outer liquid core.

Archaeomagnetic studies in Western Europe and the Middle East have led to the discovery of a new feature of geomagnetic secular variations, known as *archaeomagnetic jerks*. These events are characterized by sharp and rapid directional variations, synchronous with intensity maxima (Gallet et al., 2009).

Similarly to the contemporary geomagnetic jerks, the archaeomagnetic ones manifest themselves with an abrupt strengthening of geomagnetic field (up to 15%–30%). Their duration, however, being ~100 years, is intermediate between the duration of contemporary jerks (~1 year), and that of geomagnetic excursions (~10^3 years) (Genevey and Gallet, 2002; Gallet et al., 2003). The appearance of archaeomagnetic jerks has been compared with reliable palaeoclimatic indicators such as glacier oscillations in the Swiss Alps (Gallet et al., 2005). The authors have claimed that periods of climatic cooling in Western Europe (registered by the glacier onset) coincide fairly well with the occurrence of archaeomagnetic jerks (especially in the phase of a sharp increase in field intensity); see Fig. 2.1.

Based on archaeomagnetic data from Mesopotamia, Gallet et al. (2006) were able to expand the time range up to 5000 years, identifying archaeomagnetic jerks at ~1600, 2100, and 2700 years BC. Comparing the archaeomagnetic date with the petrologic tracers of drift ice in deep-sea sediment cores from the North Atlantic (broadly interpreted as proxies of palaeotemperature variations), Gallet et al. (2006) have found that the four cooling periods, detected in the last three millennia BC, correspond fairly well to the appearance of archaeomagnetic jerks (Fig. 2.2).

The reliably identified archaeomagnetic jerks appear at about AD 1600, 1400, 800, and 200 years and 800 years BC, while those in AD 1800 and 600 years and 350 years BC should be considered less reliable (Genevey and Gallet, 2002; Gallet et al., 2003, 2005).

Analysis of Holocene lake sediments in Sweden shows the existence of abrupt changes in the direction of geomagnetic vector between 4000 and 2000 years BC (Snowball and Sandgren, 2004). In addition, Stoner et al. (2005) reported that palaeomagnetic secular variation during the late Holocene show evidence for at least three rapid (lasting less than 100 years) changes in the position of the North magnetic pole in the Canadian High Arctic.

The origin of archaeomagnetic jerks is still not very well understood, but the broadly accepted opinion is that some of them are related to the processes in the liquid core and at the core–mantle boundary, and could be described by the low-degree spherical harmonics in Gaussian expansion.

2.3 Drift of geomagnetic poles and its influence on climate variability

Bakhmutov (2006) has pointed out that over the past 13,000 years, the alternation of cold and warm periods in palaeoclimate records of Northern Europe correspond to a movement of the *virtual geomagnetic pole* (VGP) towards or away from Northern Europe. The drift of the VGP has been reconstructed from palaeomagnetic sediments of different genesis (lake, lake-glacial, glacial-marine, and marine), on the territories of Karelia and the Kola Peninsula,

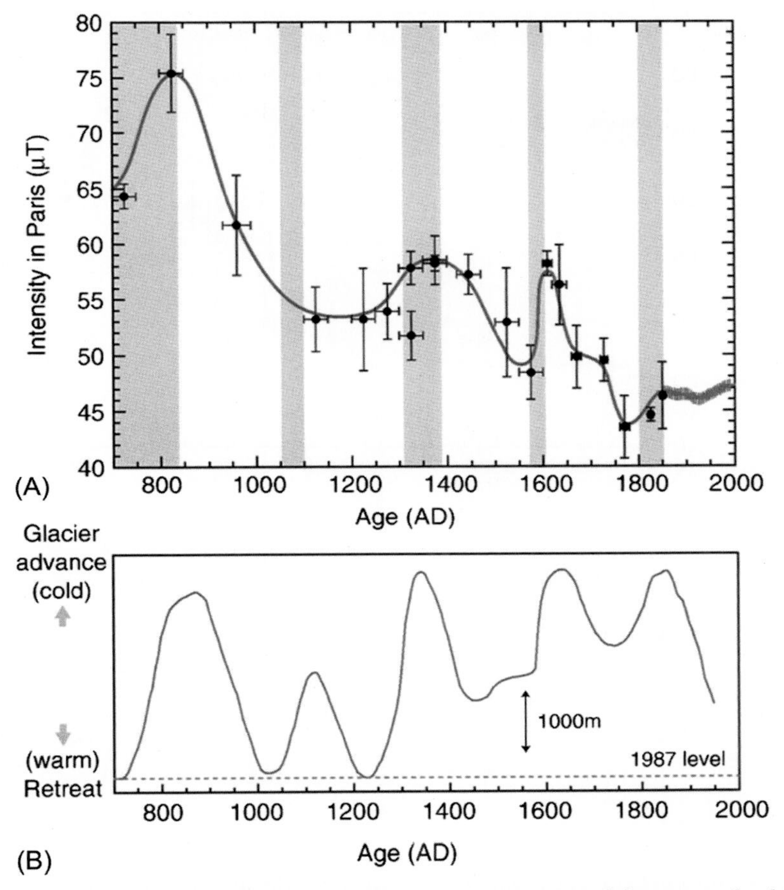

FIG. 2.1 Variations of geomagnetic field intensity in Paris obtained from French faience potsherds data for the last 1300 years (A). *Vertical and horizontal error bars* correspond to standard deviations of intensity means and age brackets of the dated sites, respectively. The geomagnetic field intensity variations in Paris, since 1850 onwards, are deduced from geomagnetic field models (Jackson et al., 2000), indicated in the figure by *small crosses*. Climatic variations during the past millennium are deduced from retreats and advances of the Alpine glaciers (B). Cooling periods are indicated in (A) by *shaded bands. Reproduced with permission from Gallet, Y., Genevey, A., Fluteau, F., 2005. Does Earth's magnetic field secular variation control centennial climate change? Earth Planet. Sci. Lett. 236, 339–347 (their Fig. 2b, c, page 344).*

as well from archaeomagnetic studies in Central and Eastern Europe, for the last several thousand years.

The results have been related to climatic changes according to the Blytt–Sernander scheme for Northern Europe, given in the left part of Fig. 2.3. Following this scheme of deglaciation of the Scandinavian Ice Sheet, six periods (chronozones) of the late Glacial period (starting ~13,000 years ago) have been identified: three *interstadials* (warm periods)—i.e. Raunis (RN), Bølling (BØ), and Allerød (AL); and three *stadials* (cold periods)—i.e. Oldest Dryas (DR-1), Older (DR-2), and Younger (DR-3) Dryas. The end of the Younger Dryas (about 11,700 years ago) marked the beginning of the Holocene geological epoch.

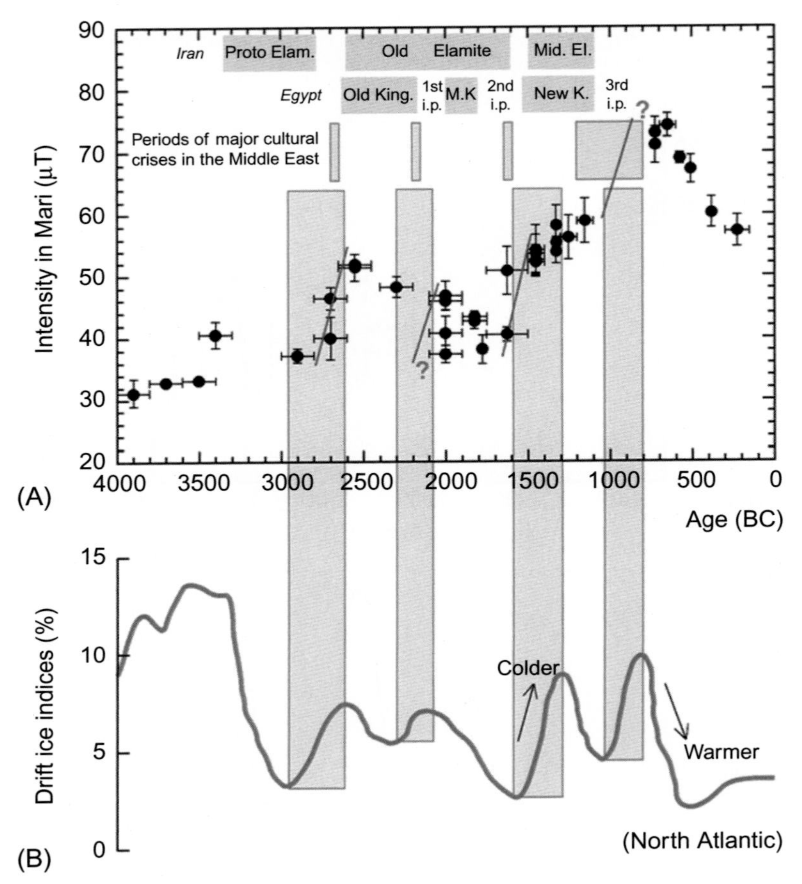

FIG. 2.2 (A) Variations of Earth's magnetic field intensity from Mesopotamia during the past four millennia BC, compared with the main cultural changes in the eastern Mediterranean and Mesopotamia. (B) Climate change in the North Atlantic is determined through the entire Holocene from petrologic tracers of drift ice in deep-sea sediment cores (after Bond et al., 2001); the four cooling periods observed over the last three millennia BC are marked as *shaded bands*. Note that periods of cooling either coincide or are relatively close to appearance of archaeomagnetic jerks. *From Gallet, Y., Genevey, A.M., Le Goff, Fluteau, F., Eshraghi, A., 2006. Possible impact of the earth's magnetic field on the history of ancient civilizations. Earth Planet. Sci. Lett. 246, 17–26 (their Fig. 2.s4, page 23). Rights holder: Elsevier.*

The Holocene epoch is characterized by five climatic stages: Preboreal (PB), Boreal (BO), Atlantic (AT), Subboreal (SB), and Subatlantic (SA). Each of these is divided into subperiods with cooling and warming phases. The discrepancies regarding the boundaries of these periods among different authors do not affect the general manifestation of palaeoclimatic variations.

Dynamics of the environmental conditions for the centre of the Russian Plain are shown in the middle of Fig. 2.3 (Velichko, 1999). The right-hand side of the figure shows the drift trajectories of the virtual geomagnetic pole (VGP) for individual time intervals.

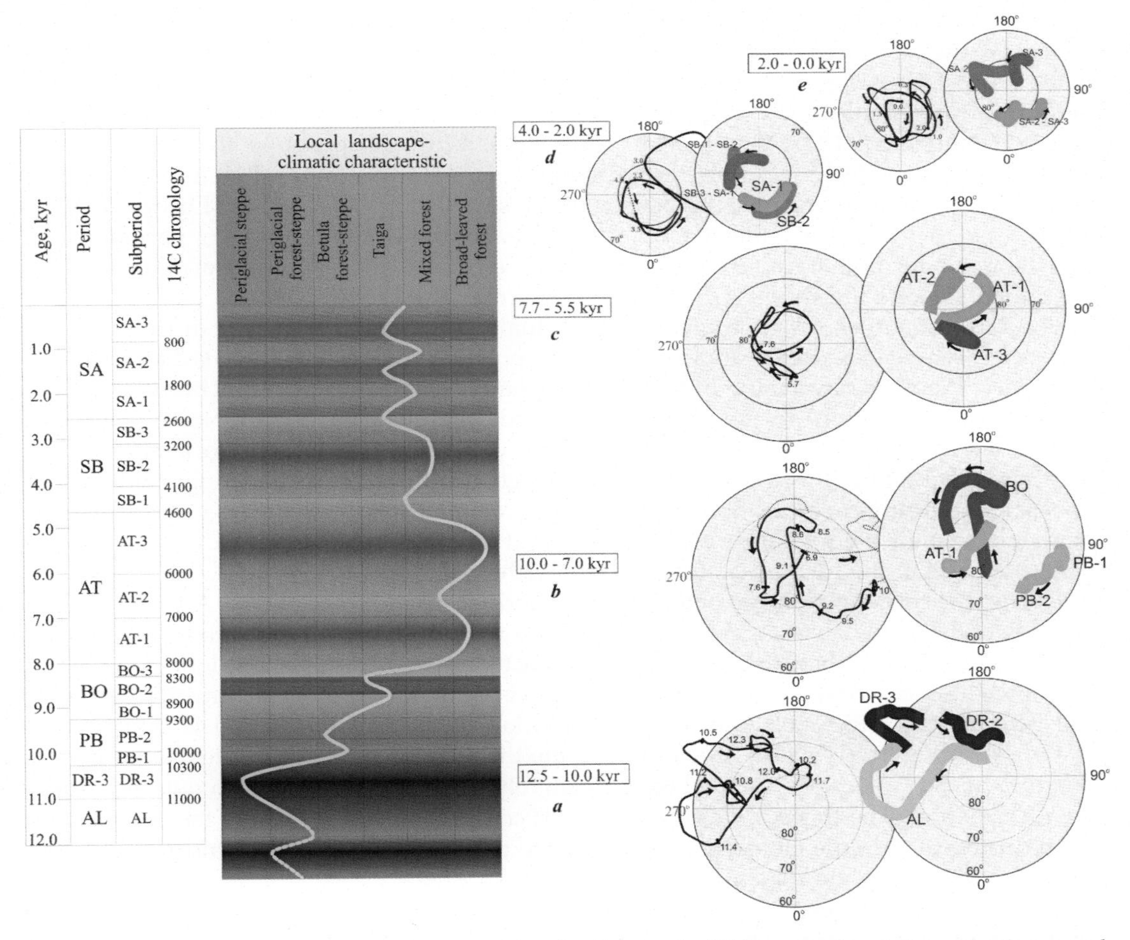

FIG. 2.3 (*left*) Age dating based on Blytt–Sernander classification; (*middle*) temporal evolution of landscape and climatic conditions in the centre of the Russian Plain, based on the materials of N.A. Khotinsky and V.A. Klimanov; (*right*) trajectory of the virtual geomagnetic pole is calculated by palaeomagnetic data, corresponding to the Late Weichselian (a), early and middle Holocene (b, c). The drift paths during the Late Holocene (d, e) are derived by archaeomagnetic data. The *numbers* on the curves indicate the radiocarbon age (in thousands of years); the *arrows* show the direction of drift motion. The *dotted line* in (b) panel illustrates the geomagnetic drift path at the Late Weichselian–Holocene boundary. The stadials (cold) and interstadials (warm) stages are coloured in *blue* and *red*, respectively. (middle) *Courtesy of Velichko, A. (Ed.), 1999. The Climate and Landscape Changes During the Last 65 Myr. GEOS, Moscow, p. 260 (in Russian);* (right) *Modified from Rusakov, O.M., Zagniy, G.F., 1973. Archaeomagnetic secular variation study in the Ukraine and Moldavia. Archaeometry 15, 153–157.*

Comparison of the drift trajectories and changes in climatic conditions during the late Glacial and Holocene epochs allows detection of the following relationship (Bakhmutov, 2006). When the trajectory of the VGP is close to Northern Europe, cold *stadials* (blue tones in Fig. 2.3) are replaced by warm *interstadials* (red tones in Fig. 2.3). In contrast, when the pole moves far away from Northern Europe, a consecutive cooling occurs, with some delay in time.

Simplified fragments of the virtual geomagnetic pole trajectory are shown on the right side of Fig. 2.3 (the cold *stadials* are marked in blue, and the warm *interstadials* in red). For example, in the stages DR-2 (12.3–11.8 ky) and DR-3 (10.8–10.3 ky), the trajectory of the virtual geomagnetic pole was significantly far away from Europe, compared to its position during the Allerød interstadial (AL, 11.8–10.8 ky) (Fig. 2.3a). With the onset of the Preboreal (PB, 10.3 ky) (i.e. the period preceding the subsequent general warming of climate), the trajectory of the virtual geomagnetic pole has been at the closest distance to Northern Europe. The pole has crossed the Kola Peninsula at about 9.7–9.5 ky. For the whole subsequent period, the virtual geomagnetic pole trajectory has never come so close to Europe. The warming trend has continued, despite the short-term alternating warmer and colder periods.

Fig. 2.3b illustrates the trajectory of the virtual geomagnetic pole drift during the early Holocene, where the dotted line indicates its position at the late Weichselian–Holocene boundary. Polovtsian warming in the Russian plain is associated with this border. The pole has drifted furthermore to the northwest towards the Svalbard Island–North geographic pole–Chukotka–Alaska. The Pereslavl cold period corresponds to this temporal interval, during the second half of PB-2. Up to about 7.6 ky ago, the climate was generally cold with the lowest temperatures near the Boreal–Atlantic boundary. This extreme (8.3 ky ago) is fixed on many global palaeotemperature curves and at that time, the trajectory of the virtual geomagnetic pole was at the furthest distance from Northern Europe.

During the subsequent onset of the warmest periods in the Holocene (i.e. Atlantic, AT), the position of the virtual geomagnetic pole correlates with the Northern Greenland region. During the AT period, the pole moved close to Europe twice (with the later time loop passing near the north-eastern coast of Greenland), and once went far away, reaching the longitude 180–240 degrees (Fig. 2.3c). The climatic condition of the Atlantic period is characterized by two warming (in AT-1 and AT-3), and during the second one the absolute maximum temperature of the Holocene was reached. A relatively cooler climate has been detected during the AT-2 period. The trajectory of the VGP drift corresponds fairly well to the climate alteration during the middle Holocene, being close to the Europe in the AT-1 and AT-3 periods, and away from it in the AT-2 period.

In the late Holocene, the trajectory of the virtual geomagnetic pole, obtained from archaeomagnetic data from Ukraine, is shown in Fig. 2.3d, e. The data confirm the alternation of stadials (cold) and interstadials (warm) periods, depending on the relative position of the virtual geomagnetic pole to Northern Europe.

The exception is the time interval of about 4.5 ky (at the beginning of the Subboreal SB-1), when the virtual pole was close to Europe, but climatic conditions in the centre of the Russian plain remained cold. This exception from the general pattern may be due to insufficient information about this period, or to regional changes in the landscape and climatic conditions due to any other reasons. For the late Holocene, three cooling and two warming phases are clearly correlated with the movement of geomagnetic virtual pole away from or close to Northern Europe.

Consequently, the analysis of the trajectory of virtual geomagnetic pole movement, juxtaposed to the climatic changes during the late Glacial and the Holocene epochs, reveals a relationship between alternations of cold and warm periods in Northern Europe, and the location of the geomagnetic pole. This result suggests that changes in the position of the geomagnetic pole somehow influences the temperature fluctuations in the Northern Hemisphere.

Putting the question about the mechanism for such a relation, it is worth remembering the earlier works of Bucha and Rein (1976), Bucha (1977), and Bucha (1988), which suggested that the drift of geomagnetic poles could affect the displacements of a large low-pressure region in the Northern Hemisphere, and correspondingly the atmospheric dynamics. The latter, in turn, is associated with an increase of cyclonic activity and changes in climate. According to the authors' hypothesis, processes in the auroral oval, initiated by increased geomagnetic activity, have a significant effect on the tropospheric circulation. Consequently, any changes in the location of geomagnetic pole (on timescales $\sim10^2$–10^3 years) should affect the long-term changes in the climatic conditions (Bucha and Bucha, 1998; Bakhmutov, 2006).

2.4 Centennial synchronization between palaeomagnetic and palaeoclimatic changes in North America and Europe

The long-scale variability ($\sim10^3$ years) of the positions of geomagnetic poles (discussed in Section 2.3) is related mainly to the variation in the dipolar component of geomagnetic field. According to recent knowledge, this could be attributed to the nonaxisymmetric nature of the field near the poles—an essential characteristic of geodynamo. More specifically, the projection of the radial (i.e. vertical) component of geomagnetic field on the core–mantle boundary reveals that the vertical field is strongest not at the poles (as should be expected if the geomagnetic field is an axial dipole), but outside a circle with radius $\sim$20 degrees. The low field intensity near the poles, related possibly to the weaker convection inside the tangent cylinder (see Gubbins and Herrero-Bervera, 2007; *Role of the Inner core, in Geodynamo*), could well explain the drift of the virtual palaeomagnetic poles around the geographic ones (Gubbins and Herrero-Bervera, 2007; *Inner core tangent cylinder*).

This section shows another evidence for existing temporal-spatial synchronization between Earth's magnetic field and climate, derived from a comparison of their centennial variation during four consecutive centuries (i.e. 15th–18th). According to our understanding, the secular variations on a multidecadal timescale could be attributed to the inner core oscillation with a period of $\sim$1 year (Gubbins, 1981; Glatzmaier and Roberts, 1996), or to torsional oscillations, imposed on the core by the upper level mantle (Braginsky, 1970; Zatman and Bloxham, 1997).

Fig. 2.4 shows a comparison of the nondipolar component of geomagnetic field, derived by the model CALS7K (Korte and Constable, 2005), and palaeoclimate data from the standardized tree-rings width chronology, during the preindustrial period 1500–1800.

It is easily seen that the persistently stronger geomagnetic field over North America, lasting for more than four centuries, is accompanied by reduced tree growth, which is indicative of cooler and drier climate conditions in moderate climatic zones (Fritts, 1965). Although this rule gives a very rough estimation of the climate, because the width of tree rings could be influenced by a wider variety of environmental factors, such as temperature, soil moisture, sunshine, etc. (García-Suárez et al., 2009), it gives a rough idea about the prevailing climate during the examined period.

Unlike North America, the temporal evolution of geomagnetic field in the European region is more dynamical (see the right-hand column in Fig. 2.4). For example, in 1500, a strong

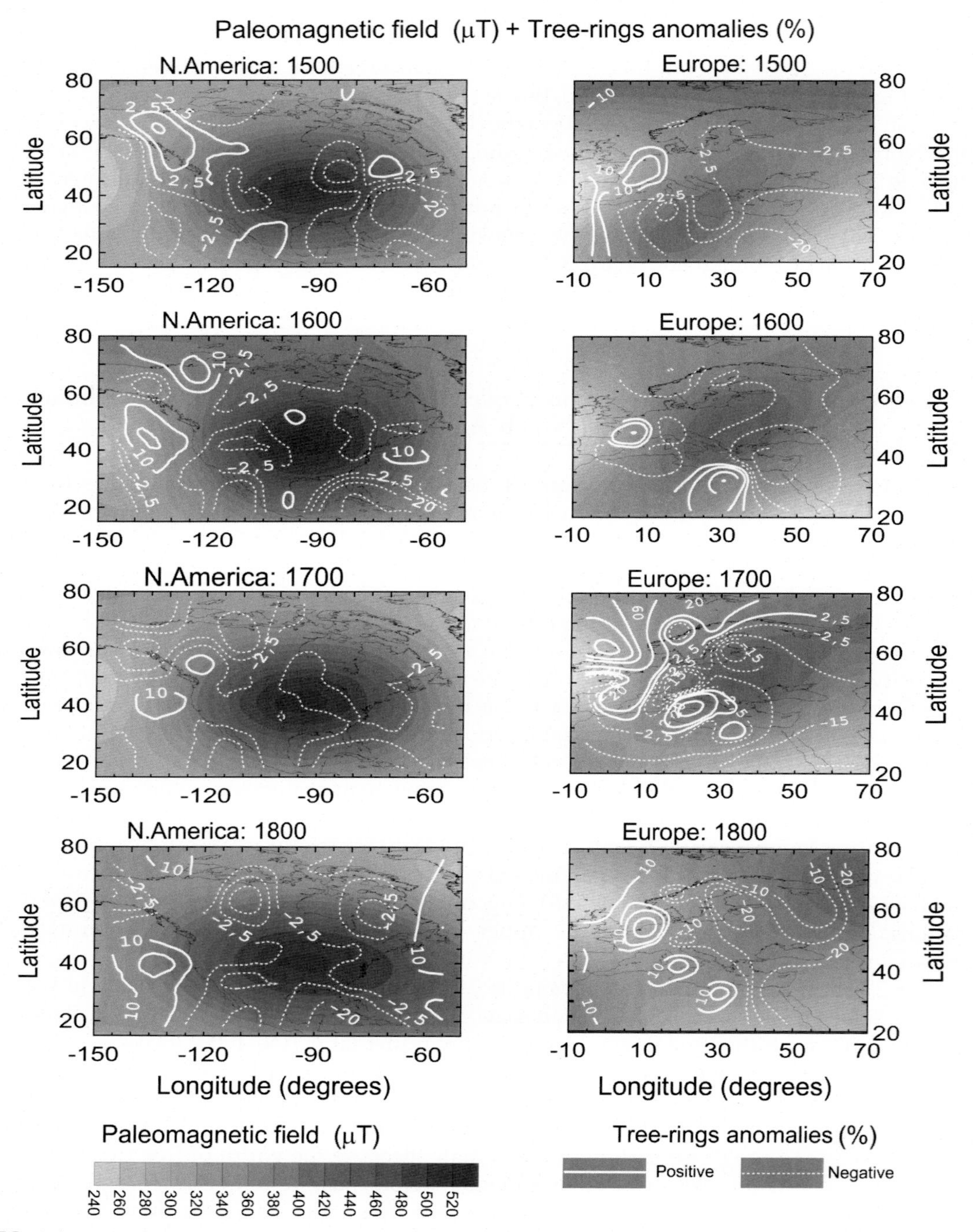

FIG. 2.4 Centennial snapshots of nondipolar component of geomagnetic field in (μT), at the core–mantle boundary, over North America and Europe, during the period 1500–1800 *(coloured shading)*. Overdrawn are contours of percentage deviation of trees growth from the standard one; *positive* values indicate faster trees growth (interpreted as a mild weather); *negative* values correspond to a slower trees growth (interpreted as a severe climatic conditions). *Data sources: For palaeoclimate conditions https://www.ncdc.noaa.gov/data-access/paleoclimatology-data/datasets/tree-ring; for palaeomagnetic field, CALS7K model (Korte, M., Constable, C.G., 2005. Continuous geomagnetic field models for the past 7 millennia: 2. CALS7K. Geochem. Geophys. Geosyst. 6. https://doi.org/10.1029/2004GC000801 (web archive link)).*

geomagnetic field was placed in the region of Central Mediterranean-North Africa, which gradually weakened and moved north-eastwards during the examined period. The climate responded in accordance with a subsequent replacement of cold and dry conditions by warmer and wetter ones. These synchronized changes of geomagnetic field and climate on centennial timescales give another indication for a possible relation between them.

In resume, palaeomagnetic and palaeoclimate records provide a lot of evidence suggesting that variations of climate parameters and geomagnetic field intensity are synchronized in space and time. The temporal covariations are observed at different timescales from decadal to millennial. However, the existing controversy in the manifestation of this relation requires a reasonably sound physical mechanism, which could explain the nonstationary character of such a relation, as well as the heterogeneity of its spatial distribution. Unfortunately, the hypotheses currently proposed are unable to meet these requirements.

References

Alexandrescu, M., Gibert, D., Hulot, G., Mouël, J.L.L., Saracco, G., 1996. Worldwide wavelet analysis of geomagnetic jerks. J. Geophys. Res. Solid Earth 101, 21975–21994. https://doi.org/10.1029/96JB01648.

Bakhmutov, V., 2006. The connection between geomagnetic secular variation and long-range development of climate changes for the last 13,000 years: the data from NNE Europe. Quat. Int. The Ukraine Quaternary Explored: the Middle and Upper Pleistocene of the Middle Dnieper Area and its Importance for East-West European Correlation. 149, 4–11. https://doi.org/10.1016/j.quaint.2005.11.013.

Bakhmutov, V.G., Martazinova, V.F., Kilifarska, N.A., Melnyk, G.V., Ivanova, E.K., 2014. Geomagnetic field and climate variability. 1. Spatial-temporal distribution of geomagnetic field and climatic parameters during XX century. Geofiz. Zh. 36, 81–104. https://doi.org/10.24028/gzh.0203-3100.v36i1.2014.116153.

Beer, J., Blinov, A., Bonani, G., Finkel, R.C., Hofmann, H.J., Lehmann, B., Oeschger, H., Sigg, A., Schwander, J., Staffelbach, T., Stauffer, B., Suter, M., Wötfli, W., 1990. Use of 10 Be in polar ice to trace the 11-year cycle of solar activity. Nature 347, 164–166. https://doi.org/10.1038/347164a0.

Bond, G., Kromer, B., Beer, J., Muscheler, R., Evans, M.N., Showers, W., Hoffmann, S., Lotti-Bond, R., Hajdas, I., Bonani, G., 2001. Persistent solar influence on North Atlantic climate during the Holocene. Science 294, 2130–2136. https://doi.org/10.1126/science.1065680.

Braginsky, S.I., 1970. Torsional magnetohydrodynamic vibrations inthe Earth's core and variations in day length. Geomagn. Aeron. 10, 1–10.

Bucha, V., 1977. Mechanism of solar – terrestrial relations and changes of atmospheric circulation. Stud. Geophys. Geod. 21, 351–360.

Bucha, V., 1988. Influence of solar activity on atmospheric circulation types. Ann. Geophys. 6, 513–524.

Bucha, V., Bucha, V., 1998. Geomagnetic forcing of changes in climate and in the atmospheric circulation. J. Atmos. Sol. Terr. Phys. 60, 145–169. https://doi.org/10.1016/S1364-6826(97)00119-3.

Bucha, V., Rein, F., 1976. Changes in the geomagnetic field and solar wind – causes of changes of climate and atmospheric circulation. Stud. Geophys. Geod. 20 (4), 346–365.

Bucha, V., Zikmunda, O., 1976. Variations of the geomagnetic field, the climate and weather. Stud. Geophys. Geod. 20, 149–167. https://doi.org/10.1007/BF01626048.

Butchvarova, V., Kovacheva, M., 1993. European changes in the paleotemperature and Bulgarian archaeomagnetic data. Bulg. Geophys. J. 19, 19–23.

Carcaillet, J.T., Thouveny, N., Bourlès, D.L., 2003. Geomagnetic moment instability between 0.6 and 1.3 Ma from cosmonuclide evidence. Geophys. Res. Lett. 30. https://doi.org/10.1029/2003GL017550.

Carcaillet, J., Bourlès, D.L., Thouveny, N., Arnold, M., 2004. A high resolution authigenic 10Be/9Be record of geomagnetic moment variations over the last 300 ka from sedimentary cores of the Portuguese margin. Earth Planet. Sci. Lett. 219, 397–412. https://doi.org/10.1016/S0012-821X(03)00702-7.

Carslaw, K.S., Harrison, R.G., Kirkby, J., 2002. Cosmic rays, clouds, and climate. Science 298, 1732–1737. https://doi.org/10.1126/science.1076964.

Channell, J.E.T., 1999. Geomagnetic paleointensity and directional secular variation at Ocean Drilling Program (ODP) Site 984 (Bjorn Drift) since 500 ka: comparisons with ODP Site 983 (Gardar Drift). J. Geophys. Res. Solid Earth 104, 22937–22951. https://doi.org/10.1029/1999JB900223.

Channell, J.E.T., Hodell, D.A., McManus, J., Lehman, B., 1998. Orbital modulation of the Earth's magnetic field intensity. Nature 394, 464. https://doi.org/10.1038/28833.

Cliver, E.W., Boriakoff, V., Feynman, J., 1998. Solar variability and climate change: geomagnetic aa index and global surface temperature. Geophys. Res. Lett. 25, 1035–1038. https://doi.org/10.1029/98GL00499.

Courtillot, V., Ducruix, J., Le Mouël, J.L.L., 1978. Série D. Sur une accélération récente de la variation séculaire du champ magnétique terrestre. C. R. Acad. Sci. 287, 1095–1098.

Courtillot, V., Mouel, J.L.L., Ducruix, J., Cazenave, A., 1982. Geomagnetic secular variation as a precursor of climatic change. Nature 297, 386–387. https://doi.org/10.1038/297386a0.

Courtillot, V., Gallet, Y., Mouël, J.L.L., Fluteau, F., Genevey, A., 2007. Are there connections between the Earth's magnetic field and climate? Earth Planet. Sci. Lett. 253, 328–339. https://doi.org/10.1016/j.epsl.2006.10.032.

Dickinson, R.E., 1975. Solar variability and the lower atmosphere. Bull. Am. Meteorol. Soc. 56, 1240–1248. https://doi.org/10.1175/1520-0477(1975)056<1240:SVATLA>2.0.CO;2.

Doake, C.S.M., 1978. Climatic change and geomagnetic field reversals: a statistical correlation. Earth Planet. Sci. Lett. 38, 313–318. https://doi.org/10.1016/0012-821X(78)90105-X.

El-Borie, M.A., Al-Thoyaib, S.S., 2006. Can we use geomagnetic activity index to predict partially the variability in global mean temperature? Int. J. Phys. Sci. 1 (2), 67–74.

Elsasser, W., Ney, E.P., Winckler, J.R., 1956. Cosmic-ray intensity and geomagnetism. Nature 178, 1226. https://doi.org/10.1038/1781226a0.

Frank, M., 2000. Comparison of cosmogenic radionuclide production and geomagnetic field intensity over the last 200 000 years. Philos. Trans. R. Soc. Lond. Ser. A Math. Phys. Eng. Sci 358, 1089–1107

Fritts, H.C., 1965. Tree-ring evidence for climatic changes in western North America. Mon. Weather Rev. 93, 421–443. https://doi.org/10.1175/1520-0493(1965)093<0421:TREFCC>2.3.CO;2.

Fuller, M., 2006. Geomagnetic field intensity, excursions, reversals and the 41,000-yr obliquity signal. Earth Planet. Sci. Lett. 245, 605–615. https://doi.org/10.1016/j.epsl.2006.03.022.

Gallet, Y., Genevey, A., Courtillot, V., 2003. On the possible occurrence of 'archaeomagnetic jerks' in the geomagnetic field over the past three millennia. Earth Planet. Sci. Lett. 214, 237–242. https://doi.org/10.1016/S0012-821X(03)00362-5.

Gallet, Y., Genevey, A., Fluteau, F., 2005. Does Earth's magnetic field secular variation control centennial climate change? Earth Planet. Sci. Lett. 236, 339–347. https://doi.org/10.1016/j.epsl.2005.04.045.

Gallet, Y., Genevey, A., Le Goff, M., Fluteau, F., Eshraghi, A., 2006. Possible impact of the Earth's magnetic field on the history of ancient civilizations. Earth Planet. Sci. Lett. 246, 17–26.

Gallet, Y., Genevey, A., Le Goff, M., Warmé, N., Gran-Aymerich, J., Lefèvre, A., 2009. On the use of archeology in geomagnetism, and vice-versa: recent developments in archeomagnetism. C. R. Phys. 10, 630–648. https://doi.org/10.1016/j.crhy.2009.08.005.

García-Suárez, A.M., Butler, C.J., Baillie, M.G.L., 2009. Climate signal in tree-ring chronologies in a temperate climate: a multi-species approach. Dendrochronologia 27, 183–198. https://doi.org/10.1016/j.dendro.2009.05.003.

Genevey, A., Gallet, Y., 2002. Intensity of the geomagnetic field in Western Europe over the past 2000 years: new data from ancient French pottery. J. Geophys. Res. Solid Earth 107, EPM 1-1–EPM 1-18. https://doi.org/10.1029/2001JB000701.

Glatzmaier, G.A., Roberts, P.H., 1996. On the magnetic sounding of planetary interiors. Phys. Earth Planet. In. 98, 207–220. https://doi.org/10.1016/S0031-9201(96)03188-3.

Gray, L.J., Beer, J., Geller, M., Haigh, J.D., Lockwood, M., Matthes, K., Cubasch, U., Fleitmann, D., Harrison, G., Hood, L., Luterbacher, J., Meehl, G.A., Shindell, D., van Geel, B., White, W., 2010. Solar influences on climate. Rev. Geophys. 48. https://doi.org/10.1029/2009RG000282.

Gubbins, D., 1981. Rotation of the inner core. J. Geophys. Res. Solid Earth 86, 11695–11699. https://doi.org/10.1029/JB086iB12p11695.

Gubbins, D., Herrero-Bervera, E. (Eds.), 2007. Encyclopedia of Geomagnetism and Paleomagnetism. Encyclopedia of Earth Sciences Series, Springer Netherlands.

Guyodo, Y., Valet, J.-P., 1999. Global changes in intensity of the Earth's magnetic field during the past 800 kyr. Nature 399, 249–252. https://doi.org/10.1038/20420.

Guyodo, Y., Gaillot, P., Channell, J.E.T., 2000. Wavelet analysis of relative geomagnetic paleointensity at ODP Site 983. Earth Planet. Sci. Lett. 184, 109–123. https://doi.org/10.1016/S0012-821X(00)00313-7.

Harrison, R.G., 2004. The global atmospheric electrical circuit and climate. Surv. Geophys. 25, 441–484. https://doi.org/10.1007/s10712-004-5439-8.

Hauchecorne, A., Bertaux, J.-L., Lallement, R., 2007. Impact of solar activity on stratospheric ozone and No2 observed by GOMOS/ENVISAT. In: Calisesi, Y., Bonnet, R.-M., Gray, L., Langen, J., Lockwood, M. (Eds.), Solar Variability and Planetary Climates, Space Sciences Series of ISSI. Springer New York, New York, NY, pp. 393–402. https://doi.org/10.1007/978-0-387-48341-2_31.

Heikkilä, U., Beer, J., Abreu, J.A., Steinhilber, F., 2013. On the atmospheric transport and deposition of the cosmogenic radionuclides (10Be): a review. Space Sci. Rev. 176, 321–332. https://doi.org/10.1007/s11214-011-9838-0.

Jackman, C.H., Marsh, D.R., Kinnison, D.E., Mertens, C.J., Fleming, E.L., 2016. Atmospheric changes caused by galactic cosmic rays over the period 1960–2010. Atmos. Chem. Phys. 16, 5853–5866. https://doi.org/10.5194/acp-16-5853-2016.

Jackson, A., Jonkers, A., Walker, M., 2000. Four centuries of geomagnetic secular variation from historical record. Philos. Trans. R. Soc. Lond. Ser. A 358, 957–990.

Kent, D.V., 1982. Apparent correlation of palaeomagnetic intensity and climatic records in deep-sea sediments. Nature 299, 538. https://doi.org/10.1038/299538a0.

Kent, D.V., Carlut, J., 2001. A negative test of orbital control of geomagnetic reversals and excursions. Geophys. Res. Lett. 28, 3561–3564. https://doi.org/10.1029/2001GL013118.

Kent, D.V., Opdyke, N.D., 1977. Palaeomagnetic field intensity variation recorded in a Brunhes epoch deep-sea sediment core. Nature 266, 156. https://doi.org/10.1038/266156a0.

Kitaba, I., Hyodo, M., Katoh, S., Dettman, D.L., Sato, H., 2013. Midlatitude cooling caused by geomagnetic field minimum during polarity reversal. Proc. Natl. Acad. Sci. U. S. A. 110, 1215–1220. https://doi.org/10.1073/pnas.1213389110.

Knudsen, M.F., Riisager, P., 2009. Is there a link between Earth's magnetic field and low-latitude precipitation? Geology 37, 71–74. https://doi.org/10.1130/G25238A.1.

Korte, M., Constable, C.G., 2005. Continuous geomagnetic field models for the past 7 millennia: 2. CALS7K. Geochem. Geophys. Geosyst. 6https://doi.org/10.1029/2004GC000801.

Kovaltsov, G.A., Usoskin, I.G., 2007. Regional cosmic ray induced ionization and geomagnetic field changes. In: Advances in Geosciences. Presented at the Solar, Heliospheric and External Geophysical Effects on the Earth's Environment: Scientific and Educational Initiatives—EGU General Assembly 2006, Vienna, Austria, 2–7 April 2006. Copernicus GmbH, pp. 31–35. https://doi.org/10.5194/adgeo-13-31-2007.

Kovaltsov, G.A., Usoskin, I.G., 2010. A new 3D numerical model of cosmogenic nuclide 10Be production in the atmosphere. Earth Planet. Sci. Lett. 291, 182–188. https://doi.org/10.1016/j.epsl.2010.01.011.

Krivolutsky, A.A., Klyuchnikova, A.V., Zakharov, G.R., Vyushkova, T.Y., Kuminov, A.A., 2006. Dynamical response of the middle atmosphere to solar proton event of July 2000: three-dimensional model simulations. Adv. Space Res. 37, 1602–1613. https://doi.org/10.1016/j.asr.2005.05.115 Particle Acceleration; Space Plasma Physics; Solar Radiation and the Earth's Atmosphere and Climate.

Le Mouël, J.-L., Kossobokov, V., Courtillot, V., 2005. On long-term variations of simple geomagnetic indices and slow changes in magnetospheric currents: the emergence of anthropogenic global warming after 1990? Earth Planet. Sci. Lett. 232, 273–286. https://doi.org/10.1016/j.epsl.2004.07.046.

Lucas, G.M., Baumgaertner, A.J.G., Thayer, J.P., 2015. A global electric circuit model within a community climate model. J. Geophys. Res. Atmos. 120, 12,054–12,066. https://doi.org/10.1002/2015JD023562.

Macmillan, S., 1996. A geomagnetic jerk for the early 1990's. Earth Planet. Sci. Lett. 137, 189–192. https://doi.org/10.1016/0012-821X(95)00214-W.

Malkus, W.V.R., 1968. Precession of the Earth as the cause of geomagnetism. Science 160, 259–264. https://doi.org/10.1126/science.160.3825.259.

Mandea, M., Bellanger, E., Le Mouël, J.-L., 2000. A geomagnetic jerk for the end of the 20th century? Earth Planet. Sci. Lett. 183, 369–373. https://doi.org/10.1016/S0012-821X(00)00284-3.

Masarik, J., Beer, J., 1999. Simulation of particle fluxes and cosmogenic nuclide production in the Earth's atmosphere. J. Geophys. Res. Atmos. 104, 12099–12111. https://doi.org/10.1029/1998JD200091.

Masarik, J., Beer, J., 2009. An updated simulation of particle fluxes and cosmogenic nuclide production in the Earth's atmosphere. J. Geophys. Res. Atmos. 114. https://doi.org/10.1029/2008JD010557.

Ménabréaz, L., Thouveny, N., Bourlès, D.L., Vidal, L., 2014. The geomagnetic dipole moment variation between 250 and 800 ka BP reconstructed from the authigenic 10Be/9Be signature in West Equatorial Pacific sediments. Earth Planet. Sci. Lett. 385, 190–205. https://doi.org/10.1016/j.epsl.2013.10.037.

Mufti, S., Shah, G.N., 2011. Solar-geomagnetic activity influence on Earth's climate. J. Atmos. Sol. Terr. Phys. 73, 1607–1615. https://doi.org/10.1016/j.jastp.2010.12.012.

Nurgaliev, D.K., 1991. Solar activity, geomagnetic variations, and climate changes. Geomagn. Aeron. 31 (1), 14–18.

Petrova, G.N., Raspopov, O.M., 1998. Relation between the changes in the geomagnetic moment and paleoclimate for the past 12 ky. Geomagn. Aeron. 38 (5), 652–658.

Petrova, G.N., Nechaeva, T.B., Pospelova, G.A., 1992. Characteristic Changes in the Geomagnetic Field in the Past. Nauka, Moscow (in Russian).

Pudovkin, M.I., Babushkina, S.V., 1991. The influence of electromagnetic and corpuscular radiation of solar flares on the intensity of zonal atmosphere circulation. Geomagn. Aeron. 31 (3), 493–499.

Pudovkin, M.I., Raspopov, O.M., 1992. Physical mechanism of solar activity influence on the lower atmosphere and meteoparameters. Geomagn. Aeron. 32 (5), 1–22.

Pudovkin, M.I., Veretenenko, S.V., 1992. Variations of the meridional profile of atmospheric pressure during a geomagnetic disturbance. Geomagn. Aeron. 32, 118–122.

Raisbeck, G.M., Yiou, F., Cattani, O., Jouzel, J., 2006. 10 Be evidence for the Matuyama–Brunhes geomagnetic reversal in the EPICA Dome C ice core. Nature 444, 82–84. https://doi.org/10.1038/nature05266.

Raspopov, O.M., Dergachev, V.S., Shumilov, O.I., Crir, K.M., Petrova, G.N., 2000. The effect of cosmic ray flux variations caused by changes in the geomagnetic dipole moment on climate variability. Geomagn. Aeron. 40 (1), 97–108.

Roberts, A.P., Winklhofer, M., Liang, W.-T., Horng, C.-S., 2003. Testing the hypothesis of orbital (eccentricity) influence on Earth's magnetic field. Earth Planet. Sci. Lett. 216, 187–192. https://doi.org/10.1016/S0012-821X(03)00480-1.

Roberts, A.P., Tauxe, L., Heslop, D., 2013. Magnetic paleointensity stratigraphy and high-resolution Quaternary geochronology: successes and future challenges. Quat. Sci. Rev. 61, 1–16. https://doi.org/10.1016/j.quascirev.2012.10.036.

Rossi, C., Mertz-Kraus, R., Osete, M.-L., 2014. Paleoclimate variability during the Blake geomagnetic excursion (MIS 5d) deduced from a speleothem record. Quat. Sci. Rev. 102, 166–180. https://doi.org/10.1016/j.quascirev.2014.08.007.

Rusch, D.W., Gérard, J.-C., Solomon, S., Crutzen, P.J., Reid, G.C., 1981. The effect of particle precipitation events on the neutral and ion chemistry of the middle atmosphere—I. Odd nitrogen. Planet. Space Sci. 29, 767–774. https://doi.org/10.1016/0032-0633(81)90048-9.

Rycroft, M.J., Israelsson, S., Price, C., 2000. The global atmospheric electric circuit, solar activity and climate change. J. Atmos. Sol. Terr. Phys. 62, 1563–1576. https://doi.org/10.1016/S1364-6826(00)00112-7.

Seppälä, A., Clilverd, M.A., Rodger, C.J., Verronen, P.T., Turunen, E., 2008. The effects of hard-spectra solar proton events on the middle atmosphere. J. Geophys. Res. Space Physics. 113https://doi.org/10.1029/2008JA013517.

Seppälä, A., Lu, H., Clilverd, M.A., Rodger, C.J., 2013. Geomagnetic activity signatures in wintertime stratosphere wind, temperature, and wave response. J. Geophys. Res. Atmos. 118, 2169–2183. https://doi.org/10.1002/jgrd.50236.

Shea, M.A., Smart, D.F., 2004. Solar Variability and Climate Change. Preliminary study of cosmic rays, geomagnetic field changes and possible climate changes. Adv. Space Res. 34, 420–425. https://doi.org/10.1016/j.asr.2004.02.008.

Simon, Q., Thouveny, N., Bourlès, D.L., Valet, J.-P., Bassinot, F., Ménabréaz, L., Guillou, V., Choy, S., Beaufort, L., 2016. Authigenic 10Be/9Be ratio signatures of the cosmogenic nuclide production linked to geomagnetic dipole moment variation since the Brunhes/Matuyama boundary. J. Geophys. Res. Solid Earth 121, 7716–7741. https://doi.org/10.1002/2016JB013335.

Simon, Q., Bourlès, D.L., Bassinot, F., Nomade, S., Marino, M., Ciaranfi, N., Girone, A., Maiorano, P., Thouveny, N., Choy, S., Dewilde, F., Scao, V., Isguder, G., Blamart, D., 2017. Authigenic 10Be/9Be ratio signature of the Matuyama–Brunhes boundary in the Montalbano Jonico marine succession. Earth Planet. Sci. Lett. 460, 255–267. https://doi.org/10.1016/j.epsl.2016.11.052.

Simon, Q., Bourlès, D.L., Thouveny, N., Horng, C.-S., Valet, J.-P., Bassinot, F., Choy, S., 2018. Cosmogenic signature of geomagnetic reversals and excursions from the Réunion event to the Matuyama–Brunhes transition (0.7–2.14 Ma interval). Earth Planet. Sci. Lett. 482, 510–524. https://doi.org/10.1016/j.epsl.2017.11.021.

Snowball, I., Sandgren, P., 2004. Geomagnetic field intensity changes in Sweden between 9000 and 450 cal BP: extending the record of "archaeomagnetic jerks" by means of lake sediments and the pseudo-Thellier technique. Earth Planet. Sci. Lett. 227, 361–376. https://doi.org/10.1016/j.epsl.2004.09.017.

Stauning, P., 2011. Solar activity–climate relations: a different approach. J. Atmos. Sol. Terr. Phys. 73, 1999–2012. https://doi.org/10.1016/j.jastp.2011.06.011.

Stoner, J., Francus, P., Bradley, R., Patridge, W., Abbott, M., Retelle, M., Lamoureux, S., Channell, J., 2005. Abrupt shifts in the position of the north magnetic pole from Arctic lake sediments: relationship to archeomagnetic jerks. In: Eos Trans. AGU Fall Meeting. 86 (52), pap. GP44A-02.

Suganuma, Y., Yokoyama, Y., Yamazaki, T., Kawamura, K., Horng, C.-S., Matsuzaki, H., 2010. 10Be evidence for delayed acquisition of remanent magnetization in marine sediments: implication for a new age for the Matuyama–Brunhes boundary. Earth Planet. Sci. Lett. 296, 443–450. https://doi.org/10.1016/j.epsl.2010.05.031.

Svensmark, H., 1998. Influence of cosmic rays on Earth's climate. Phys. Rev. Lett. 81, 5027–5030. https://doi.org/10.1103/PhysRevLett.81.5027.

Svensmark, H., 2000. Cosmic rays and Earth's climate. Space Sci. Rev. 93, 175–185. https://doi.org/10.1023/A:1026592411634.

Svensmark, H., Friis-Christensen, E., 1997. Variation of cosmic ray flux and global cloud coverage—a missing link in solar-climate relationships. J. Atmos. Sol. Terr. Phys. 59, 1225–1232. https://doi.org/10.1016/S1364-6826(97)00001-1.

Tauxe, L., Steindorf, J.L., Harris, A., 2006. Depositional remanent magnetization: toward an improved theoretical and experimental foundation. Earth Planet. Sci. Lett. 244, 515–529. https://doi.org/10.1016/j.epsl.2006.02.003.

Thouveny, N., Carcaillet, J., Moreno, E., Leduc, G., Nérini, D., 2004. Geomagnetic moment variations and paleomagnetic excursions since 400 ka BP: a stacked record from sedimentary sequences of the Portuguese margin. Earth Planet. Sci. Lett. 219, 377–396.

Thouveny, N., Bourlès, D.L., Saracco, G., Carcaillet, J.T., Bassinot, F., 2008. Paleoclimatic context of geomagnetic dipole lows and excursions in the Brunhes, clue for an orbital influence on the geodynamo? Earth Planet. Sci. Lett. 275, 269–284. https://doi.org/10.1016/j.epsl.2008.08.020.

Tinslev, B.A., 1996. Solar wind modulation of the global electric circuit and apparent effects on cloud microphysics, latent heat release, and tropospheric dynamics. J. Geomag. Geoelec. 48, 165–175. https://doi.org/10.5636/jgg.48.165.

Usoskin, I.G., Gladysheva, O.G., Kovaltsov, G.A., 2004. Cosmic ray-induced ionization in the atmosphere: spatial and temporal changes. J. Atmos. Sol. Terr. Phys. 66, 1791–1796. https://doi.org/10.1016/j.jastp.2004.07.037.

Valet, J., Meynadier, L., 1993. Geomagnetic field intensity and reversals during the past four million years. Nature 366, 234. https://doi.org/10.1038/366234a0.

Valet, J.-P., Bassinot, F., Bouilloux, A., Bourlès, D., Nomade, S., Guillou, V., Lopes, F., Thouveny, N., Dewilde, F., 2014. Geomagnetic, cosmogenic and climatic changes across the last geomagnetic reversal from Equatorial Indian Ocean sediments. Earth Planet. Sci. Lett. 397, 67–79. https://doi.org/10.1016/j.epsl.2014.03.053.

Valev, D., 2006. Statistical relationships between the surface air temperature anomalies and the solar and geomagnetic activity indices. Phys. Chem. Earth A/B/C, Long Term Changes and Trends in the Atmosphere. 31, 109–112. https://doi.org/10.1016/j.pce.2005.03.005.

Vasiliev, S.S., Dergachev, V.A., Raspopov, O.M., Jungner, H., 2012. Long-term variations in the flux of cosmogenic isotope 10Be over the last 10000 years: variations in the geomagnetic field and climate. Geomagn. Aeron. 52, 121–128. https://doi.org/10.1134/S001679321201015X.

Velichko, A. (Ed.), 1999. The Climate and Landscape Changes During the Last 65 Myr. GEOS, Moscow, p. 260 (in Russian).

Veretenenko, S., Ogurtsov, M., 2012. Regional and temporal variability of solar activity and galactic cosmic ray effects on the lower atmosphere circulation. Adv. Space Res. 49, 770–783. https://doi.org/10.1016/j.asr.2011.11.020.

Vieira, L.E.A., da Silva, L.A., 2006. Geomagnetic modulation of clouds effects in the Southern Hemisphere Magnetic Anomaly through lower atmosphere cosmic ray effects. Geophys. Res. Lett. 33. https://doi.org/10.1029/2006GL026389.

Vitinskii, U.I., Oll, A.I., Sazonov, B.I., 1976. Sun and Earth's Atmosphere. Hydrometeoizdat, Saint Petersburg 351 pp. (in Russian).

Wollin, G., Ericson, D.B., Ryan, W.B.F., 1971a. Variations in magnetic intensity and climatic changes. Nature 232, 549. https://doi.org/10.1038/232549a0.

Wollin, G., Ericson, D.B., Ryan, W.B.F., Foster, J.H., 1971b. Magnetism of the Earth and climatic changes. Earth Planet. Sci. Lett. 12, 175–183. https://doi.org/10.1016/0012-821X(71)90075-6.

Worm, H.-U., 1997. A link between geomagnetic reversals and events and glaciations. Earth Planet. Sci. Lett. 147, 55–67. https://doi.org/10.1016/S0012-821X(97)00008-3.

Yamazaki, T., Oda, H., 2002. Orbital influence on Earth's magnetic field: 100,000-year periodicity in inclination. Science 295, 2435–2438. https://doi.org/10.1126/science.1068541.

Zatman, S., Bloxham, J., 1997. Torsional oscillations and the magnetic field within the Earth's core. Nature 388, 760–763. https://doi.org/10.1038/41987.

CHAPTER

3

Current understanding about the factors driving climate variability

OUTLINE

The concept of *climate*, in a narrow sense, is understood as prevailing meteorological conditions (including temperature, direction, and speed of wind, amount of precipitation, clouds, etc.) over a long-term period—i.e. 30 years or more. In addition to the atmosphere, the meteorological conditions are also influenced by the other components of *climate system*, which include the hydrosphere, the cryosphere, the land surface with its different absorption and emissivity, and the biosphere (IPCC, 2007). In a broader sense, *climate* refers to the state of climatic system, in terms of the mean and its variability, over a certain time span and a certain area.

The elements of the climate system continuously interact with each other, but they all depend on the amount of solar radiation received. Therefore, by the beginning of the 19th

The Hidden Link Between Earth's Magnetic Field and Climate
https://doi.org/10.1016/B978-0-12-819346-4.00003-6

© 2020 Elsevier Inc. All rights reserved.

century, the historically observed climate variations had been suggestively attributed to changes of solar radiation intensity (Herschel, 1801).

The accumulation of satellite measurements of the total solar irradiance (for 40 years or so) reveals, however, that variations of electromagnetic radiation reaching Earth's atmosphere are negligibly small, despite the well-established 11-year cycle in solar magnetic activity. At the same time, continuously rising emissions of greenhouse gases coincide fairly well with the positive trend in the surface temperature, observed in the 20th century. This fact motivates many scientists to conclude that the anthropogenically increased concentration of CO_2 and other greenhouse gases plays a dominant role in the contemporary climate warming (IPCC, 2013, chapter 2). On the other hand, investigations of the palaeoclimate variations (IPCC, 2013, chapter 5) suggest that the role of the main driver of the long-term climate variability belongs to the Earth's orbital parameters. Accordingly, the role of CO_2 density in the overall picture is to amplify the orbital variations. Moreover, part of climate variability is attributed to internal modes of climate system (e.g. Atlantic Multidecadal Oscillation; thermohaline circulation; northern and southern annular modes etc., IPCC, 2013), although the driving forces of these internal modes are not clearly understood.

This ambiguity, and to some extent controversy in attribution of drivers of climate variability, has been the motivation for writing this chapter. It offers a brief synthesis and critical reassessment of the main factors, and the most popular mechanisms of their influence on the climate system.

3.1 External forcings and mechanisms of their influence on climate

All environmental processes influencing Earth's climate are called *forcing factors*. They may be *external* to Earth or generated in some of its climate system components (atmosphere, hydrosphere, cryosphere, etc.). The existing differentiation between *external* and *internal* forcings is relatively controversial and depends on the definition of the concept of the climate system. Relying on the IPCC definition, *external* forcings are the forcings related to: (i) variations of magnetic and radiative activity of the Sun; (ii) variations of particles flux intensity (of solar, galactic or extragalactic origin) continuously bombarding Earth's atmosphere; (iii) lithospheric activity (i.e. volcanoes; plate tectonics and related movements of continents and changes of their relief and/or climatic zones; regional changes of geomagnetic field, etc.); and (iv) changes in the orbital parameters of the planet. The *internal* forcings, in turn, include: (i) ocean circulation; (ii) planetary albedo; (iii) atmospheric composition; and (iv) life systems and particularly human activity.

As pointed out above, all components of climatic system receive their energy from the Sun, so the following sections put the stress on various manifestations of solar variability and their possible influence on climate.

3.1.1 Variations in solar luminosity

The variability of the Sun is driven by: (i) its evolution as a star (controlled by the conditions in the solar core); and (ii) variations of solar magnetic field, generated by the solar dynamo in the convection zone and the atmosphere of the Sun (Solanki, 2002). Among different manifestations of solar activity, its luminosity (i.e. the radiative power to emit electromagnetic radiation) is the most likely factor influencing Earth's climate. The luminosity depends mainly on the mass of the star and changes slowly, over the periods of hundreds of millions of years (Beer et al., 2000; Nikolov, 2011). Since the Sun is presently in the middle of its ~10-milliard-year-old evolution along the main sequence, the regular enhancement of solar diameter, as well as the brightness of the star within the last several million years, is relatively small.

In addition to the nuclear fusion in the core of the Sun, the solar insolation (i.e. its electromagnetic radiation, integrated over all wavelengths) depends on: (i) transportation of radiated energy through radiative and convective zone of the Sun; (ii) emission of solar radiation by the photosphere; and (iii) the Sun–Earth distance. The transport of energy through the radiative zone (0.3–0.7 solar radii) is probably very stable on a million-year timescale and does not generate any measurable variability. However, the heat transport through the convective zone (0.7–1.0 solar radii) could be a significant source of variability of the solar radiation, on timescales of years to 100 kyr (Nesme-Ribes et al., 1993). The greatest part of the total solar irradiance (TSI, i.e. the integral value over the whole emission spectra) variations are related to the individual features of the photosphere (i.e. sunspots, plages, magnetic network, etc.—particularly by the net effect of dark and bright magnetic contribution) and magnetic or thermal excitations near the bottom of the convective zone (Beer et al., 2000).

Instrumental measurements of TSI show that its variation within the most common ~11-year solar cycle is only ~0.1% (Fröhlich, 2006). This variability is additionally modulated by aperiodic episodes of reduced solar activity, with characteristic timescales of 150–220 years (e.g. Weiss, 1990; Rozelot, 1995). The initial estimations of the amplitude of such fluctuations have been in the range of ~0.2% from the Maunder minimum (observed in the 17th century) to the recent solar minimums (Lean, 2000). Recent recalibration of the historical records reveals, however, that the amplitude of these irregular variations is much smaller than previously expected (Foukal et al., 2006; Kopp et al., 2016). Thus the solar energy received by our planet seems to vary very little with time, and according to Foukal et al. (2006), it is "insufficient to drive climate variability at centennial, millennial or even million-year time scales."

Despite the small amplitude of the 11-year variations of TSI, its 'fingerprint' on the sea surface temperature, and tropospheric temperature and dynamics has been detected (Labitzke and Loon, 1995; White et al., 1997; van Loon and Shea, 2000; Haigh, 2003; Gleisner and Thejll, 2003; Kodera, 2004; Brönnimann et al., 2006; etc.). The consistent explanation of all these results becomes very difficult, not only because of some controversy between them, but also because the variability of solar radiation reaching the troposphere and Earth's surface is negligibly small (according to the fifth IPCC report, less than 0.1%). Consequently, if quasidecadal variability of climate parameters is somehow related to variations in TSI, this suggests the existence of amplifying mechanism(s). One such mechanism has been proposed by Meehl et al. (2009), who suggested that the solar signal in the tropical Pacific Ocean is regionally amplified. Based on the ocean–atmosphere interaction and the intensification of the Walker circulation within the Pacific basin, this mechanism could not explain, however, the global character of quasidecadal oscillations found in the sea surface and the upper ocean temperature (Tourre et al., 2001; White et al., 2003).

Observational evidence shows that the amplitude of quasidecadal variations in atmospheric temperature decreases from the lower stratosphere towards the surface and the upper ocean levels (e.g. White, 2006; Frame and Gray, 2009). For this reason, several mechanisms have been suggested in the scientific literature, attempting to explain the translation of the solar signal from the upper atmosphere (where the amplitude is larger) down to the troposphere and surface. The stronger response of the upper atmosphere is due to the noticeable difference in solar emission in the extreme ultraviolet (EUV) and UV bands between solar maximum and minimum conditions (i.e. up to 6% at wavelengths near 200 nm and 2%–4% in the band 240–320 nm; Lean and Rind, 1998). Reaching the middle atmosphere, the solar UV radiation alters the photochemical production of ozone, which in turn is one of its strongest absorbers, thus affecting the stratospheric temperature and dynamics (Gray et al., 2010 and references therein). The ozone, in turn, also varies with solar cycle and the amplitude of its variation is ~2%–4% in the upper–middle stratosphere (Soukharev and Hood, 2006).

Therefore, it is not surprising that this well-established relation between variations of solar short wave radiation, stratospheric O_3, temperature, and circulation is actually the foundation of most of the proposed mechanisms for downward transmission of solar influence. Regression analyses of atmospheric temperature reveal that besides in the upper stratosphere, where the maximum of solar signal has been found, the secondary maximum has been discovered near the tropopause (see Fig. 3.1). Similar results are also reported by Frame and Gray (2009), i.e. their Fig. 1, based on the ERA 40 reanalysis data.

Meanwhile, some authors have deduced that a warmer lower stratosphere-upper troposphere should reduce the vertical tropospheric temperature gradient in periods of high solar activity (White et al., 2003; White, 2006). As a result, the upward thermal flux transported by Hadley circulation should be reduced, followed by an increase of the tropical upper ocean temperature, when the Sun is more active. This suggestion has been further confirmed by the numerical experiments of Haigh et al. (2005), which show (even with a highly idealized model) that warming of the tropical lower stratosphere causes weakening and expansion of the Hadley and poleward movement of the Ferrel circulation cells. This result motivated the authors to conclude that solar heating of the lower stratosphere may influence the tropospheric circulation, without any direct forcing beneath the tropopause. The weaker point in this hypothesis, however, is the unclear mechanism of the lower stratosphere extra-warming in periods of high solar activity. The intensification of the positive Arctic Oscillation mode (as proposed by Hood and Soukharev, 2012), and corresponding weakening of the Brewer-Dobson circulation, could not solve the problem of causality, because the factors driving this internal climatic mode are still unclear.

Another hypothesis, which is well accepted by the climatological community over the world, is suggested by Kodera and Kuroda (2002). According to this hypothesis, the stronger upper stratospheric subtropical jet (in periods of solar maximum) should deflect the winter planetary waves, directing them towards the pole. Thus instead of 'breaking' in the tropical upper stratosphere and 'pumping' the Brewer-Dobson circulation (e.g. through the deposition of their easterly momentum there, as described by Holton et al., 1995), the planetary waves are reflected back to the pole (Kodera and Kuroda, 2002). The authors hypothesized that this will weaken the Brewer-Dobson circulation, due to the suggestive reduction of upward motion in equatorial region, accompanied by an enhancement of the lower stratospheric temperature.

Although the warmer (in solar maximum) tropical lower stratosphere supports this hypothesis, the warmer subtropics and higher ozone density found in the lower stratosphere

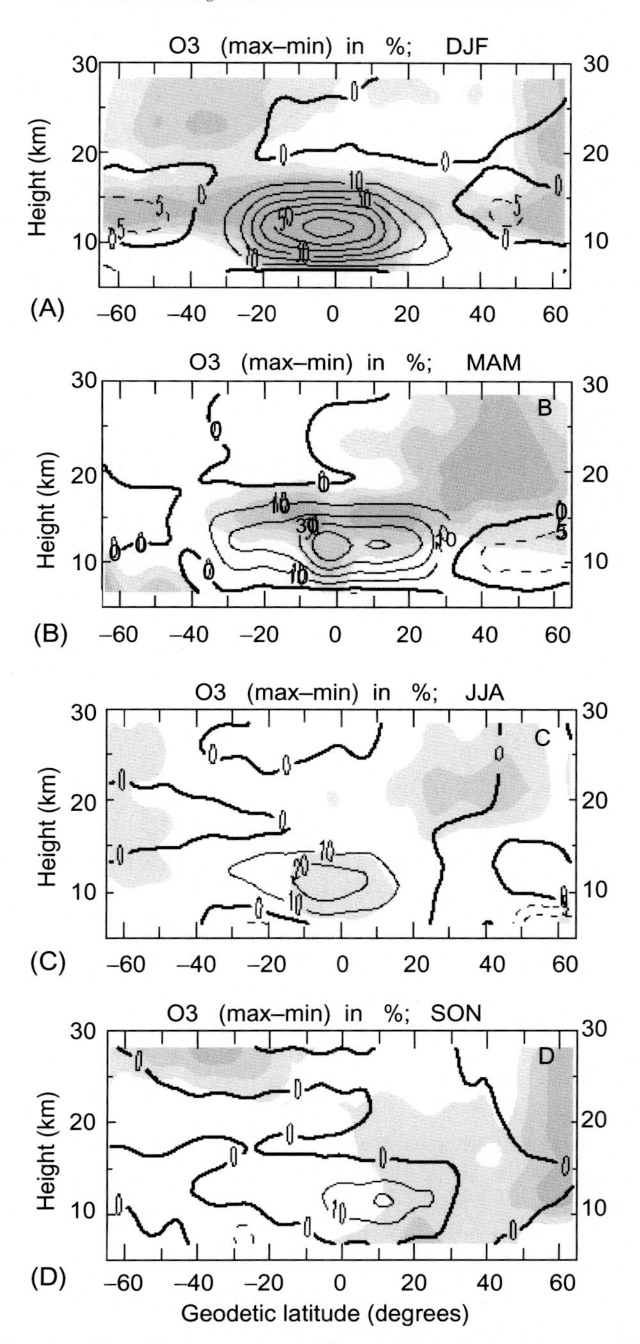

FIG. 3.1 Estimated solar signal found in temperature difference (i.e. solar maximum–solar minimum, in [K]) derived from a multiple regression analysis of HALOE data for the period October 1991–March 2003; the four panels illustrate regression coefficients between temperature and solar radio emission at 10.7 cm, derived for: (A) boreal winter (December, January, February); (B) boreal spring equinox (March, April, May); (C) boreal summer (June, July, August); and (D) boreal autumn equinox (September, October, November). Shading denotes regions where values are significant at the 80% *(lightest gray)*, 90% *(medium gray)*, and 95% *(dark gray)* levels, respectively, using a student's *t*-test. *From Kilifarska, N.A., 2005. Solar modulation of temperature, water vapour and ozone seasonal cycle in the upper troposphere/lower stratosphere. C. R. Acad. Bulg. Sci. 58 (10), 1151–1158.*

(Hood and Soukharev, 2012) require acceleration (but not deceleration) and/or poleward expansion of Hadley circulation. The Kodera and Kuroda (2002) mechanism contradicts also to the results of White et al. (2003), White (2006), and numerical simulations of Haigh et al. (2005), reporting a weakening of Hadley cells in periods of active Sun.

Moreover, there are some theoretically based objections against the validity of Kodera and Kuroda's (2002) hypothesis. Thus Plumb and Eluszkiewicz (1999) and Seviour et al. (2012) pointed out that wave forcing maximizes in the extra-tropical winter hemisphere, and is unable to explain the strongest upwelling observed in the summer hemisphere—even if the wave-driven circulation from the winter hemisphere could penetrate in the summer one, as argued by Tung and Kinnersley (2001). Plumb and Semeniuk (2003) have shown in addition, that the influence of the wave-driven circulation on the tropical lower stratosphere is limited by the rapid attenuation of the upwelling beneath the level of wave dissipation. Further analyses of the problem show that the year-round upwelling in the tropics should be attributed to the equatorial waves and diabatic heating, and to a lesser extent to the weak penetration of extra-tropical waves closer to the equator (Semeniuk and Shepherd, 2001; Scott, 2002; Chen and Sun, 2011).

Another highly debated problem is the mechanism of excitation of quasidecadal variability of the upper ocean temperature. Some authors insist that it originates in the ocean oscillations (Farneti and Vallis, 2011), while others like White et al. (1997, 2003, 2006) suggest a solar influence, mediated by the lower stratosphere. The impact of the ocean in climate variability, as well as the roles of a 'master' and a 'slave' in atmospheric-ocean interactions, will be briefly discussed in Section 3.2.

3.1.2 Variations of the solar magnetic field

Unlike the electromagnetic radiation, which directly influences the composition and atmospheric thermodynamics, as well as the surface temperature, the well-pronounced variations in solar magnetic field do not have a direct influence on Earth's climate. Nevertheless, many correlations between solar magnetic field and climate variables have been reported from many authors. For example, the covariance between the length of quasiperiodic sunspot cycle and the Northern Hemisphere land air temperature, found by Friis-Christensen and Lassen (1991), excited the solar and climate communities during the last decade of the 20th century, asking a question about the possible mechanism of solar magnetic field influence on climate. Interestingly, the relation is better pronounced for historical records (up to five centuries ago; Lassen and Friis-Christensen, 1995) and has become worse since around 1990 (Thejll and Lassen, 2000). Some authors interpret this result as a conformation of the dominant role of anthropogenic warming in the recent climate change. The most probable explanation of this unsteady relation should be sought, however, in the long-term oscillations (with variable periods and amplitudes) of the nonlinear magneto-hydrodynamic solar dynamo (Weiss, 1990; Rozelot, 1995).

Although the relation between the variable length of sunspot cycle and climate variability remains unexplained, the important message of the above studies is that the long-term variations of solar magnetic field and climate variations are somehow interrelated. The delayed response of the sea surface temperature to the variations of the sunspots cycle length

(Lassen and Friis-Christensen, 1995) is a hint that the effectiveness of the mechanism of their synchronization is suppressed by the greater heat capacity of the ocean, which leads to a time delay of the oceanic response.

Some authors suggest that the mechanism of solar influence on climate operates through excitation of the North Atlantic Oscillation (NAO) (Kodera, 2002; Ruzmaikin and Feynman, 2002). It is worth noting that the NAO is the key and primary source of variability of the North Atlantic climate on many timescales (Marshall et al., 2001). In this context, more recent studies deserve an attention with their claim that centennial variations of solar magnetic dynamo and the interplay between zonal and meridional components of solar magnetic field determine the long-term variations of the internal atmospheric mode NAO (Georgieva et al., 2007, 2012). These authors show that variations of solar activity initiated by the zonal (toroidal) component of the Sun's magnetic field decreases the Northern Hemisphere zonal circulation. This corresponds to a more negative phase of NAO mode. In contrast, solar activity related to the meridional (poloidal) component of the solar magnetic field increases zonal atmospheric circulation, a state corresponding to more positive NAO. Although the authors have not provided a physical mechanism for such an influence, their results deserve further consideration. General circulation models are still unable to simulate the relation between solar forcing and NAO during the past 1000 years (IPCC, 2013, chapter 5). This fact shows that there is no mechanism of such an influence built in the recent climate models.

The main quasiperiodicity of solar magnetic field variations is a bi-decadal cycle of magnetic field polarity (i.e. change of magnetic dipole orientation from N → S, to S → N direction), with an average period of ~22 years. The fingerprints of this periodicity on Earth's climate parameters have been found in the surface temperature (White et al., 1997; Miyahara et al., 2008; Souza Echer et al., 2012; Qu et al., 2012; Barlyaeva, 2013), precipitation (Almeida et al., 2004), lower stratospheric ozone (Peters et al., 2008; Kilifarska, 2011), Earth rotation (Scafetta, 2010; Chapanov et al., 2009; Zotov et al., 2016), etc. The bi-decadal signal found in the dynamical anomalies of surface pressure, averaged over 40–70°N latitudes, during the period 1900–2010, is illustrated in Fig. 3.2. Comparison with the Hovmoller diagram of

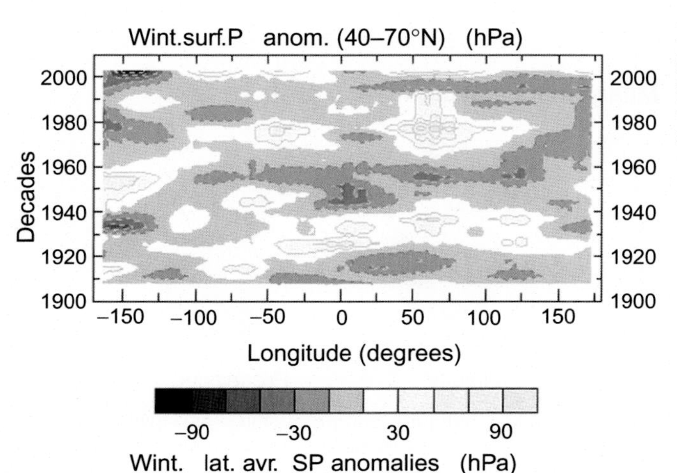

FIG. 3.2 Hovmoller diagram of surface pressure deviations from dynamically varying decadal mean, averaged over the 40–70°N latitudes. *Data source is ERA-20C reanalysis.*

lower stratospheric ozone anomalies, shown in Fig. 4.16, implies that quasi bi-decadal variations of the extratropical surface pressure could be related to the similar variations in ozone density.

3.1.3 Changes of Earth orbital parameters (Milankovitch cycles)

The orbital parameters of Earth planet vary continuously with time, although very slowly. The shape of Earth's revolution around the Sun (called *eccentricity*, with a main period of 413.10^3 years), the tilt of its axis with respect to the orbital plane (called *obliquity*, with a period of 41.10^3 years), and its slow wobble (known as *precession*, with a period of 25,771 years) result from the gravitational disturbances of the Sun and other planets in the solar system (mainly the biggest ones: Jupiter and Saturn). In the early 1900s, the Serbian astronomer Milutin Milankovitch suggested that these quasiperiodic changes in Earth's orbit are the primary drivers of the recurrent periods of glaciation and interstadials, during the 4.6 milliard years of Earth's history (Fig. 3.3).

Similar hypotheses were developed much earlier, by James Croll and Joseph Adhémar during the 19th century. However, at that time, geological dating methods were not sufficiently advanced to provide reliable dated evidence in support of the theory. In fact, the Milankovitch hypothesis came into focus of attention of the scientific community only after the publication of the pioneering work of Hays et al. (1976). The latter authors have provided the first, high-resolution, high-precision record of long-term global climatic change using deep ocean cores.

Among all the orbital cycles, Milankovitch believed that obliquity had the greatest effect on climate, by varying the summer insolation in northern high latitudes. Therefore, he deduced a 41,000-year periodicity in the Ice Ages appearance. Milankovitch emphasized the importance of changes experienced at 65°N latitudes, because of the large land surface, which responds much more quickly to the amount of solar heating received than the ocean does.

Subsequent research (e.g. Shackleton et al., 1990; Abe-Ouchi et al., 2013) has shown, however, that the Ice Age cycles of the Quaternary glaciation during the last million years have a

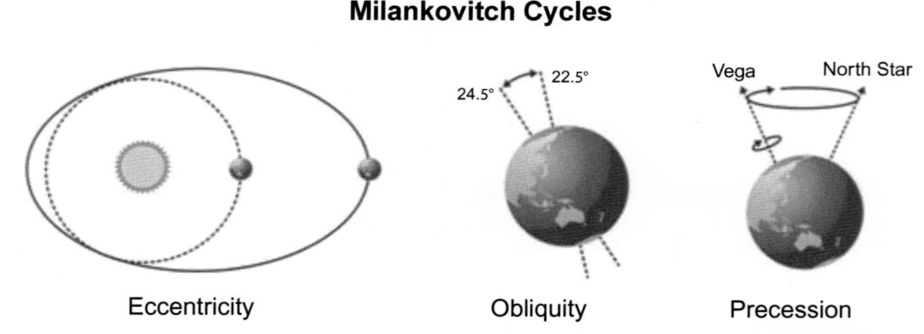

FIG. 3.3 Illustration of the effects of Earth's orbit eccentricity, obliquity, and precession suggestively responsible for the periodical glaciation of the planet. *From Skeptical Science, https://skepticalscience.com/print.php?n=116.*

periodicity of 100.10^3 years, which is closer to the eccentricity cycle than to the obliquity one (with a period of 41.10^3 years). Moreover, palaeoclimate records show that variation of Earth's climate is much more extreme than the variation in solar radiation intensity imposed by the orbital changes. Recent modelling of the effect of orbital inclination on the amount of received total solar irradiance (TSI) reveals that inclination modulation is not enough to explain Earth's glaciation cycle of 100.10^3 years (Vieira et al., 2012). These model simulations have been verified by the measurement of TSI by TIM/SORCE (Kopp et al., 2005) and VIRGO/SOHO (Fröhlich, 2006) instruments, and show that at timescales of 10^5 years, the amplitude of TSI variations is ~0.003 W/m^2, which is much smaller than the variations due to orbital eccentricity (1.5 W/m^2) or solar magnetic activity (1–8 W/m^2) (Viera et al., 2012).

These and some other discrepancies with Milankovitch predictions, as well as the negligible long-term variations of solar luminosity (Foukal et al., 2006), suggest either the existence of more effective external forcing or the presence of amplifying mechanism(s) of solar variability.

3.1.4 Galactic cosmic rays

The intensity of energetic particles of galactic or extragalactic origin, which continuously bombard Earth's atmosphere, is modulated by the heliospheric interplanetary magnetic field (IMF). The main (and well-known) modulation of GCR consists of a simple scattering and deflection of less energetic particles by the magnetic irregularities, propagating with solar plasma within the heliosphere. The effect is much stronger in periods of active Sun, because the turbulent IMF deflects cosmic rays much more efficiently than in periods of low solar activity, when the IMF is more regular. The heterogeneous heliospheric plasma not only reduces the number of cosmic rays reaching the inner heliosphere and Earth, but also changes the energy spectrum and the direction of its propagation.

The simple analysis of the time evolution of GCR flux, as measured at Earth's surface, shows that it is different in two successive sunspot number cycles (see Fig. 3.4). The consecutive alteration of cycles with sharp (i.e. 1960s, 1980s, and 2000s) and flatter maximums (i.e. 1970s, 1990s) is easily seen in Fig. 3.4.

The main reason for this difference is the periodic reversal of solar magnetic field (from solar north to solar south pole and vice versa, with a period of ~22 years). The orientation of the heliomagnetic field, in turn, affects the charged particles' propagation; this effect is known as charge-sign dependence of GCR modulation (e.g. Potgieter, 2014 and references therein). Thus, in the case of N $\rightarrow$ S orientation of the solar magnetic field (positive polarity), galactic protons enter the heliosphere mainly in the small regions near the poles, where the solar magnetic field decreases rapidly with a distance (i.e. as $1/r^2$, Jokipii, 1989). In this case, the direction of gradient and curvature drifts of positively charged particles (induced during their spiralling along the magnetic field lines) coincides with their diffusion motion along the magnetic field lines, bringing the particles into the heliosphere. When entering the heliosphere, they are rapidly ejected along the magnetic lines in the equatorial neutral current sheet, and the neutron monitors on the Earth (orbiting the Sun near its ecliptic) encounter a prolonged period of high GCR flux (flatter maximums in Fig. 3.4).

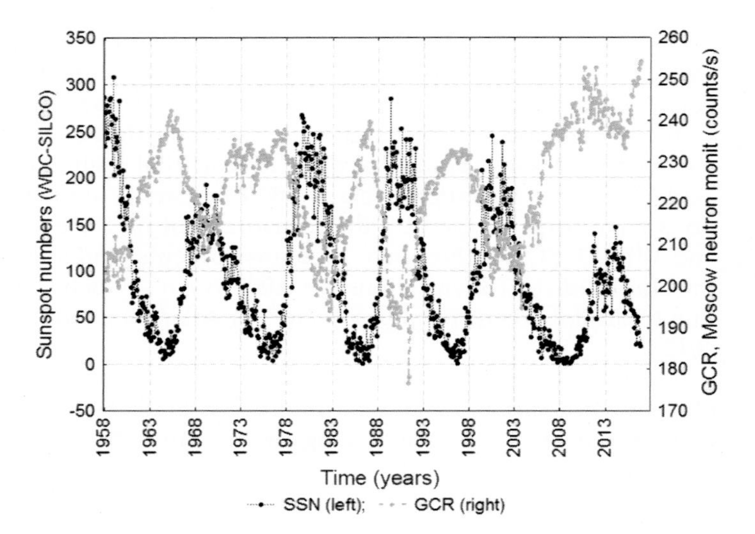

FIG. 3.4 Monthly values of sunspot numbers and GCR measured by Moscow neutron monitor. *Data sources: Solar data are from WDC-SILCO, while GCR are from the neutron monitors database NMBD.*

Oppositely, in S → N orientation of the solar magnetic field (negative polarity), galactic protons drift is directed out of the solar poles (Jokipii, 1989). Guided by heliospheric magnetic lines, protons reach the equatorial current sheet from where they are able to enter the heliosphere, crossing the turbulent interplanetary magnetic field. On their way towards the Sun, protons experience waviness of the heliomagnetic field and get progressively reduced as solar activity surges. For this reason, their flux density increases slowly when approaching the solar minimum, and then rapidly decreases again (the sharp cycles in Fig. 3.4).

The idea that cosmic rays could influence Earth's climate dates back to Ney (1959), who pointed out that if there is a relation between ionization created by GCR near the tropopause and weather events (i.e. thunderstorm activity), one should expect that at longer timescales, a solar-cycle modulation of such a relation might be observed (due to the solar wind modulation of GCR intensity). This imprint of solar activity onto the intensity of GCR flux entering Earth's atmosphere is used nowadays for palaeoreconstruction of ancient solar activity, based on the ^{14}C and ^{10}Be isotopes found in tree rings, sediments, ice core, etc.

In the atmosphere, the ^{10}Be appears as a product of spallation of larger atomic nuclei that have collided with cosmic rays. It becomes attached to aerosols and is removed from the atmosphere within 1–2 years, mainly by wet precipitation. ^{14}C, on the other hand, is oxidized to $^{14}CO_2$ and enters the carbon cycle, where it exchanges between the atmosphere, biosphere, and ocean (Scherer et al., 2006). As a consequence, analysis of these nuclides provides information not only about the production history, but also about the atmospheric transport and mixing processes, before their storage in the archive. Cosmogenic radionuclides records are influenced by both solar activity and variations in the geomagnetic dipole moment. However, ^{10}Be records are used mainly as a proxy for the past variations of solar cycle. Additional information can be gained from the ^{14}C records, sampled from tree rings or other organics, which anticorrelate with Earth's magnetic moment. This information allows deconvolution

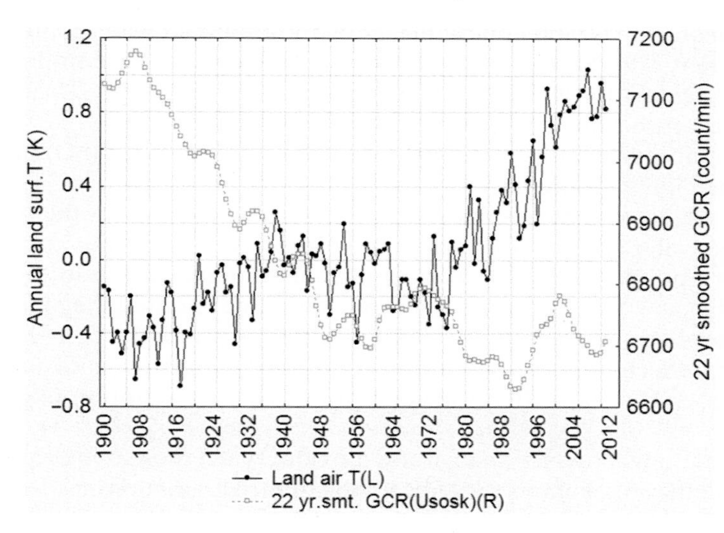

FIG. 3.5 Time series of annual mean global land surface temperature anomalies (i.e. deviations from the average over the period 1961–90) and galactic cosmic ray intensity during the period 1850–2010. *Data source: CRUTEM 4v data set, and reconstructed galactic cosmic rays smoothed by a 22-year running window (Usoskin, I.G., et al., 2008. Cosmic Ray Intensity Reconstruction. IGBP PAGES/World Data Center for Paleoclimatology, Data Contribution Series # 2008-013. NOAA/NCDC Paleoclimatology Program, Boulder, CO, USA).*

of the 11-year sunspot cycle from the variations of the geomagnetic field. However, due to the much shorter half-life of ^{14}C isotopes, at $\approx$5730 year, compared to that of ^{10}Be, at $\approx 1.5 \times 10^6$ year, the utility of ^{14}C is limited to $\approx 4 \times 10^4$ years, while that of ^{10}Be is limited to $\approx 10 \times 10^6$ years (Frank, 2000).

The ratio of oxygen isotopes $^{18}O/^{16}O$ (derived from marine or lake sediments, caves stalagmites, deep ice core, etc.) is used as a proxy for Earth's palaeo-temperature. Thorough analysis of the oxygen and cosmogenic isotopes at different timescales—from million years to centuries and decades—is presented in the review by Scherer et al. (2006). The results of many authors, gathered there, show convincingly the synchronous variation of GCR and temperature, in alternating episodes of warming and cooling in Earth's climate history. In addition, the simple examination of the global land surface air temperature and GCR time series shown in Fig. 3.5 confirms fairly well their anticorrelation for the last 150 years. More evidence for imprint of GCR on the surface T is provided in Chapter 4, while a suggestive mechanism of their temporal covariance is described in Chapter 7.

3.1.5 Lithospheric activity and its influence on climate variations

3.1.5.1 Volcanoes

The effect of volcanic gases and dust may warm or cool the Earth's surface, depending on how sunlight interacts with the volcanic material. Volcanic dust blasted into the atmosphere causes temporary cooling. The amount of cooling depends on the amount of dust put into the air, and the duration of the cooling depends on the size of the dust particles. Particles the size of sand grains fall out of the air within a few minutes and stay close to the volcano. These particles have little effect on climate. Tiny dust-size ash particles thrown into the lower atmosphere will float around for hours or days, causing darkness and cooling directly beneath the ash cloud, but these particles are quickly washed out of the air by the abundant water in the lower atmosphere, and subsequent rain. However, dust tossed into the dry upper troposphere and the stratosphere can remain for weeks to months before it finally settles. These dust particles block sunlight and cause some cooling over large areas of the Earth.

Volcanoes that release large amounts of sulphur compounds like sulphur oxide (SO) or sulphur dioxide (SO_2) affect the climate more strongly than those volcanoes ejecting just dust. The sulphur compounds are gases that rise easily into the stratosphere. Once there, they combine with the available water and form a haze of tiny droplets of sulphuric acid. These tiny droplets are very bright and reflect a great amount of sunlight back into the space. These droplets eventually grow with time and once they are large enough, they fall down to the surface. The stratosphere, however, is so dry that this takes months or even years to happen. Consequently, reflective hazes of sulphur droplets can cause significant cooling of the Earth for as long as 2–3 years after a major sulphur-bearing eruption (e.g. Shindell et al., 2003).

Volcanoes also release large amounts of water and carbon dioxide. When these two compounds are in the form of gases in the atmosphere, they absorb the infrared radiation emitted by the planet and hold it in the troposphere. This causes warming of the air and surface below. Over long periods of time (thousands or millions of years), multiple eruptions of giant volcanoes, such as flood basalt volcanoes, can raise carbon dioxide levels enough to cause significant global warming.

A similar effect could produce the large volumes of SO_2 that erupt frequently (every year or two, even several times per year for decades). The higher SO_2 density overdrives the oxidizing capacity of the atmosphere, and prevents the formation of sulphuric acid (aerosols), which cools the atmosphere (Ward, 2009). The effect of such volcanic activity is global warming of the climate (Ward, 2009).

3.1.5.2 Plate tectonics

Earth's lithosphere consists of a series of rigid plates that are in constant motion, although very slowly. Having different masses and speeds, these plates interact with each other; this is the subject of investigation of scientific branch *plate tectonics*. One consequence of plate tectonics is continuously changing distribution of continents over the surface of the planet. In Earth's history can be found times when Greenland straddled the Equator and times when the Sahara Desert was at the South Pole (Boron, 1996). An example of continents' distribution during the late Triassic geological epoch is shown in Fig. 3.6.

The topography or elevation of the continents is also governed by plate tectonics. Due to the continuous motion of the plates, collision between continents is inevitable. Usually, the thinner and denser oceanic lithosphere is subducted beneath the thicker and less dense continental lithosphere, a process leading to an active orogenesis (i.e. the process of mountain formation by deformation of the Earth's crust) (see Fig. 3.7). This is the mechanism of formation of the highest mountains, such as the Andes mountain range or the Himalayan-Tibetan Plateau.

Plate tectonics also influences the altitude of the sea level. At boundaries where new crust is created (predominately at middle-ocean ridges), hot, thick lithosphere is accreted to the margins of diverging plates. The process is illustrated in Fig. 3.7. Over tens of millions of years, the thick lithosphere cools and contracts, which reduces the holding capacity of the

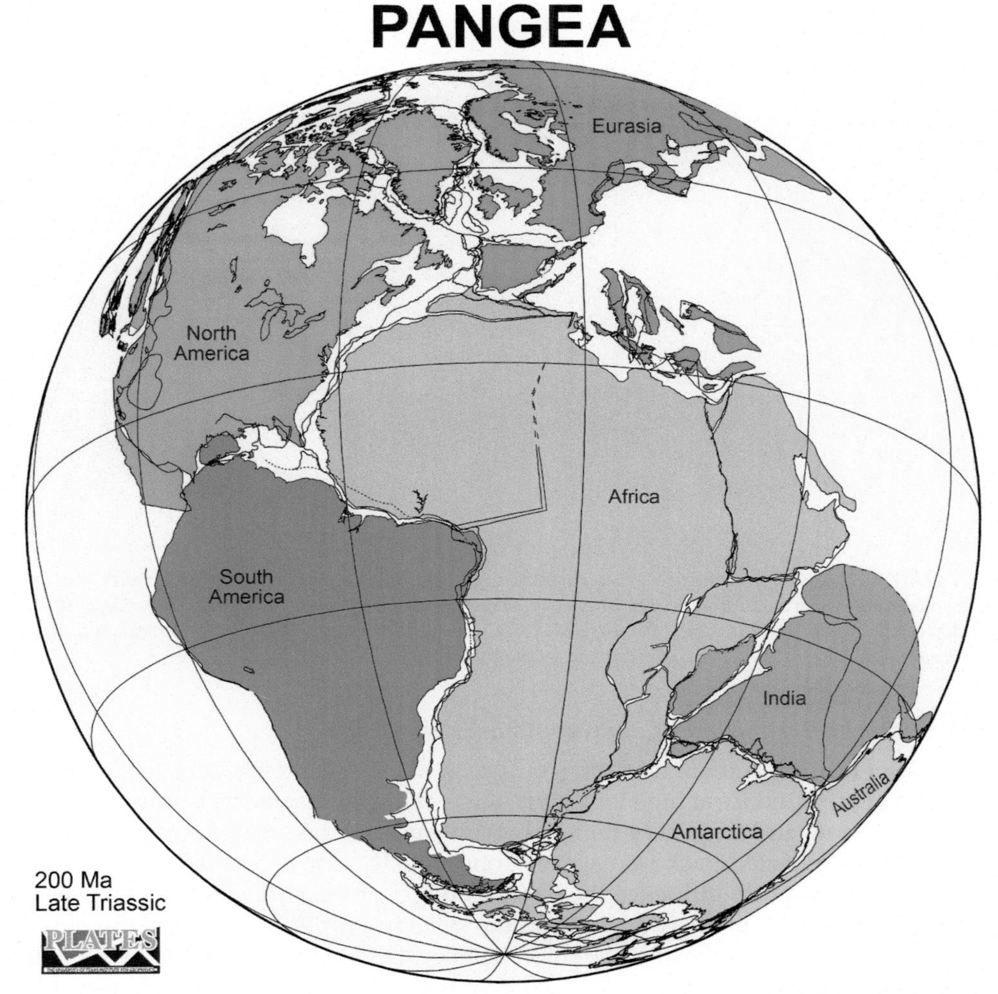

FIG. 3.6 Late Triassic (200 million years ago) continental geography, reconstructed by the Institute of Geophysics, University of Texas at Austin. With permission of Lawver, L.A., Dalziel, I.W.D., Norton, I.O., Cahagan, L.M., 2009. The PLATES 2009, Atlas of Plate reconstruction (750 Ma to Present Day). Plates Progress Report No. 325-0509, University of Texas, Technical Rep. No. 196, p. 57.

ocean basin and leads to the occurrence of some continental flooding. Geological reconstructions show that almost 20% of the modern continents were covered by ocean in the middle Cretaceous period (Boron, 1996).

In short, plate tectonic processes affect not only the distribution and elevation of the continents, but also the area of continents above sea level. However, this raises the question: how could tectonic processes in lithosphere influence Earth's climate? There are several channels of such an influence, which will be briefly described here.

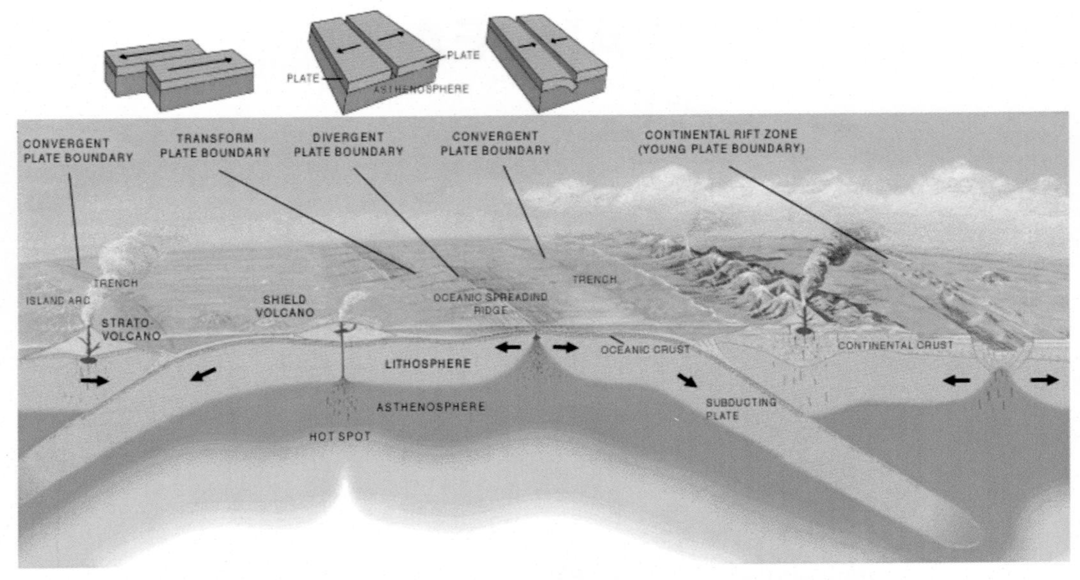

FIG. 3.7 Artistic illustration of the processes of subduction, mountain orogenesis, volcanic activity, and mid-ocean ridge: the place of a new crust formation. *From Kious, W. Jacquelyne, Tilling, Robert I., 1996. This Dynamic Earth: the Story of Plate Tectonics (Online ed.). U.S. Geological Survey. ISBN 0-16-048220-8. Retrieved 29 January 2008 (originally drawn by Jose F. Vigil). https://upload.wikimedia.org/wikipedia/commons/4/40/Tectonic_plate_boundaries.png.*

3.1.5.3 Spatial distribution of continents over the globe

The smaller heat capacity of the land (compared to the oceanic one) ensures accumulation of snow and year-round ice at high latitudes, where the solar insolation is substantially lower. The albedo of snow and ice is very high, which means that the greater part of the solar radiation received is reflected back into space. Consequently, snow-covered continents at high latitudes can significantly influence the energy budget of the atmosphere-ocean system, increasing the proportion of energy loss from the system. Tropical oceans, on the other hand, have the opposite effect on Earth's radiation balance. Due to the much smaller albedo of the ocean surface, a greater portion of solar radiation is accumulated and later on re-emitted into the atmosphere in the form of latent and sensible heat fluxes.

Another manifestation of continental influence on climate is observed in the case when they serve as a mechanical barrier for the oceanic currents. The ocean currents transport between one-third and one-half of the bulk of heat in the tropical atmosphere-ocean system. In this sense, the spatial distribution of continents can block the poleward heat transfer by ocean currents. Moreover, the higher thermal inertia of the oceans (conditioned by their higher heat capacity) tends to moderate the existing seasonal cycle of incoming solar radiation.

3.1.5.4 Continental elevation

Due to the negative temperature gradient in the troposphere, the mountains' temperature is much lower, even when they are placed at tropical latitudes. The colder temperatures there promote snow accumulation, and due to the very high albedo of snow and ice, the highly elevated lands have a cooling effect on the planetary energy balance.

Differential heating of continents and oceans (depending on their geographical position and topography) generates excitation of planetary waves in the troposphere, which influences the atmospheric circulation. Moreover, the variations in topography may influence the distribution of cold air masses or the track of winter storms—additionally cooling some regions or warming others—thus affecting the regional climate.

3.1.5.5 *Carbon dioxide (CO_2)*

Plate tectonic processes also modify the carbonate-silicate geochemical cycle. Tectonic processes related to subduction or enhanced sea-floor spreading rates initiate higher volcanic activity and consequently increased amount of carbon dioxide emissions in the atmosphere. Although the volcanic eruptions do not consist solely of CO_2, on timescales of tens of millions of years, a substantial amount of carbon dioxide could be accumulated, despite the balancing effect of igneous and metamorphic rocks weathering (Boron, 1996). Experiments with climate models suggest that quadrupling of atmospheric CO_2 concentration in the Cretaceous period, in addition to the differences in geography, could produce an 8°C increase of temperature compared to the present (Boron, 1996).

Despite the strong in-phase synchronization between carbon dioxide obtained during the past several hundred thousand years, palaeoclimate data reveal that climate change is not just about temperature. Thus changes of CO_2 density are accompanied by changes of other aspects of climate. For example, during glacial episodes the snow lines moved equatorward and towards the valleys, continents became drier, tropical monsoons were weaker, etc. Understanding of the causal relations between all these changes is the focus of contemporary climatic research.

3.1.5.6 *Variations in the geomagnetic field intensity*

Statistical relations between the geomagnetic field and Earth's climate have been noticed for many years, either in their synchronous temporal variations (Elsasser et al., 1956; Harrison and Funnell, 1964; Bucha et al., 1970; Wollin et al., 1971a,b; Courtillot et al., 1982, 2007; Gallet et al., 2005; Kilifarska, 2015) or in the similarity in their spatial distribution (King, 1974; Vieira et al., 2008; Bakhmutov et al., 2011; Kilifarska et al., 2013). These studies address the relationship between the long-term climate variability (at multidecadal and centennial timescales) and the changes in Earth's magnetic field, caused by different sources: internal (located in the Earth's core) and external (located in the ionosphere and magnetosphere). The sources of internal and external geomagnetic variability have a fundamentally different nature and different frequency range. Consequently, the timescales of obtained geomagnetic-climate relations vary from decades to millions of years, and are based on the analysis of palaeo-data and instrumental observations.

Fig. 3.8 illustrates the previously identified tendency that cooled climates are related to the stronger geomagnetic field (Wollin et al., 1971a,b; Bucha and Zikmunda, 1976; Nurgaliev, 1991; Butchvarova and Kovacheva, 1993; Gallet et al., 2005). The amplitude of longitudinal variations of latitudinally averaged (over 40–70° latitudes) geomagnetic field intensity and surface temperature for the first decades of the 20th and 21st centuries are compared in Fig. 3.8A and B. Note that the highest amplitude of surface temperature increase, since the beginning of the 20th century, corresponds to the strongest geomagnetic weakening over North America.

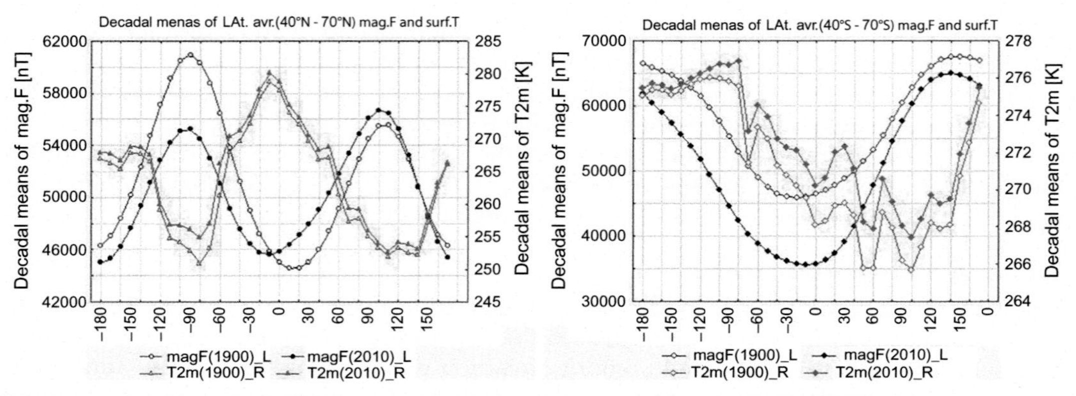

FIG. 3.8 Centennial changes of winter air surface temperature *(red/grey symbols)* and geomagnetic field intensity *(black symbols)*, latitudinally averaged (over the 40–70°N latitudes), given for the first decades of the 20th and 21st centuries: (A) in the Northern Hemisphere; and (B) in the Southern Hemisphere. *Data sources: International Geomagnetic Reference Field model (https://www.ngdc.noaa.gov/geomag/geomag.shtml) and ERA 20 century reanalysis.*

Similarly, the highest centennial warming in the Southern Hemisphere corresponds fairly well to the strongest geomagnetic weakening over the South Atlantic region (see Fig. 3.8A and B).

More evidence for the relation between these two variables is given in Chapter 4. The suggested relation between the geomagnetic field and climate seems quite an exotic idea, due to the electrical neutrality of the lower atmosphere, making it independent of the geomagnetic field. However, the numerous reports for detected statistical relations between Earth's magnetic field and some climate parameters indicate the need for a convincing mechanism(s) for such a relationship, which was our motivation for writing this book.

3.2 Internal variations of climate system within a framework of an average climate status

The Sun is the main source of energy for Earth, which means that internal climate variability obviously depends on the available energy in the climate system. The response time of various climatic components to the impact of solar electromagnetic radiation is very different. For example, the response time of the troposphere is relatively short, from days to weeks, whereas the stratosphere comes into equilibrium on a timescale of several months. Due to their large heat capacity, the oceans have a much longer response time—typically decades but potentially up to centuries or millennia. The response time of the coupled surface-troposphere system is much slower compared to that of the stratosphere, due to the higher inertia of the oceans. The response of the biosphere also varies within a broad range, from fast to very slow. As a result, the complex response of climate system to the variations of external solar forcing occurs on different space scales and timescales.

The relatively weak variances of total solar radiation reaching Earth's surface has provoked some scientists to suggest that variability of climate system is caused entirely by its internal variability. The theory of internal causation of climate change has been developed by Lorenz (1968, 1970, 1976). He suggested that climate change might just be natural variations, due to the complex nonlinear interactions among the various components of the climate system.

Without an external forcing, however, the climate would fluctuate around some stationary mean state. But palaeoclimatic records show evidence for sharp changes in the palaeoclimate conditions (Scherer et al., 2006) during Earth's history. In conjunction with the hotly debated problem for the detected hiatus in global warming (since the beginning of the 21st century), many authors have returned to the idea of internal climate variability as a possible explanation of the missing or strongly reduced positive trend in surface air temperature (Cohen et al., 2012; Li et al., 2015; Kug et al., 2015; Sun et al., 2018). The problem with this interpretation is that the authors have focused on quite a short period of 10–15 years, which reflects most probably a short time variation of current mean climate state (Mann et al., 2014; Dai et al., 2015). To estimate whether this is a turnaround point in long-term climate variability, much longer periods of observations are required.

3.2.1 The oceans

The atmosphere does not have much capability to store heat. Moreover, heat penetration deeper into the soil is limited by the low thermal conductivity of the land surface. As a result, only the top 2 m or so of the land typically play an active role in heat storage and release. Accordingly, the role of the land as a storage of the heat received from the Sun, and consequently as a memory for the climate system, is significantly reduced compared to that of the ocean. The heat capacity of the global atmosphere corresponds to the heat capacity of the top 3.5 m of the ocean (Trenberth and Stepaniak, 2003). However, the depth of the ocean, which is actively involved in climate variability, is of course much greater than that.

The oceans cover about 71% of Earth's surface and contain 97% of its water resource (Trenberth, 2001). Consequently, the ocean plays an important role in shaping Earth's climate and its variability, due to its fluid motions, high heat capacity, and ecosystems. Water vapour, evaporated from the ocean surface, provides latent heat energy to the atmosphere during the precipitation process. In units of 10^3 km^3 per year, evaporation over the oceans (413 units) exceeds precipitation (373 units), leaving a net of 40 units of moisture transported onto land as water vapour. On average, this flow must be balanced by a return flow over and beneath the ground through rivers and stream flows, and subsurface ground water flow.

The seasonal variations of the heat received from the Sun penetrate into the ocean through a combination of radiation, convective overturning (in which cooled surface waters sink, while warmer more buoyant waters below rise up), and mechanical stirring by winds. These processes mix heat through the ocean's mixed layer, which involves the top ~90 m of the ocean. The thermal inertia of a 90 m layer can add a delay of about 6 years to the temperature response to an instantaneous forcing. As a result, actual changes in climate tend to be gradual. With its mean depth of about 3800 m, the total ocean would add a delay of 230 years to the response if rapidly mixed. However, mixing is not a rapid process for most of the ocean, meaning that in reality, the response depends on the rate of ventilation of water between the well-mixed upper layers of the ocean and the deeper, more isolated layers that are separated by the thermocline (the ocean layer exhibiting a strong vertical temperature gradient). The rate of such mixing is not well-established and it varies greatly geographically. An overall estimate of the delay in surface temperature response caused by the oceans is 10–100 years.

The slowest response should be in high latitudes, where deep mixing and convection occur, while the fastest response is expected in the tropics. Consequently, the oceans have a great moderating effect on climate changes.

3.2.2 Planetary albedo

The amount of radiation reflected by a surface is called albedo. Albedo can range from a value of 0 (no reflection) to a value of 1 (100% reflection). Analysis of the energy balance equation requires knowledge about the total amount of received energy, and energy reflected, absorbed, or radiated back to space.

Earth's global temperature is relatively stable when the sum of energy gains is approximately equal to the sum of energy losses. Any imbalance between the incoming or outgoing energy disturbs Earth's radiative equilibrium, leading to a rise or fall of global temperature.

The amount of reflected solar energy depends on the properties of the reflecting surface. The average Earth's albedo is 0.31, meaning that the Earth reflects nearly a third of the incoming solar radiation back into space. Forests, oceans, cities, and deserts all have different albedos. Forests' albedo varies between 0.08 and 0.15, while that of deserts is about 0.30. The bare surface (e.g. after deforestation or agricultural burning) reflects more sunlight back to space, and has a cooling effect in Earth's radiation balance. The reflectance of bright snow and ice is even higher and their albedo is between 0.6 and 0.9.

Besides Earth's surface, the clouds also have a substantial impact on radiation balance. Clouds' albedo depends on their height, size, and the number and size of droplets inside the cloud. A large cumulonimbus cloud casts a dark shadow because light does not go through it easily. From space, the same cloud looks bright white because of its high albedo. A cirrus cloud, on the other hand, is nearly transparent but seems greyer from space, because its albedo is lower.

3.2.3 Sea ice and climate

Sea ice is an active component of the climate system and varies greatly with the seasons, but only at higher latitudes. In the Arctic, where sea ice is confined by the surrounding continents, mean sea ice thickness is 3–4 m and multiyear ice can exist. Around Antarctica, the sea ice is unimpeded and spreads out extensively, but the mean thickness of the ice there is typically much less: 1–2 m. Sea ice caps the ocean and interferes with ocean-atmosphere exchange of heat, moisture, and other gases.

Melting sea ice diminishes salinity and density of the surface water, which impedes the heat exchange in the upper ocean mixing layer. It has been suggested that this could affect the convective overturning of ocean waters and formation of deep waters, which could weaken the thermohaline circulation (Bryden et al., 2005; Curry and Mauritzen, 2005). The greatest effect of the reduced snow and ice in polar regions, however, is related to the reduced albedo of the warmer polar ocean. For example, the appearance of darker ocean surfaces that are free of ice, which absorb more solar radiation, further warms the ocean, thus strengthening the ice-albedo positive feedback that amplifies initial warming. Diminished sea ice also increases moisture fluxes into the atmosphere, which may increase fog and low clouds, adding further complexity to the net albedo change.

3.2.4 Atmospheric composition

The equilibrium between incoming solar radiation and the long-wave radiation re-emitted by Earth's surface and atmosphere back into space determine the radiation balance of the planet. About 50% of all solar energy is absorbed by the land and ocean, ~20% is absorbed by the atmosphere and clouds, and the remaining ~30% is reflected back to space by clouds, Earth's surface, and different gases and particles in the atmosphere. Without the atmosphere, Earth would be a cold and uninhabitable place. The existence of atmosphere increases the average planetary temperature by more than 15°C, due to the greenhouse effect of some atmospheric compounds. The most important of them are water vapour (H_2O), ozone (O_3), carbon dioxide (CO_2), and methane (CH_4). The most effective among these is H_2O vapour, which adsorbs between 50% and 67% of all long-wave radiation emitted from the planet (Schmidt et al., 2010).

All radiatively active gases forces Earth's energy budget out of balance, mainly by absorbing thermal infrared energy radiated by the surface. Their efficiency in trapping long wave radiation is maximal at the cold tropopause, due to the lower effective temperature of emission. Kept in the troposphere, Earth's infrared radiation warms the atmosphere and surface. Unlike water vapour, CO_2 has a much longer lifetime, allowing its homogeneous distribution over the globe. For this reason, most scientists believe that the enhanced concentration of CO_2, due to human activities like cement production, deforestation as well as the burning of fossil fuels like coal, oil, and natural gas, is the main driver of contemporary global warming (e.g. IPCC, 2007, 2013). The main problem of this hypothesis, however, is the much lower efficiency of CO_2 greenhouse warming. Model experiments of Schmidt et al. (2010) show that the impact of CO_2 in planetary warming varies between 19% and 24%, for cloudy and clear sky conditions, respectively. This means that without H_2O vapour feedback, the increased density of CO_2 is not enough to explain the climate warming observed during the second half of the 20th century.

For this reason, the water vapour feedback effect is supposed to be the main reinforcing factor of the global warming, triggered by the enhanced concentration of greenhouse gases (IPCC, 2007, 2013 and references therein). However, the rise of lower tropospheric humidity, along with the progressing increase of the surface temperature, accounts for only about 10% of the total greenhouse power of the water vapour. In fact, the strongest influence on Earth's radiation balance has a tiny amount of humidity in the upper troposphere. Numerical experiments with general circulation models (Spencer and Braswell, 1997) and satellite measurements (Inamdar et al., 2004) reveal that about 90% of the whole greenhouse effect of the atmospheric water vapour is accounted for by upper tropospheric humidity.

Atmospheric ozone is another radiatively active gas with a well-determined influence on the stratospheric temperature. However, its integral radiative effect on the surface temperature is quite small, due to the high cancelation between the negative forcing of contemporary reduced stratospheric O_3, and positive forcing of the enhanced density of tropospheric O_3 (Gauss et al., 2006). The lower stratospheric ozone, however, could affect the tropopause temperature significantly, and consequently the near tropopause humidity (Kilifarska, 2012, 2013). This could explain the increased climate sensitivity to the near tropopause ozone density, noticed long ago by many authors (Manabe and Strickler, 1964; Manabe and Wetherald, 1967; Ramanathan et al., 1976; de Forster and Shine, 1997; Stuber et al., 2001; etc.).

3.2.5 Climate modes

Historical data indicates that changes of Earth's climate take place with a distinctive spatial pattern(s) that may be characterized by several modes (repeating patterns of time-space variability) of the climate system. In meteorology and climatology, the term 'mode' is often used to describe a spatial structure with at least two strongly coupled centres of action (Wang and Schimel, 2003). Its polarity and amplitude are represented by the index of the mode. The choice of such a simplified description of the climate system is argued by the fact that the temporal variation and changes of the mode index are more predictable than climate anomalies at individual stations.

There are several naturally occurring dynamical modes detected within Earth's climate system. These include the North Atlantic Oscillation/Northern Annular Mode (NAO/NAM) and Southern Annular Mode (SAM), in the Northern and Southern hemisphere extra-tropics, respectively, El Niño-Southern Oscillation (ENSO), and various other well-known modes such as the Pacific North American (PNA) pattern, Pacific Decadal Oscillation (PDO), Atlantic Multidecadal Oscillation (AMO), etc. According to recent knowledge, the dynamical modes of Earth's climate system result from fundamental physical processes such as "instabilities of the climatological mean flow, large-scale atmosphere-ocean interaction, or interactions between the climatological mean flow and transients" (Wallace, 2000). Large-scale climate variability is often determined by interactions between several climate modes, resulting in synchronized behaviour in regions that are geographically far apart. The rapid warming in the last three decades of the 20th century coincides with a trend or phase shift in several leading modes of climate variability. In addition, various data analyses have indicated that observed climate changes in the past several decades are statistically related to trends in some of the leading modes, with modal changes explaining a major part of the recent temperature warming. These results tempt some researchers to explain climate changes simply as a phase shift or structural changes of natural climate modes. However, the experts' conclusion of the fifth IPCC report is that "internal variability alone cannot account for the observed warming since 1951." But much of the future manifestation of the climate warming is expected to occur as a distinctive spatial pattern and changes in the polarity, frequency, and/or intensity of naturally occurring modes of the climate system (Wang and Schimel, 2003; IPCC, 2013).

3.2.6 Life systems

Throughout Earth's history, climate variations have been accompanied by an appearance and disappearance of different ecosystems and species. A particular climate can be favourable to one species and devastating to another. As the climate changes, species and ecosystems respond by adapting, migrating, or reducing their population. Gradual shifts in the climate are easier to adapt to. Rapid changes, however, affect ecosystems' and species' ability for adaptation, and in such periods biodiversity is usually decreased. Studies of Earth's climatic history indicate that abrupt changes in the past climate have resulted in dramatic shifts in ecosystems.

On the other hand, life systems affect the composition of the atmosphere, and therefore the climate, because they take in and release gases like oxygen, carbon dioxide, methane, etc. The human impact on the environment is quite substantial. The enhanced emission of greenhouse gases (through burning of fossil fuels and industrial activities) and the human-driven changes in land use and land cover (i.e. deforestation, urbanization and shifts in vegetation patterns) also alter the climate, resulting in changes in the reflectivity of the Earth's surface (albedo), emissions from burning forests, urban heat island effects, and changes in the natural water cycle.

The abundance of greenhouse gases in the atmosphere is controlled by biogeochemical and carbonate-silicate geochemical cycles that continually move these components between their ocean, land, life, and atmospheric reservoirs. The abundance of carbon in the atmosphere is reduced through seafloor accumulation of marine sediments and accumulation of plant biomass. On long timescales, carbon dioxide production is balanced by the consumption during weathering of igneous and metamorphic rocks. Consequently, there is a dynamical balance between different components of the climate system. Their interconnectivity means that a significant change in any one component can influence the equilibrium of Earth's entire system.

3.3 Regional character of climate variations—Challenge for all existing hypotheses

The variations of Earth's climate system—with amplitude exceeding its internal fluctuations—should inevitably be attributed to variations of external forcings. The explanation of the well-defined positive trend in the surface temperature since the beginning of the 20th century (although with different strength) requires an increase of the amplitude of at least one of the exerted external forcings. Examination of the factors, having centennial-scale variability, reveals that neither solar luminosity nor the volcanic activity meet this requirement. Only two of all potential forcings of climate variations encounter this requirement: (i) continuously increasing CO_2 concentration, due to the anthropogenic forcing; and (ii) the negative centennial trend in galactic cosmic rays intensity, starting at the beginning of the 20th century (refer to Figs 3.5 and 3.9).

The pioneering work of Svante Arrhenius, dating back to the end of the 19th century, raised the awareness of the scientific community regarding the potential global warming of the Earth planet by human-made CO_2. Since that time, it is speculated that observed climate changes—particularly the rapid surface warming during the last two or three decades of the 20th century—may have resulted from the accumulation of anthropogenic greenhouse gases in the atmosphere. Although climatic changes are often described by a trend of the global or hemispherical means of the climate system, the observed centennial changes have distinctive spatial patterns (IPCC, 2013, chapter 14). As well as the well mixed carbon dioxide could not account for the regional specificity of climate changes, the latter are usually associated with the regional climatic modes of atmospheric and/or oceanic variability (e.g. NAO, ENSO, PDO). On the other hand, an assessment of changes in the climatic modes of variability can be problematic for several reasons, such as difficulties in separation of the natural variations and forced responses, or nonlinear relations between different climatic modes (IPCC, 2013, chapter 14). Moreover, the physical basis of climatic modes is not well understood, which means that the ambiguities in climate behaviour may be attributed to something equally vague.

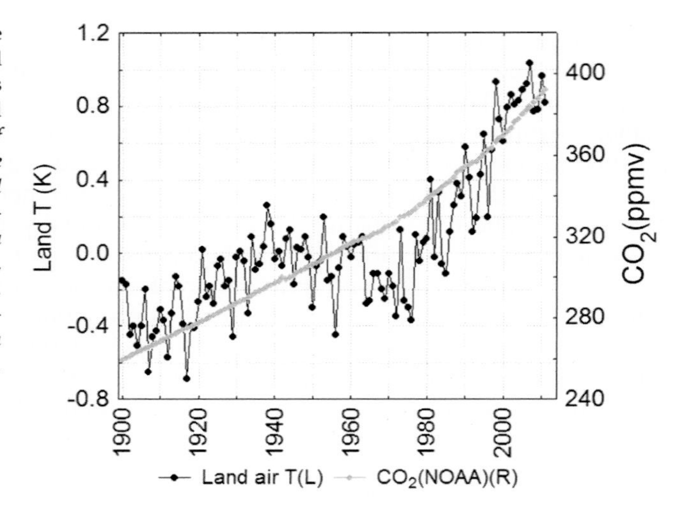

FIG. 3.9 Time series of carbon dioxide (CO_2) and the Northern Hemisphere annual deviations of land air temperature from its mean value (calculated over the period 1961–90). *Data sources: In-situ measurements of CO_2 at Mauna Loa observatory, Hawaii since 1958, interpolated back to 1900 by an exponential function; CRUTEM4 database, derived by the Climatic Research Unit of University of East Anglia (Jones, P.D., Lister, D.H., Osborn, T.J., Harpham, C., Salmon, M., Morice, C.P., 2012. Hemispheric and large-scale land-surface air temperature variations: an extensive revision and an update to 2010. J. Geophys. Res. Atmos. 117. https://doi.org/10.1029/2011JD017139).*

On the other hand, the other possibility mentioned in the beginning of this section—the GCR impact to variations of climate—has not been a subject of serious consideration, due to the lack of plausible physical mechanism. Therefore, the main purpose of this book is to show evidence for such an influence, as well as to propose a physically justifiable mechanism. Unlike the homogeneously distributed CO_2, the GCR reaching the lower atmosphere is heterogeneously distributed over the globe, due to the uneven distribution of the geomagnetic field, which modulates its flux intensity and depth of penetration in the atmosphere (for more details, see Chapter 5). This book will show that the regional specificity of climate change could be reasonably attributed to this heterogeneously distributed forcing over the globe.

References

Abe-Ouchi, A., Saito, F., Kawamura, K., Raymo, M.E., Okuno, J., Takahashi, K., Blatter, H., 2013. Insolation-driven 100,000-year glacial cycles and hysteresis of ice-sheet volume. Nature 500, 190–193. https://doi.org/10.1038/nature12374.

Almeida, A., Gusev, A., Mello, M.G.S., Martin, I.M., Pugacheva, G., Pankov, V.M., Spjeldvik, W.N., Schuch, N.J., 2004. Rainfall cycles with bidecadal periods in the Brazilian region. Geofis. Int. 43 (2), 271–279.

Bakhmutov, V.G., Martazinova, V.F., Ivanova, E.K., Melnyk, G.V., 2011. Changes of the main magnetic field and the climate in the XX-th century. Rep. Natl. Acad. Sci. Ukr. 7, 90–94.

Barlyaeva, T.V., 2013. External forcing on air–surface temperature: geographical distribution of sensitive climate zones. J. Atmos. Sol. Terr. Phys. 94, 81–92. https://doi.org/10.1016/j.jastp.2012.12.014.

Beer, J., Mende, W., Stellmacher, R., 2000. The role of the sun in climate forcing. Quat. Sci. Rev. 19, 403–415. https://doi.org/10.1016/S0277-3791(99)00072-4.

Boron, E.J., 1996. Carroll, L. (Ed.), Climatic variation in earth history. University Science Books, UCAR, Sausalito, CA, p. 25. https://opensky.ucar.edu/islandora/object/research%3A40/datastream/OBJ/view.

Brönnimann, S., Ewen, T., Griesser, T., Jenne, R., 2006. Multidecadal signal of solar variability in the upper troposphere during the 20th century. Space Sci. Rev. 125, 305–317. https://doi.org/10.1007/s11214-006-9065-2.

Bryden, H.L., Longworth, H.R., Cunningham, S.A., 2005. Slowing of the Atlantic meridional overturning circulation at 25°N. Nature 438, 655–657. https://doi.org/10.1038/nature04385.

Bucha, V., Zikmunda, O., 1976. Variations of the geomagnetic field, the climate and weather. Stud. Geophys. Geod. 20, 149–167. https://doi.org/10.1007/BF01626048.

Bucha, V., Taylor, R.E., Berger, R., Haury, E.W., 1970. Geomagnetic intensity: changes during the past 3000 years in the Western Hemisphere. Science 168, 111–114. https://doi.org/10.1126/science.168.3927.111.

Butchvarova, V., Kovacheva, M., 1993. European changes in the paleotemperature and Bulgarian rchaeomagnetic data. Bulg. Geophys. J. 19, 19–23.

Chapanov, Y., Vondrák, J., Ron, C., 2009. Common 22-year cycles of Earth rotation and solar activity. Proc. Int. Astron. Union 5, 407–409. https://doi.org/10.1017/S1743921309993000.

Chen, G., Sun, L., 2011. Mechanisms of the tropical upwelling branch of the Brewer–Dobson circulation: the role of extratropical waves. J. Atmos. Sci. 68, 2878–2892. https://doi.org/10.1175/JAS-D-11-044.1.

Cohen, J.L., Furtado, J.C., Barlow, M., Alexeev, V.A., Cherry, J.E., 2012. Asymmetric seasonal temperature trends. Geophys. Res. Lett. 39. https://doi.org/10.1029/2011GL050582.

Courtillot, V., Mouel, J.L.L., Ducruix, J., Cazenave, A., 1982. Geomagnetic secular variation as a precursor of climatic change. Nature 297, 386. https://doi.org/10.1038/297386a0.

Courtillot, V., Gallet, Y., Le Mouël, J.-L., Fluteau, F., Genevey, A., 2007. Are there connections between the Earth's magnetic field and climate? Earth Planet. Sci. Lett. 253, 328–339. https://doi.org/10.1016/j.epsl.2006.10.032.

Curry, R., Mauritzen, C., 2005. Dilution of the northern North Atlantic Ocean in recent decades. Science 308, 1772–1774. https://doi.org/10.1126/science.1109477.

Dai, A., Fyfe, J.C., Xie, S.-P., Dai, X., 2015. Decadal modulation of global surface temperature by internal climate variability. Nat. Clim. Chang. 5, 555–559. https://doi.org/10.1038/nclimate2605.

de Forster, P.M.F., Shine, K.P., 1997. Radiative forcing and temperature trends from stratospheric ozone changes. J. Geophys. Res. Atmos. 102, 10841–10855. https://doi.org/10.1029/96JD03510.

Elsasser, W., Ney, E.P., Winckler, J.R., 1956. Cosmic-ray intensity and geomagnetism. Nature 178, 1226. https://doi.org/10.1038/1781226a0.

Farneti, R., Vallis, G.K., 2011. Mechanisms of interdecadal climate variability and the role of ocean–atmosphere coupling. Clim. Dyn. 36, 289–308. https://doi.org/10.1007/s00382-009-0674-9.

Foukal, P., Fröhlich, C., Spruit, H., Wigley, T.M.L., 2006. Variations in solar luminosity and their effect on the Earth's climate. Nature 443, 161–166. https://doi.org/10.1038/nature05072.

Frame, T.H.A., Gray, L.J., 2009. The 11-yr solar cycle in ERA-40 data: an update to 2008. J. Clim. 23, 2213–2222. https://doi.org/10.1175/2009JCLI3150.1.

Frank, M., 2000. Comparison of cosmogenic radionuclide production and geomagnetic field intensity over the last 200 000 years. Philos. Trans. R. Soc. Lond. Ser. A 358, 1089–1107.

Friis-Christensen, E., Lassen, K., 1991. Length of the solar cycle: an indicator of solar activity closely associated with climate. Science 254, 698–700. https://doi.org/10.1126/science.254.5032.698.

Fröhlich, C., 2006. Solar irradiance variability since 1978. Space Sci. Rev. 125, 53–65. https://doi.org/10.1007/s11214-006-9046-5.

Gallet, Y., Genevey, A., Fluteau, F., 2005. Does Earth's magnetic field secular variation control centennial climate change? Earth Planet. Sci. Lett. 236, 339–347. https://doi.org/10.1016/j.epsl.2005.04.045.

Gauss, M., Myhre, G., Isaksen, I.S.A., Grewe, V., Pitari, G., Wild, O., Collins, W.J., Dentener, F.J., Ellingsen, K., Gohar, L.K., Hauglustaine, D.A., Iachetti, D., Lamarque, F., Mancini, E., Mickley, L.J., Prather, M.J., Pyle, J.A., Sanderson, M.G., Shine, K.P., Stevenson, D.S., Sudo, K., Szopa, S., Zeng, G., 2006. Radiative forcing since preindustrial times due to ozone change in the troposphere and the lower stratosphere. Atmos. Chem. Phys. 6, 575–599. https://doi.org/10.5194/acp-6-575-2006.

Georgieva, K., Kirov, B., Tonev, P., Guineva, V., Atanasov, D., 2007. Long-term variations in the correlation between NAO and solar activity: the importance of north–south solar activity asymmetry for atmospheric circulation. Adv. Space Res. 40, 1152–1166. https://doi.org/10.1016/j.asr.2007.02.091.

Georgieva, K., Kirov, B., Koucká Knížová, P., Mošna, Z., Kouba, D., Asenovska, Y., 2012. Solar influences on atmospheric circulation. In: Recent Progress in the Vertical Coupling in the Atmosphere-Ionosphere System.J. Atmos. Sol. Terr. Phys. pp. 90–91. https://doi.org/10.1016/j.jastp.2012.05.010 15–25.

Gleisner, H., Thejll, P., 2003. Patterns of tropospheric response to solar variability. Geophys. Res. Lett. 30. https://doi.org/10.1029/2003GL017129.

Gray, L.J., Beer, J., Geller, M., Haigh, J.D., Lockwood, M., Matthes, K., Cubasch, U., Fleitmann, D., Harrison, G., Hood, L., Luterbacher, J., Meehl, G.A., Shindell, D., van Geel, B., White, W., 2010. Solar influences on climate. Rev. Geophys. 48. https://doi.org/10.1029/2009RG000282.

Haigh, J.D., 2003. The effects of solar variability on the Earth's climate. Philos. Trans. R. Soc. Lond. Ser. A 361, 95–111. https://doi.org/10.1098/rsta.2002.1111.

Haigh, J.D., Blackburn, M., Day, R., 2005. The response of tropospheric circulation to perturbations in lower-stratospheric temperature. J. Clim. 18, 3672–3685. https://doi.org/10.1175/JCLI3472.1.

Harrison, C.G.A., Funnell, B.M., 1964. Relationship of palæomagnetic reversals and micropalæontology in two late cænozoic cores from the Pacific Ocean. Nature 204, 566. https://doi.org/10.1038/204566a0.

Hays, J.D., Imbrie, J., Shackleton, N.J., 1976. Variations in the Earth's orbit: pacemaker of the Ice Ages. Science 194, 1121–1132. https://doi.org/10.1126/science.194.4270.1121.

Herschel, W., 1801. Observations tending to investigate the nature of the sun, in order to find the causes or symptoms of its variable emission of light and heat: with remarks on the use that may possibly be drawn from solar observations. Philos. Trans. R. Soc. Lond. 91, 265–318. https://doi.org/10.1098/rstl.1801.00.

Holton, J.R., Haynes, P.H., McIntyre, M.E., Douglass, A.R., Rood, R.B., Pfister, L., 1995. Stratosphere–troposphere exchange. Rev. Geophys. 33, 403–439.

Hood, L.L., Soukharev, B.E., 2012. The lower-stratospheric response to 11-yr solar forcing: coupling to the troposphere–ocean response. J. Atmos. Sci. 69, 1841–1864. https://doi.org/10.1175/JAS-D-11-086.1.

Inamdar, A.K., Ramanathan, V., Loeb, N.G., 2004. Satellite observations of the water vapor greenhouse effect and column longwave cooling rates: relative roles of the continuum and vibration-rotation to pure rotation bands. J. Geophys. Res. Atmos. 109. https://doi.org/10.1029/2003JD003980.

IPCC, 2007. In: Solomon, S. et al., (Eds.), (Intergovernmental Panel on Climate Change), Climate Change 2007: The Physical Science Basis. Cambridge University Press, New York 996. pp.

IPCC, 2013. In: Stocker, T.F. et al., (Eds.), (Intergovernmental Panel on Climate Change), Climate Change 2013: The Physical Science Basis—Contribution of Working Group I. Cambridge University Press, Cambridgeand New York, NY. 1535 pp.

Jokipii, J.R., 1989. The physics of cosmic-ray modulation. Adv. Space Res. 9, 105–119. https://doi.org/10.1016/0273-1177(89)90317-7.

Kilifarska, N.A., 2011. Long-term variations in the stratospheric winter time ozone variability—22 year cycle. C. R. Acad. Bulg. Sci. 64 (6), 867–874.

Kilifarska, N.A., 2012. Mechanism of lower stratospheric ozone influence on climate. IREPHY 6 (3), 279–290.

Kilifarska, N.A., 2013. An autocatalytic cycle for ozone production in the lower stratosphere initiated by galactic cosmic rays. C. R. Acad. Bulg. Sci. 66 (2), 243–252.

Kilifarska, N.A., 2015. Bi-decadal solar influence on climate, mediated by near tropopause ozone. J. Atmos. Sol. Terr. Phys. 136, 216–230. https://doi.org/10.1016/j.jastp.2015.08.005. SI:Vertical Coupling.

Kilifarska, N.A., Bakhmutov, V.G., Melnyk, G.V., 2013. Geomagnetic influence on Antarctic climate—evidence and mechanism. Int. Rev. Phys. 7 (3), 242–252.

King, J.W., 1974. Weather and the Earth's magnetic field. Nature 247, 131. https://doi.org/10.1038/247131a0.

Kodera, K., 2002. Solar cycle modulation of the North Atlantic Oscillation: implication in the spatial structure of the NAO. Geophys. Res. Lett. 29, 59-1–59-4. https://doi.org/10.1029/2001GL014557.

Kodera, K., 2004. Solar influence on the Indian Ocean Monsoon through dynamical processes. Geophys. Res. Lett. 31. https://doi.org/10.1029/2004GL020928.

Kodera, K., Kuroda, Y., 2002. Dynamical response to the solar cycle. J. Geophys. Res. Atmos. 107, ACL 5-1–ACL 5-12. https://doi.org/10.1029/2002JD002224.

Kopp, G., Lawrence, G., Rottman, G., 2005. The total irradiance monitor (TIM): science results. In: Rottman, G., Woods, T., George, V. (Eds.), The Solar Radiation and Climate Experiment (SORCE): Mission Description and Early Results. Springer New York, New York, NY, pp. 129–139. https://doi.org/10.1007/0-387-37625-9_8.

Kopp, G., Krivova, N., Wu, C.J., Lean, J., 2016. The impact of the revised sunspot record on solar irradiance reconstructions. Sol. Phys. 291, 2951–2965. https://doi.org/10.1007/s11207-016-0853-x.

Kug, J.-S., Jeong, J.-H., Jang, Y.-S., Kim, B.-M., Folland, C.K., Min, S.-K., Son, S.-W., 2015. Two distinct influences of Arctic warming on cold winters over North America and East Asia. Nat. Geosci. 8, 759–762. https://doi.org/10.1038/ngeo2517.

Labitzke, K., Loon, H.V., 1995. Connection between the troposphere and stratosphere on a decadal scale. Tellus A 47, 275–286. https://doi.org/10.1034/j.1600-0870.1995.t01-1-00008.x.

Lassen, K., Friis-Christensen, E., 1995. Variability of the solar cycle length during the past five centuries and the apparent association with terrestrial climate. J. Atmos. Terres. Phys. 57, 835–845. https://doi.org/10.1016/0021-9169(94)00088-6.

Lean, J., 2000. Evolution of the Sun's spectral irradiance since the maunder minimum. Geophys. Res. Lett. 27, 2425–2428. https://doi.org/10.1029/2000GL000043.

Lean, J., Rind, D., 1998. Climate forcing by changing solar radiation. J. Clim. 11, 3069–3094. https://doi.org/10.1175/1520-0442(1998)011<3069:CFBCSR>2.0.CO;2.

Li, C., Stevens, B., Marotzke, J., 2015. Eurasian winter cooling in the warming hiatus of 1998–2012. Geophys. Res. Lett. 42, 8131–8139. https://doi.org/10.1002/2015GL065327.

Lorenz, E.N., 1968. Climatic determinism. In: Causes of Climatic Change. Meteorological Monographs, No. 30, American Meteorological Society, pp. 1–3.

Lorenz, E.N., 1970. Climatic change as a mathematical problem. J. Appl. Meteorol. 9, 325–329.

Lorenz, E.N., 1976. Nondeterministic theories of climatic change. Quat. Res. 6, 495–506.

Manabe, S., Strickler, R.F., 1964. Thermal equilibrium of the atmosphere with a convective adjustment. J. Atmos. Sci. 21, 361–385. https://doi.org/10.1175/1520-0469(1964)021<0361:TEOTAW>2.0.CO;2.

Manabe, S., Wetherald, R.T., 1967. Thermal equilibrium of the atmosphere with a given distribution of relative humidity. J. Atmos. Sci. 24, 241–259. https://doi.org/10.1175/1520-0469(1967)024<0241:TEOTAW>2.0.CO;2.

Mann, M.E., Steinman, B.A., Miller, S.K., 2014. On forced temperature changes, internal variability, and the AMO. Geophys. Res. Lett. 41, 3211–3219. https://doi.org/10.1002/2014GL059233.

Marshall, J., Kushnir, Y., Battisti, D., Chang, P., Czaja, A., Dickson, R., Hurrell, J., McCartney, M., Saravanan, R., Visbeck, M., 2001. North Atlantic climate variability: phenomena, impacts and mechanisms. Int. J. Climatol. 21, 1863–1898. https://doi.org/10.1002/joc.693.

Meehl, G.A., Arblaster, J.M., Matthes, K., Sassi, F., van Loon, H., 2009. Amplifying the Pacific climate system response to a small 11-year solar cycle forcing. Science 325, 1114–1118. https://doi.org/10.1126/science.1172872.

Miyahara, H., Yokoyama, Y., Masuda, K., 2008. Possible link between multi-decadal climate cycles and periodic reversals of solar magnetic field polarity. Earth Planet. Sci. Lett. 272, 290–295. https://doi.org/10.1016/j.epsl.2008.04.050.

Nesme-Ribes, E., Ferreira, E.N., Sadourny, R., Treut, H.L., Li, Z.X., 1993. Solar dynamics and its impact on solar irradiance and the terrestrial climate. J. Geophys. Res. Space Physics 98, 18923–18935. https://doi.org/10.1029/93JA00305.

Ney, E.P., 1959. Cosmic radiation and the weather. Nature 183, 451–452. https://doi.org/10.1038/183451a0.

Nikolov, T., 2011. Global Climate Change in the History of Earth. Academic Publishing House 'Marin Drinov', Sofia.

Nurgaliev, D.K., 1991. Solar activity, geomagnetic variations, and climate changes. Geomagn. Aeron. 31, 14–18.

Peters, D.H.W., Gabriel, A., Entzian, G., 2008. Longitude-dependent decadal ozone changes and ozone trends in boreal winter months during 1962000. Ann. Geophys. 26, 1275–1286. https://doi.org/10.5194/angeo-26-1275-2008.

Plumb, R.A., Eluszkiewicz, J., 1999. The Brewer–Dobson circulation: dynamics of the tropical upwelling. J. Atmos. Sci. 56, 868–890. https://doi.org/10.1175/1520-0469(1999)056<0868:TBDCDO>2.0.CO;2.

Plumb, R.A., Semeniuk, K., 2003. Downward migration of extratropical zonal wind anomalies. J. Geophys. Res. Atmos. 108. https://doi.org/10.1029/2002JD002773.

Potgieter, M.S., 2014. The charge-sign dependent effect in the solar modulation of cosmic rays. Adv. Space Res. Cosmic Ray Origins: Viktor Hess Centennial Anniversary. 53, 1415–1425. https://doi.org/10.1016/j.asr.2013.04.015.

Qu, W., Zhao, J., Huang, F., Deng, S., 2012. Correlation between the 22-year solar magnetic cycle and the 22-year quasicycle in the earth's atmospheric temperature. Astron. J. 144, 6. https://doi.org/10.1088/0004-6256/144/1/6.

Ramanathan, V., Callis, L.B., Boughner, R.E., 1976. Sensitivity of surface temperature and atmospheric temperature to perturbations in the stratospheric concentration of ozone and nitrogen dioxide. J. Atmos. Sci. 33, 1092–1112. https://doi.org/10.1175/1520-0469(1976)033<1092:SOSTAA>2.0.CO;2.

Rozelot, J.P., 1995. On the chaotic behaviour of the solar activity. Astron. Astrophys. 297, L45.

Ruzmaikin, A., Feynman, J., 2002. Solar influence on a major mode of atmospheric variability. J. Geophys. Res. Atmos. 107, ACL 7-1–ACL 7-11. https://doi.org/10.1029/2001JD001239.

Scafetta, N., 2010. Empirical evidence for a celestial origin of the climate oscillations and its implications. J. Atmos. Sol. Terr. Phys. 72, 951–970. https://doi.org/10.1016/j.jastp.2010.04.015.

Scherer, K., Fichtner, H., Borrmann, T., Beer, J., Desorgher, L., Flükiger, E., Fahr, H.-J., Ferreira, S.E.S., Langner, U.W., Potgieter, M.S., Heber, B., Masarik, J., Shaviv, N., Veizer, J., 2006. Interstellar-terrestrial relations: variable cosmic environments, the dynamic heliosphere, and their imprints on terrestrial archives and climate. Space Sci. Rev. 127, 327–465. https://doi.org/10.1007/s11214-006-9126-6.

Schmidt, G.A., Ruedy, R.A., Miller, R.L., Lacis, A.A., 2010. Attribution of the present-day total greenhouse effect. J. Geophys. Res. Atmos. 115. https://doi.org/10.1029/2010JD014287.

Scott, R.K., 2002. Wave-driven mean tropical upwelling in the lower stratosphere. J. Atmos. Sci. 59, 2745–2759. https://doi.org/10.1175/1520-0469(2002)059<2745:WDMTUI>2.0.CO;2.

Semeniuk, K., Shepherd, T.G., 2001. Mechanisms for tropical upwelling in the stratosphere. J. Atmos. Sci. 58, 3097–3115. https://doi.org/10.1175/1520-0469(2001)058<3097:MFTUIT>2.0.CO;2.

Seviour, W.J.M., Butchart, N., Hardiman, S.C., 2012. The Brewer–Dobson circulation inferred from ERA-interim. Q. J. R. Meteorol. Soc. 138, 878–888. https://doi.org/10.1002/qj.966.

Shackleton, N.J., Berger, A., Peltier, W.R., 1990. An alternative astronomical calibration of the lower Pleistocene timescale based on ODP Site 677. Earth Environ. Sci. Trans. R. Soc. Edinb. 81, 251–261. https://doi.org/10.1017/S0263593300020782.

Shindell, D.T., Schmidt, G.A., Miller, R.L., Mann, M.E., 2003. Volcanic and solar forcing of climate change during the preindustrial era. J. Clim. 16, 4094–4107. https://doi.org/10.1175/1520-0442(2003)016<4094:VASFOC>2.0.CO;2.

Solanki, S.K., 2002. Solar variability and climate change: is there a link? Astron. Geophys. 43, 5.9–5.13. https://doi.org/10.1046/j.1468-4004.2002.43509.x.

Soukharev, B.E., Hood, L.L., 2006. Solar cycle variation of stratospheric ozone: multiple regression analysis of long-term satellite data sets and comparisons with models. J. Geophys. Res. Atmos. 111. https://doi.org/10.1029/2006JD007107.

Souza Echer, M.P., Echer, E., Rigozo, N.R., Brum, C.G.M., Nordemann, D.J.R., Gonzalez, W.D., 2012. On the relationship between global, hemispheric and latitudinal averaged air surface temperature (GISS time series) and solar activity. J. Atmos. Sol. Terr. Phys. 74, 87–93. https://doi.org/10.1016/j.jastp.2011.10.002.

Spencer, R.W., Braswell, W.D., 1997. How dry is the tropical free troposphere? Implications for global warming theory. Bull. Amer. Meteor. Soc. 78, 1097–1106. https://doi.org/10.1175/1520-0477(1997)078<1097:HDITTF>2.0.CO;2.

Stuber, N., Ponater, M., Sausen, R., 2001. Is the climate sensitivity to ozone perturbations enhanced by stratospheric water vapor feedback? Geophys. Res. Lett. 28, 2887–2890. https://doi.org/10.1029/2001GL013000.

Sun, X., Ren, G., Ren, Y., Fang, Y., Liu, Y., Xue, X., Zhang, P., 2018. A remarkable climate warming hiatus over Northeast China since 1998. Theor. Appl. Climatol. 133, 579–594. https://doi.org/10.1007/s00704-017-2205-7.

Thejll, P., Lassen, K., 2000. Solar forcing of the northern hemisphere land air temperature: new data. J. Atmos. Sol. Terr. Phys. 62, 1207–1213. https://doi.org/10.1016/S1364-6826(00)00104-8.

Tourre, Y.M., Rajagopalan, B., Kushnir, Y., Barlow, M., White, W.B., 2001. Patterns of coherent decadal and interdecadal climate signals in the Pacific Basin during the 20th century. Geophys. Res. Lett. 28, 2069–2072. https://doi.org/10.1029/2000GL012780.

Trenberth, K.E., 2001. Earth system processes. In: Munn, et al., (Eds.), Encyclopedia of Global Environmental Change. The Earth System: Physical and Chemical Dimensions of Global Environmental Change, vol. 1. John & Sons Ltd, pp. 13–30.

Trenberth, K.E., Stepaniak, D.P., 2003. Seamless poleward atmospheric energy transports and implications for the Hadley circulation. J. Clim. 16, 3706–3722. https://doi.org/10.1175/1520-0442(2003)016<3706:SPAETA>2.0.CO;2.

Tung, K.K., Kinnersley, J.S., 2001. Mechanisms by which extratropical wave forcing in the winter stratosphere induces upwelling in the summer hemisphere. J. Geophys. Res. Atmos. 106, 22781–22791. https://doi.org/10.1029/2001JD900228.

van Loon, H., Shea, D.J., 2000. The global 11-year solar signal in July–August. Geophys. Res. Lett. 27, 2965–2968. https://doi.org/10.1029/2000GL003764.

Vieira, L.E.A., da Silva, L.A., Guarnieri, F.L., 2008. Are changes of the geomagnetic field intensity related to changes of the tropical Pacific sea-level pressure during the last 50 years? J. Geophys. Res. Space Phys. 113. https://doi.org/10.1029/2008JA013052.

Vieira, L.E.A., Norton, A., Dudok de Wit, T., Kretzschmar, M., Schmidt, G.A., Cheung, M.C.M., 2012. How the inclination of Earth's orbit affects incoming solar irradiance. Geophys. Res. Lett. 39, L16104. https://doi.org/10.1029/2012GL052950.

Wallace, J.M., 2000. North Atlantic oscillatiodannular mode: two paradigms—one phenomenon. Q. J. R. Meteorol. Soc. 126, 791–805. https://doi.org/10.1002/qj.49712656402.

Wang, G., Schimel, D., 2003. Climate change, climate modes, and climate impacts. Annu. Rev. Environ. Resour. 28, 1–28. https://doi.org/10.1146/annurev.energy.28.050302.105444.

Ward, P.L., 2009. Sulfur dioxide initiates global climate change in four ways. Thin Solid Films 517, 3188–3203. https://doi.org/10.1016/j.tsf.2009.01.005.

Weiss, N.O., 1990. Periodicity and aperiodicity in solar magnetic activity. Philos. Trans. R. Soc. Lond. Ser. A Math. Phys. Sci. 330, 617–625. https://doi.org/10.1098/rsta.1990.0042.

White, W.B., 2006. Response of tropical global ocean temperature to the Sun's quasi-decadal UV radiative forcing of the stratosphere. J. Geophys. Res. Oceans 111. https://doi.org/10.1029/2004JC002552.

White, W.B., Lean, J., Cayan, D.R., Dettinger, M.D., 1997. Response of global upper ocean temperature to changing solar irradiance. J. Geophys. Res. Oceans 102, 3255–3266. https://doi.org/10.1029/96JC03549.

White, W.B., Dettinger, M.D., Cayan, D.R., 2003. Sources of global warming of the upper ocean on decadal period scales. J. Geophys. Res. Ocean 108. https://doi.org/10.1029/2002JC001396.

Wollin, G., Ericson, D.B., Ryan, W.B.F., 1971a. Variations in magnetic intensity and climatic changes. Nature 232, 549. https://doi.org/10.1038/232549a0.

Wollin, G., Ericson, D.B., Ryan, W.B.F., Foster, J.H., 1971b. Magnetism of the earth and climatic changes. Earth Planet. Sci. Lett. 12, 175–183. https://doi.org/10.1016/0012-821X(71)90075-6.

Zotov, L., Bizouard, C., Shum, C.K., 2016. A possible interrelation between Earth rotation and climatic variability at decadal time-scale. Geodesy Geodyn. Special Issue: Geodetic and Geophysical Observations and Applications and Implications. 7, 216–222. https://doi.org/10.1016/j.geog.2016.05.005.

CHAPTER

4

Contemporary evidence for existing relation between geomagnetic and climatic parameters

OUTLINE

4.1 Choosing the methods for climate analysis

The use of linear statistical methods is common practice in recent climate studies. In the case of linear (or quasilinear) and stationary relations between forcing factors and climate response, linear methods provide valuable information about the effectiveness of the applied impact. These methods should be used dominantly for relatively short periods, for which the interactions between forcing factors and climate system are approximately stationary and linear. However, they should be applied with caution, because their reliability depends on three basic assumptions, which are frequently violated in climatic records. Thus the choice of using linear statistics is justifiable only if: (i) time series are stationary (i.e. their mean and standard

https://doi.org/10.1016/B978-0-12-819346-4.00004-8

© 2020 Elsevier Inc. All rights reserved.

deviation are independent of time); (ii) the relations between forcing factors and climate response are linear; and (iii) these relations are stationary with time.

Climate records actually appear to be distinctly nonlinear (King, 1996; Mikšovský and Raidl, 2006; Wu et al., 2007; Halko and Tsilika, 2014; etc.), especially the longer ones. Consequently, if we want to estimate the impact of the examined factors on climate variability as accurately as possible, the nonlinear functions should be fitted to the climate data records. Climate series, in addition, are nonstationary, meaning that functional dependence between climate and forcing factors changes with time. Consequently, each approximately stationary period will be characterized by different linear functional relations between forcing factors and the climate system's response. For this reason, linear statistical methods are not suitable for predictions, especially long-term ones. However, they could help us to understand the magnitude and nature of interactions in the climate system, giving insights into the mechanisms responsible for climate change. Understanding of mechanisms can add to understanding of climate dynamics on all timescales (King, 1996), which allow physical modelling.

4.1.1 Linear statistics as an option for time series analyses

Determination of the linear trend, in order to get rid of it in the analysed time series, is one of the first goals of linear statistical methods. In the most common understanding, the trend (i.e. the tendency over the examined data record) consists of the straight-line best fit, which is assumed to continue into the future, when new observations will be available. However, examination of sufficiently long data records reveals that the latter assumption is usually not warranted. Fig. 4.1 illustrates this point, presenting the zonally averaged air surface temperature at 40°N latitude, for the period 1957–2015 (continuous grey line with circles), its 11-year running average (black squares), and trend calculated over the periods 1957–99 and 2000–15 (thick, grey long dashes). Data used are from ERA 40 and ERA Interim reanalyses, merged at January 2000. The nonstationarity of the climate is well demonstrated by the different trends observed until 1999 and after that. It should be noted that the changed slope of the trend after 1999 is not an artefact of merging, because the mean values of both reanalyses have been adjusted before merging. The nonlinear response of the climate system to forcing factors is also clear, and is especially pronounced during the first period (note that climate variability is interpolated by the quadratic trend function). Consequently, estimations of the linear trend for a given time interval make little sense in terms of uncovering the underlying mechanisms driving observed changes, which are likely nonlinear and possibly nonstationary over the examined time window. Fig. 4.1 shows also that temperature variations are better described by a running average curve (black squares) than by the piecewise trends.

Consequently, to reveal the meandering of Earth's climate between warmer and cooler periods (evident in palaeoclimatic time records), it is better to work with dynamically varying means. The problem with this approach is that the researcher must predefine the averaging window. Their choice, however, could be facilitated by: (i) the WMO definition of climate 'norms' (used for climate reference) as 30-year data averages; and (ii) the timescales of variability in which the researcher is interested. Thus for multidecadal and centennial timescales, an averaging window of 11 or 22 years is quite reasonable and has its physical basis; the

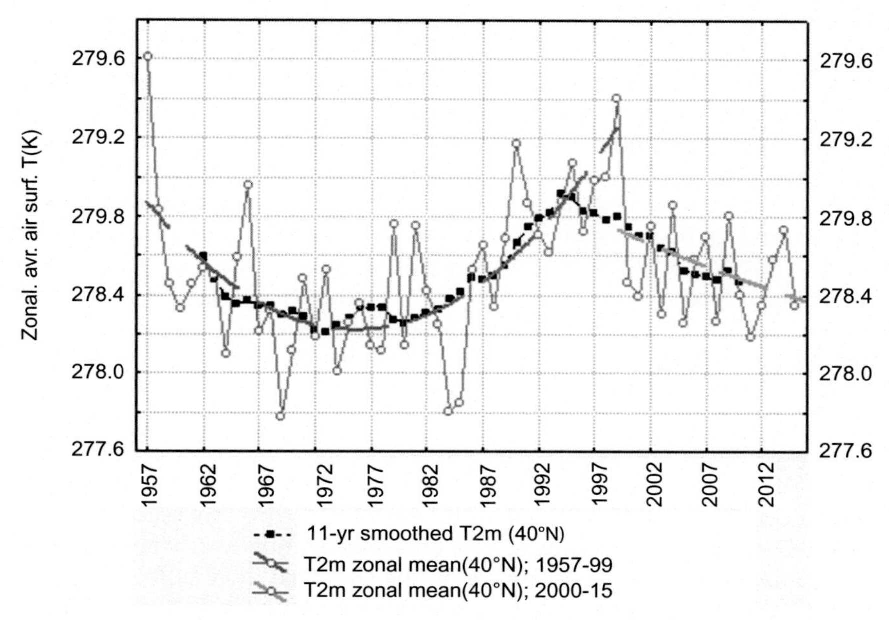

FIG. 4.1 Zonally averaged time record of the winter air surface temperature at 40°N latitude *(grey, continuous line with circles)* and its trends for the periods 1957–99 and 2000–15 *(thick, dashed curves). Data source: ERA40 and ERA Interim merged in January 2000 (long-term averages of two reanalyses have been equalized before merging). From Kilifarska, N., 2017. Mechanism of Relation Between Cosmic Rays, Geomagnetic Field and Earth's Climate, (DSc thesis), Sofia, Bulgaria.*

Schwabe and Hale cycles of solar variability are eliminated or strongly reduced. Another argument for choosing such a time window is the necessity of time for the surface T to achieve its thermal equilibrium. According to Beltrami et al. (2011) the equilibrium surface T at centennial timescales is approached at ~500–600 m below the surface. Assuming the value of $10^{-6}\,m^2\,s^{-1}$ for thermal diffusivity (Beltrami et al., 2011), it takes 15–19 years for the surface temperature anomaly to propagate downward to 500–600 m beneath the surface. For a higher coefficient of diffusivity, i.e. $1.5 \times 10^{-6}\,m^2\,s^{-1}$, the time for achieving climatic equilibrium temperature is shortened to 11–13 years. Bearing in mind the different thermal diffusivity of the ocean and land from one side, and different rock properties from the other, the various delays of the surface T response (to the applied external forcing) become easier to understand.

Another widely used linear technique is multiple regression, intended to attribute the observed climate variations to different impact factors. The accuracy of differentiation between examined variables, however, cannot be ascertained, which makes the separation of the impacts elusive to some extent. Moreover, the list of explanatory (independent) variables is frequently a mixture of different timescales of variability. This means that the long-term variations, which are usually nonlinear, are not adequately detected and attributed by the linear multiple regression method. In addition, climate response to some forcing factors could be delayed, which usually is not taken into account in multiple regression models.

An alternative linear method, overcoming some of multiple regression's deficiencies, is lagged cross-correlation analysis. This allows detection of delayed climate response to a given forcing, which in turn can add to the understanding of climate system dynamics. Below, an estimation is provided of the potential impacts of different factors on the near-surface temperature variations during the past 111 years. Temperature data are from the HadCRUT3v data set for the Northern Hemisphere, which are zonally and latitudinally averaged deviations from the climate 'norm' for the period 1961–90. The examined explanatory variables include carbon dioxide (CO_2), geomagnetic field total intensity in Arosa (Switzerland), multidecadal variations of Sunspot numbers (SSN), galactic cosmic rays (GCRs), and total ozone density from Arosa station. Details for explanatory (independent) variables can be found in Kilifarska (2012a,b) and Kilifarska et al. (2015).

The cross-covariance coefficients at lag m have been calculated by moving the axis of independent variable (i.e. the cause, or forcing parameter) backward in time, i.e.:

$$c_{xy}(k) = \frac{1}{N-1}\sum_{m}(Y_t - \overline{Y})(X_{t+m} - \overline{X}) \quad for\ t = 1\ to\ N;\ \ m = 0\ to\ -k \tag{4.1}$$

where N is the number of observations in the time series. For this reason, the time delay is given in our tables as negative values. The cross-correlation coefficients are calculated as usual by normalization of cross-covariance on the standard deviations of both time series.

Results of the applied lagged correlation analysis, presented in Table 4.1, confirm the IPCC conclusions for the leading role of CO_2 in the contemporary global warming of Earth. The contribution of solar irradiance (presented here by the SSN) and GCRs' variability are either statistically nonsignificant, or significant but very small. The only factors (besides CO_2) having a relatively substantial influence on the temperature during the period 1990–2010 are the total ozone density (TOZ) and, surprisingly, Earth's magnetic field.

Examination of the variables time series clarifies the reason for the very high correlation of carbon dioxide and geomagnetic field with surface temperature (see Fig. 4.2A). It is easily noticeable that both forcing factors evolve quasilinearly with time, especially in the second half of the 20th century, when global warming occurred. Consequently, the applied linear statistical method detects the positive linear trends of independent (e.g. CO_2 or geomagnetic field intensity) and dependent (surface temperature) time series. As a result, the calculated cross-correlation coefficients is high.

The other factors, like solar electromagnetic radiation (represented by SSN) and galactic cosmic rays, have well-pronounced decadal variability, which are not clearly defined and persistent in the surface temperature (see Fig. 4.3). Moreover, SSN, GCRs, and total ozone (TOZ) have a well-defined nonlinear multidecadal variability (see Fig. 4.2B). Thus, the

TABLE 4.1 Lagged correlation coefficients of land air temperature (T) with different forcing factors, statistically significant at 2σ level.

	CO_2	Geomag. field	GCR	TOZ	11-yr smt. SSN	22-yr smt. GCR	11-yr smt. TOZ
R	**0.82**	**0.65**	**−0.3**	**−0.53**	**0.38**	**−0.52**	**−0.68**
Lag	(0)	(0)	(−20)	(0)	(−16)	(−12)	(0)

The 2σ significance (bold) indicates that there is a 5% probability that the relation between the variables found in our samples is a "fluke." The numbers in brackets show the time delay of the T response in years.

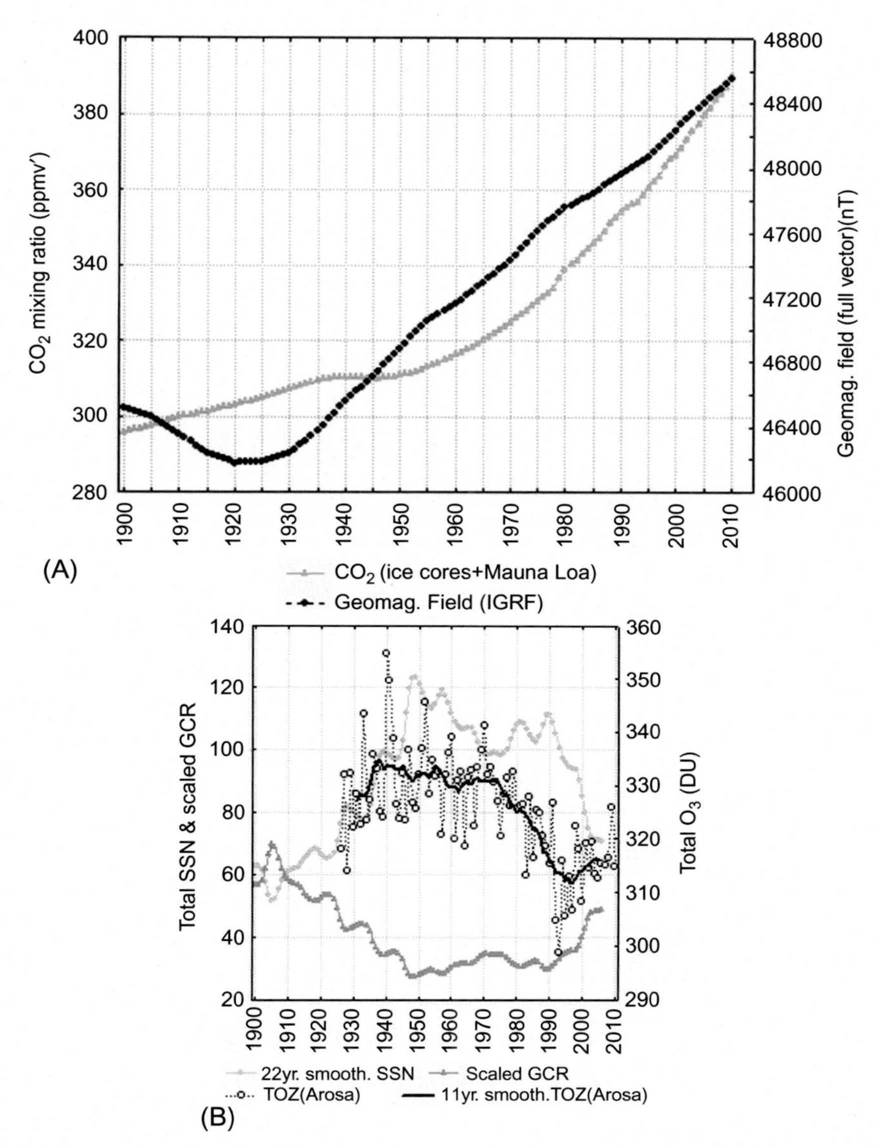

FIG. 4.2 (A) Time series of yearly CO_2 values *(grey curve)* and the geomagnetic field intensity *(black curve)*; (B) 22-year smoothed sunspot numbers (SSN; *grey dots*) and galactic cosmic rays (GCRs; *dark grey triangles*), with annual values of total ozone density (TOZ; *black circles*) from Arosa, Switzerland and its centennial nonlinear trend derived by 11-year running window *(continuous black curve). From Kilifarska, N., 2017. Mechanism of Relation Between Cosmic Rays, Geomagnetic Field and Earth's Climate, (DSc thesis), Sofia, Bulgaria.*

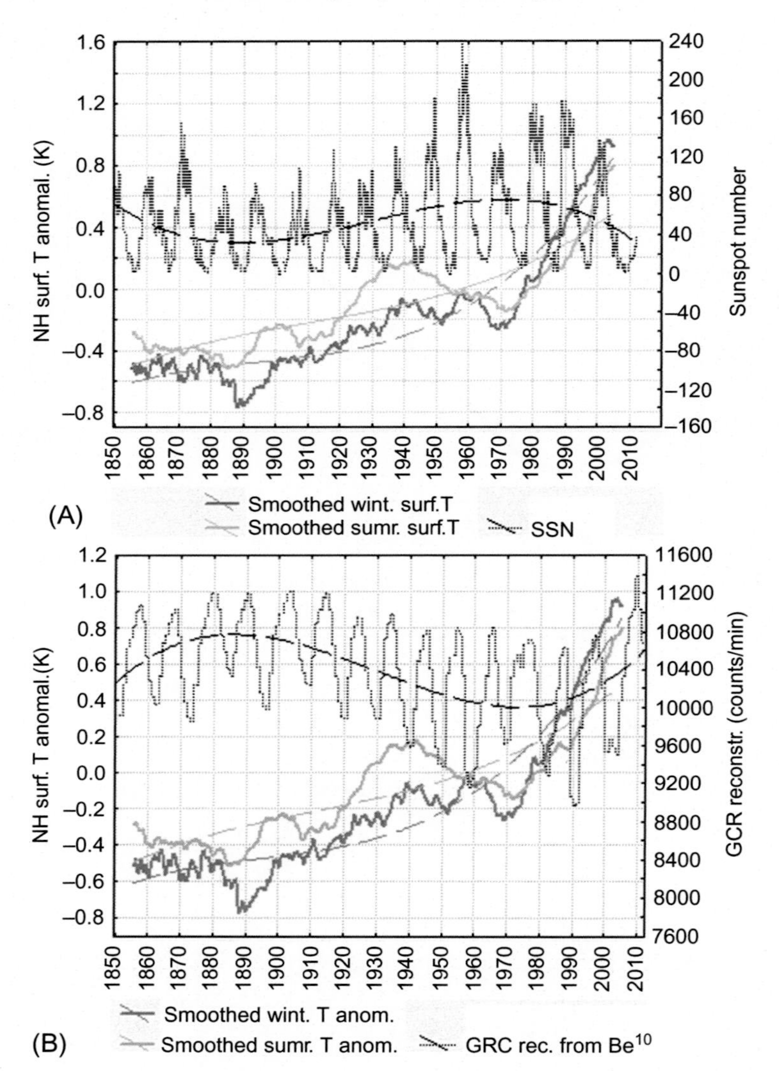

FIG. 4.3 Time series of land surface air temperature *(winter = dark grey; summer = pale grey)* compared with: (panel A) sunspot numbers (SSN) = *black dots*, and (panel B) galactic cosmic rays (GCR) = *black dots*. The polynomial fits of each curve are shown with *long dashes. Data source: Climate Research Unit, University of East Anglia, CRUTEM4 data for temperature; centennial reconstruction of GCR is from Usoskin, I.G., et al., 2008. Cosmic Ray Intensity Reconstruction. IGBP PAGES/World Data Center for Paleoclimatology, Data Contribution Series # 2008-013, NOAA/NCDC Paleoclimatology Program, Boulder, CO; data for SSN are from WDC-SILCO, Royal Observatory of Belgium, Brussels. From Kilifarska, N., 2017. Mechanism of Relation Between Cosmic Rays, Geomagnetic Field and Earth's Climate, (DSc thesis), Sofia, Bulgaria.*

correlation analysis has found much less similarity (or none at all) with smoothly increasing surface temperature. Note, however, that correlation with smoothed SSN, GCRs, and TOZ is much higher (see Table 4.1).

4.1.2 Some results from nonlinear statistics

Climate nonlinearity is frequently identified with chaos or randomness and consequently with its unpredictability (e.g. IPCC, 2007). This point of view erroneously implies that nonlinear climate variability cannot be discriminated from its random fluctuations. This section aims to show that the correct attribution of given factors' impact on the overall climate variability depends on the knowledge of its temporal evolution, and on the functional dependence between applied impact and climate response. The latter, however, is usually unknown. That is why determination of the nonlinear best fit is not an easy task. The higher regression coefficient in nonlinear fitting does not always correspond to the best functional dependence between dependent and explanatory variables. Consequently the final decision must rely on the whole range of diagnostics provided by statistical packages. For example, residuals (i.e. the differences between measured and modelled values of dependent variable) of a good statistical model should be normally distributed and must not autocorrelate. Moreover, all points of scatterplots presenting *calculated versus normally distributed* residuals or *observed versus predicted* values should be grouped along the regression line, and nonlinear "patterns" (like S shape) should not exist. Additional diagnostic for the good model is the random distribution along the zero line of the points in a *predicted versus residual* scatterplot, etc.

An illustration of one nonlinear technique, a nonlinear estimation, is given below. Some nonlinear functions of analysed forcing factors have been fitted to the variation of the land surface temperature anomalies, determined for the period 1900–2010 (CRUTEM3v data). In addition to the above-mentioned requirements (which should be obeyed by every satisfactory statistical model), a comparison of graphical best fit and modelled fitting function has been used as an additional diagnostics for model accuracy. For this reason, we have restricted the list of explanatory variables to no more than two, because we are able to visualize and compare up to 3D fitting surfaces. The nonlinear fits used are given below:

$$\mathrm{LandT} = b_0 + b_1 CO_2 + b_2 (CO_2)^2$$

$$\mathrm{LandT} = b_0 + b_1 \mathrm{magF} + b_2 (\mathrm{magF})^2$$

$$\mathrm{LandT} = b_0 + b_1 \mathrm{SSN}_{22}^{1.5} + b_2 \mathrm{SSN}_{11}^{0.5} + b_3 \cdot \mathrm{hypot}(\mathrm{SSN}_{11}; \mathrm{SSN}_{22}) + b_4 \mathrm{hypot}(\mathrm{SSN}_{11}; \mathrm{SSN}_{22})^2$$

$$\mathrm{LandT} = b_0 + b_1 \mathrm{TOZ}^{-1} + b_2 (\mathrm{TOZ}_{11})^{-1} + b_3 (\mathrm{TOZ} \cdot \mathrm{TOZ}_{11})^2 + b_4 (\mathrm{TOZ}_{11})^{-2}$$

where $\mathbf{b_i}$ are regression coefficients, function hypot(x,y) is defined as $\mathrm{hypot} = \sqrt{x^2 + y^2}$, and nonlinear estimations have been performed with STATISTICA 8. Nonlinear regression coefficients are given in Table 4.2. Comparison with linear correlation coefficients in Table 4.1 reveals that quasilinearly evolving forcings like CO_2, and to a lesser extent geomagnetic field, gain little by the nonlinear fitted function between their own and surface temperature variations. The nonlinearly evolving explanatory variables like solar irradiance and ozone, on the other hand, reveal better their potential impact in climate variability. Thus Table 4.2 shows

TABLE 4.2 Statistically significant (at 95%) nonlinear regression coefficients of land air temperature (T).

	CO_2	Geomag. field	ann + 11-yr smt. SSN	ann + 11-yr smt. TOZ
R	**0.86**	**0.84**	**0.56**	**0.78**
R^2	74%	71%	31%	61%

The statistical significance at 95% level (bold) indicates that there is a 5% probability that the relation between the variables found in our samples is a "fluke." Percentage variability of surface T, described by each of the variables, is given in the second row.

that the anthropogenic increase of CO_2 density has at least two potential competitors for the leading role in the last global warming period: geomagnetic field and total ozone density. Note that the same functions applied in another version of the software could lead to some differences in calculated regression coefficients.

The above experiments with linear and nonlinear statistical methods illustrate that correct attribution of climate variations to various explanatory variables critically depends on their temporal evolution, and on the functional dependence between applied forcing and climate system response. For analysis of the long-term variations, nonlinear methods are more suitable, because they describe adequately both linear and nonlinear impacts. Unfortunately, they are time-consuming and experience-demanding, and for these reasons are rarely used. Of course, as any statistical method, the nonlinear ones could not be used for prediction, because of continuously changing connectivity between forcing factors and climate response (known as nonstationarity of climate variation). Therefore the main goal of the nonlinear statistic is to understand the mechanisms, which could be built further into the physical models.

4.2 Convolution of palaeomagnetic and palaeoclimate variations

The problem of the relationship between Earth's magnetic field and climate has been explored on the timescales of thousands and millions of years—based on the palaeo-data, and on centennial and multidecadal timescales—using instrumental observations of climate variations and solar and magnetic activity. Numerous studies have addressed the relationship between the climatic trends and the changes in Earth's magnetic field caused by different sources: internal (located in Earth's core, e.g. Elsasser et al., 1956; Harrison and Funnell, 1964; Bucha et al., 1970; Wollin et al., 1971; King, 1974; Courtillot et al., 2007; Vieira et al., 2008; Kilifarska et al., 2015) and external (located on the Sun, ionosphere, or magnetosphere, e.g. Bucha and Bucha, 1998; Bucha, 2002; Palamara and Bryant, 2004; Lu et al., 2008; Paluš and Novotná, 2009; Seppälä et al., 2009; Dobrica et al., 2009; Li et al., 2011; Seppälä et al., 2013; Okeke and Audu, 2017; etc.). These sources have fundamentally different natures and different frequency ranges, and the information about them is obtained by different methods. The emphasis in this book is put on the long-term variations of the geomagnetic field (originating in the core–mantle boundary) and their imprint on the climate variability.

The statistical evidence for existing synchronicity between geomagnetic field and climate time series is highly controversial. For example, up to now there have been two mutually exclusive suggestions about the hypothesized connection between geomagnetic field and climate. One of them relates the periods of geomagnetic field weakening and/or magnetic poles reversals to periods of climate cooling (e.g. Valet and Meynadier, 1993; Worm, 1997;

Knudsen and Riisager, 2009; Kitaba et al., 2013). The other states just the opposite: that the geomagnetic strengthening is followed by the climate cooling and vice versa (e.g. Wollin et al., 1971; Bucha and Zikmunda, 1976; Nurgaliev, 1991; Butchvarova and Kovacheva, 1993; Gallet et al., 2005; Bakhmutov et al., 2011). This controversy is not supportive of the idea of any relation between the geomagnetic field and climate variability. The existence of such scepticism is well formulated by Worm (1997), who states that obviously two mutually exclusive statements cannot be true at the same time.

One reason for this controversial picture of the geomagnetic–climate relations could be the fact that palaeomagnetic field intensity reconstructions, based on the marine and lacustrine sediments, could be contaminated by climatic/lithologic processes (Kent, 1982; Guyodo et al., 2000; Roberts et al., 2003). The natural remanent magnetization (NRM) possessed by the sediments can be attributed to the alignment of detrital magnetic grains in Earth's magnetic field at the time of their deposition, or shortly after that (Kent and Opdyke, 1977). In favourable circumstances, this remanent magnetization can be interpreted as a record of palaeomagnetic field behaviour. Disturbances in the NRM of sediments can be induced by many factors, including the climatologically defined depositional conditions. This means that the long-term climate periodicities (including those ascribed to the variations of Earth's orbital parameters) could be projected on the palaeomagnetic records.

According to other hypothesis, the covariance between temporal variations of geomagnetic field and climate could be related to the variations in Earth's orbital parameters, affecting both climatic conditions and geomagnetic field. The idea of climate dependence on the long-term periodicity of orbital parameters was suggested by Milutin Milankovitch in 1920, while that of precession as a driver of geomagnetic dynamo can be traced back to Blackett's experiments in the 1950s (Blackett, 1952). Orbital periods in sedimentary relative palaeointensity records have been considered evidence for orbital influence on the geodynamo (Kent and Opdyke, 1977; Channell et al., 1998; Yamazaki, 1999; Yamazaki and Oda, 2002).

Exploring the presence of climate signal in palaeointensity records, Frank (2000) concluded furthermore that it is a consequence of inadequate normalization and correction of the geomagnetic data, and that there is no clear evidence for orbital forcing and any other residual climatic influence in palaeomagnetic data. In line with these conclusions, the parallel analysis of palaeointensity and oxygen isotope data for 1.5 million years (Channell et al., 2009) reveals that there is no tendency "for relative palaeointensity minima in the stack, to occur at particular phases of orbital variations such as lows or decreasing values of obliquity[a]." Previous analyses of Roberts et al. (2003) and Xuan and Channell (2008) also reveal that palaeomagnetic records show no statistically significant coherency with the orbital signals.

The other reason for existing controversy in climate–geomagnetic relations is the idealized expectation of the most of the researchers that the spatial geomagnetic variations are zonally symmetric, based on the idea for the quasidipole character of Earth's magnetic field. Indeed this expectation appears to be valid at distances far from the Earth's surface. Close to it, however, the deviations from the magnetic dipole become substantial—not only in their spatial distribution, but also in their temporal variability (e.g. Freon and McCracken, 1962; Shea et al., 1965; Bakhmutov, 2006b; Muxworthy, 2017; Cai et al., 2017; Driscoll and Wilson, 2018).

[a] *obliquity* or the axial tilt, is the angle between Earth's rotational axis and its orbital axis, or, equivalently, the angle between its equatorial plane and orbital plane, varying between ~22.1 and 24.5 degrees.

Consequently, one could suggest that spatial–temporal heterogeneity of geomagnetic variations could be projected on the climatic variability. In line with this suggestion, the study of Pavón-Carrasco et al. (2018) reveals that during the Holocene, the production rates of cosmogenic radionuclides in the atmosphere were controlled not only by the strength of geomagnetic field (as determined by its dipole magnetic moment) but also by its nondipole part.

Recent investigation of the problem of causality in geomagnetic–climate relations reveals that the driver of their coherent variability is "the system producing geomagnetic field" (Campuzano et al., 2018). For this reason, the contemporary evidence for existing relations between the geomagnetic field and climate (provided in the following sections) will be interpreted in the light of the changing geomagnetic field, which induces the corresponding changes in the climatic system.

4.3 Geomagnetic signal in recent climate records

4.3.1 Similarities in spatial–temporal variations of the geomagnetic field and some climatic parameters

Many authors (e.g. Vieira and da Silva, 2006; Vieira et al., 2008; Winkler et al., 2008; Kerton, 2009; De Santis et al., 2012; Bakhmutov et al., 2014; Cnossen et al., 2016) have pointed out that changes of geomagnetic intensity, or the geomagnetic poles location, could be projected on the spatial distribution of climate system variables. An illustration of this idea is shown in the Hovmöller diagram presented in Fig. 4.4, which reveals the persisting similarity (for more than a century) in the longitudinal distribution of latitudinally averaged (over 40–80°N latitudes) geomagnetic field intensity, and the mean January surface temperature and pressure, during the period 1900–2010. Note that longitudinal sectors with higher geomagnetic field intensity coincide very well with the cooler, high surface pressure sectors in the winter Northern Hemisphere, during the 111 years examined. This result could be considered as a contemporary confirmation of the hypothesis that geomagnetic strengthening is accompanied by climate cooling, reported by Wollin et al. (1971), Bucha and Zikmunda (1976), Nurgaliev (1991), Butchvarova and Kovacheva (1993), Gallet et al. (2005), Bakhmutov et al. (2011), etc. In addition, it is worth pointing out the regional character and seasonality of the magnetic–climate coupling. For example, in summer the geomagnetic–climate connectivity is less pronounced.

Another illustration of this relation, based on the seasonal winter data from ERA 20th-century re-analysis for the period 1900–2010, is shown in Fig. 4.5. The almost perfect anticorrelation between Northern Hemisphere geomagnetic field intensity and surface temperature is impressive (Fig. 4.5A). Moreover, a closer look reveals that in the North American sector, where the centennial geomagnetic weakening is stronger, the rise in the mean winter temperature since 1900 is highest (compare the darkest and the palest grey curves). Over Europe and Central Asia, the geomagnetic field is slightly strengthened during the past century, and temperature changes are very small or negligible. In the Southern Hemisphere, the geomagnetic field decreases at all longitudes, but more strongly in the South Atlantic sector. Respectively, the centennial temperature rise in this sector is stronger (see Fig. 4.5B).

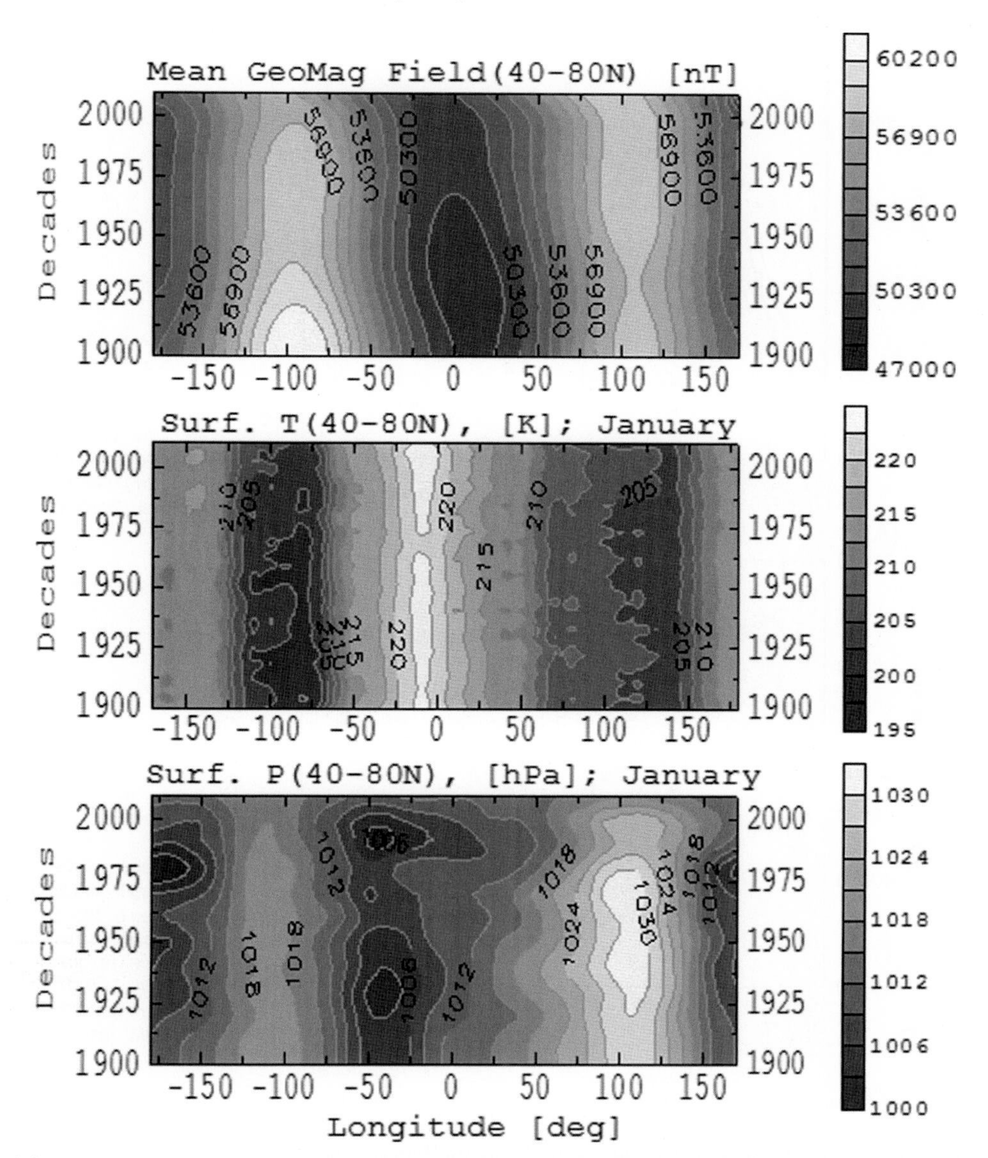

FIG. 4.4 (*top*) Longitude-time variations of latitudinally averaged (over the 40–80°N latitudes) geomagnetic field total intensity in nT; (*middle*) winter near-surface air temperature dynamical anomalies in °C, defined as deviations from the temporary varying decadal means; (*bottom*) winter surface pressure dynamical anomalies in hPa for the period 1900–2010. Note the good correspondence between geomagnetic, temperature, and pressure extrema. *Data source: Temperature and pressure are taken from ERA 20th-century reanalysis and geomagnetic field data, from the IGRF (International Reference Geomagnetic field), https://www.ngdc.noaa.gov/geomag/geomag.shtml.*

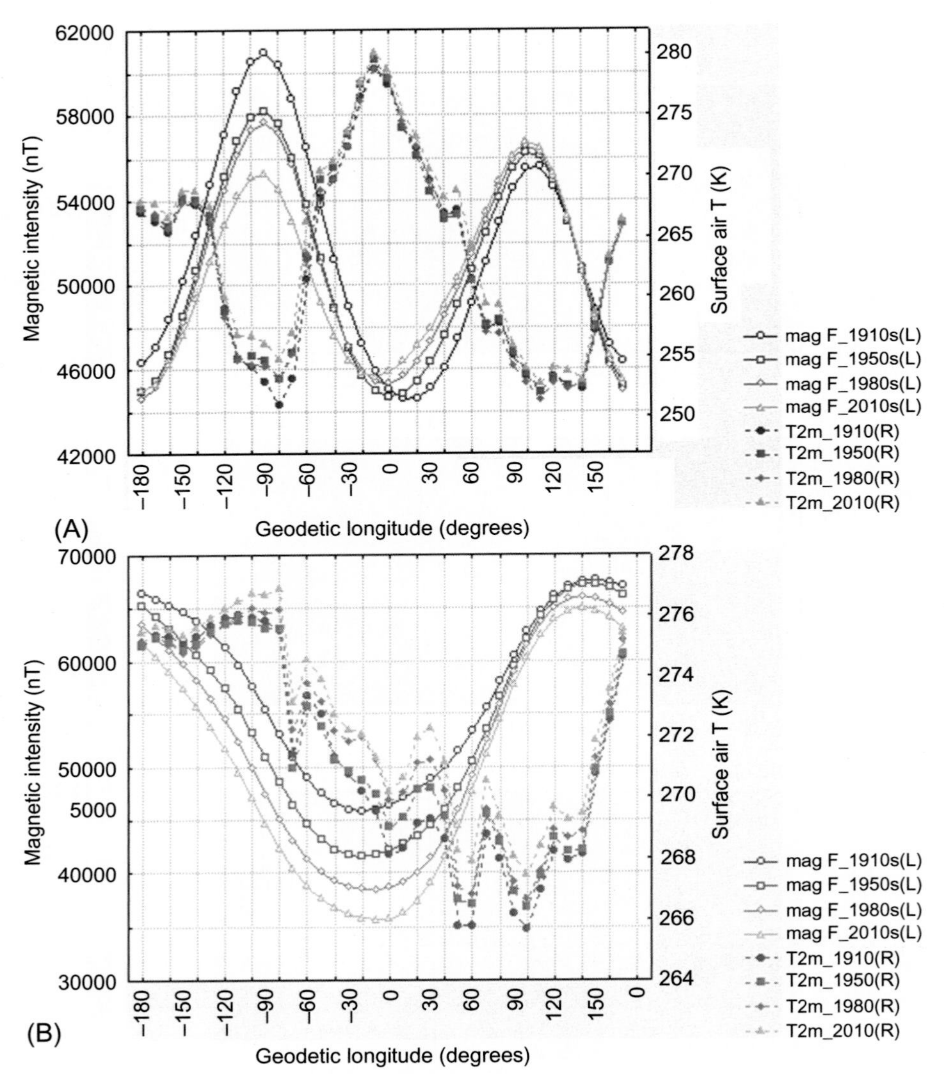

FIG. 4.5 (A) Decadal variations of the Northern Hemisphere winter (Dec–Mar) air surface temperature *(full symbols)* and geomagnetic field intensity *(open symbols)*, latitudinally averaged (over the 40–70°N latitudes), during the 20th and 21st centuries. (B) Decadal variations of the Southern Hemisphere winter (Jun–Sep) air surface temperature *(full symbols)* and geomagnetic field intensity *(open symbols)*, latitudinally averaged (over the 40–70°S latitudes), during the 20th and 21st centuries. *(A) Data sources: International Geomagnetic Reference Field model (https://www.ngdc.noaa.gov/geomag/geomag.shtml) and ERA 20th-century reanalysis for air temperature at 2 m above the surface (T2m) (https://apps.ecmwf.int/datasets/data/era20c-daily/levtype=sfc/type=an/). (B): Data sources: International Geomagnetic Reference Field model (https://www.ngdc.noaa.gov/geomag/geomag.shtml) and ERA 20th-century reanalysis for air temperature at 2 m above the surface (T2m) (https://apps.ecmwf.int/datasets/data/era20c-daily/levtype=sfc/type=an/).*

However, the persistence of this antiphased spatial distribution could be coincidental. For example, one possible explanation for the double-wave distribution of the Northern Hemisphere surface temperature is the alteration of continents and oceans (the two continents in the Northern Hemisphere are divided by two oceans) and corresponding differential heating rates of the near surface air. Scepticism about the geomagnetic temperature relation is additionally fed by the fact that in the Southern Hemisphere, the phase shift between the regions of the strongest geomagnetic intensity and the surface temperature minimum is less than 90 degrees (see Fig. 4.5B).

Another way to assess the real existence of geomagnetic–climate connectivity is to examine whether their temporal variations are synchronous or not. To do this, the time series of winter surface temperature (smoothed by an 11-year moving average), magnetic field intensity, and geomagnetic secular variations—i.e. the first time derivative of geomagnetic intensity (smoothed by a 5-year moving window)—have been analysed by the use of the lagged cross-correlation analysis in each node of our grid (with 10-degree steps in latitude and longitude). From the highest, statistically significant at 2σ level, correlation coefficients have been created four different correlation maps, for the Northern and Southern hemispheres correspondingly, which are presented in Figs 4.6 and 4.7. The time series have been smoothed by different windows due to their different roughness; e.g. the smooth geomagnetic intensity records are used without smoothing, while magnetic secular variations are smoothed by a 5-point window. The strong interannual variability of the air surface temperature, on the other hand, is smoothed by a longer (11-point) window. The 111-year climatology of dynamical temperature anomalies (i.e. the deviations of the mean winter temperature from its centennially varying decadal means) is given in Figs 4.6 and 4.7 as a background (coloured shading). The contours in Fig. 4.6 show that synchronization between the temporary varying geomagnetic field and surface temperature is better pronounced in the Southern Hemisphere than in the Northern Hemisphere (contours in Fig. 4.6). Comparison with the centennial climatology of temperature deviations from its 10-year climatic norms shows no clear connection with regions of stronger geomagnetic–temperature correlation (except the warmer spot in north-eastern Canada, which coincides with the negatively correlated geomagnetic field and temperature). Obviously this result is predefined by the slow and smooth changes of geomagnetic field intensity, in contrast to the relatively variable near-surface air temperature.

On the other hand, Fig. 4.7 illustrates fairly well that the areas with prevailing negative temperature anomalies in the Northern Hemisphere are in good correspondence with the antiphase covariance between geomagnetic secular variations and near-surface temperature. In contrast, the positive correlation coefficients seem partially 'attracted' by the positive temperature anomalies. In the Southern Hemisphere, where the oceans dominate, the dynamical temperature anomalies are much smaller and any relation with the areas of strong geomagnetic–temperature connections could hardly be detected.

The time lag of the surface temperature response to the geomagnetic forcing is given in the bottom panels of Figs 4.6 and 4.7. Generally, the near-surface temperature responds to changes of geomagnetic intensity without, or with a small time delay (1–2 years). The temperature response to the geomagnetic secular variations, however, is substantially delayed, being in the order of 10–20 years in regions with maximal correlation in the Northern Hemisphere. In the Southern Hemisphere, the time lag is slightly smaller, possibly due to the

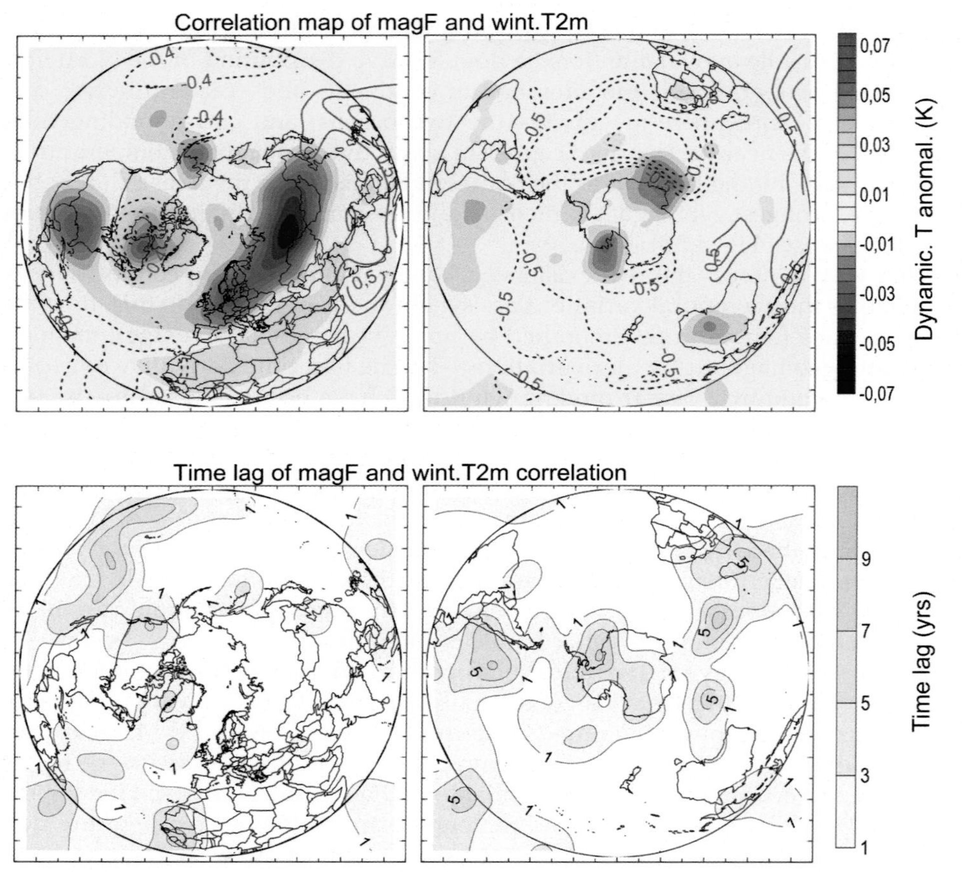

FIG. 4.6 (*top*) Correlation maps of geomagnetic field intensity and winter temperature (i.e. Dec–Mar in the NH and Jun–Sep in the SH) at 2 m above the surface—T2m *(continuous contours = positive correlation; dashed ones = negative correlation)*. The *background coloured shading* presents the 111-year climatology of T2m dynamical anomalies (i.e. temperature deviations from its dynamically varying decadal mean); Northern Hemisphere = left, Southern Hemisphere = right panel. (*bottom*) Time delay of the surface temperature response to geomagnetic forcing. *Data sources: Geomagnetic field intensity is from the IGRF model (https://www.ngdc.noaa.gov/geomag/geomag.shtml); T2m are taken from ERA 20th-century reanalysis (https://apps.ecmwf.int/datasets/data/era20c-daily/levtype=sfc/type=an/).*

higher coefficient of diffusivity of the ocean and correspondingly shorter time needed to achieve thermal equilibrium (see Section 4.1.1).

Fig. 4.6 explains why Sternberg and Damon (1979) did not find any significant relation between surface temperature and geomagnetic variations, when analysing the instantaneous (with zero time lag) correlations in 85 geomagnetic stations. In fact their analysis did not contain any station in the region of strongest connectivity between geomagnetic field and temperature, i.e. the region between South Africa and Antarctica. Fig. 4.7 shows that surface temperature covariates better with geomagnetic secular variations instead with field intensity. Despite that the above examples are not sufficient, they support the existence of a relation

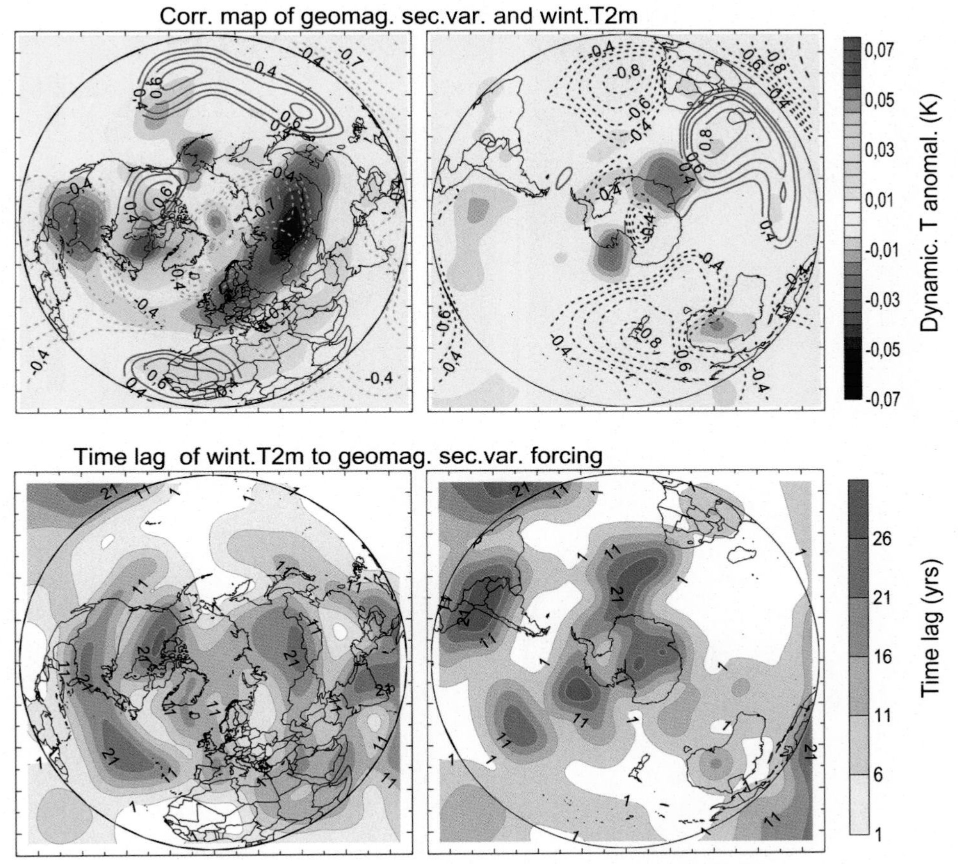

FIG. 4.7 The same as Fig. 4.6, but for correlation between geomagnetic secular variations and winter near-surface air temperature (T2m). *Data sources: Geomagnetic secular variations are from the IGRF model (https://www.ngdc.noaa.gov/geomag/geomag.shtml); T2m are taken from ERA 20th-century reanalysis (https://apps.ecmwf.int/datasets/data/era20c-daily/levtype=sfc/type=an/).*

between geomagnetic field and surface temperature, which is possibly mediated by some other factor(s).

One of the potential mediators of geomagnetic influence on the surface temperature is the ozone in the lower stratosphere (e.g. Kilifarska, 2012b; Kilifarska et al., 2015). Although the suggested geomagnetic influence on the ozone density is inserted through the help of galactic cosmic rays, the correlation between the geomagnetic field and ozone at 70 hPa shows consistent patterns in both hemispheres, i.e. circular belts of increased correlation coefficients (see Fig. 4.8). The different sign of correlation over the mid-latitude Atlantic Ocean and Europe on one side, and the Pacific Ocean on the other, is a result of the different centennial evolution of geomagnetic field in these regions (see Chapter 1). It is worth noting that before the creation of correlation maps, the lagged correlation coefficients were weighted by the autocorrelation function of geomagnetic field intensity, corresponding to a particular time lag of

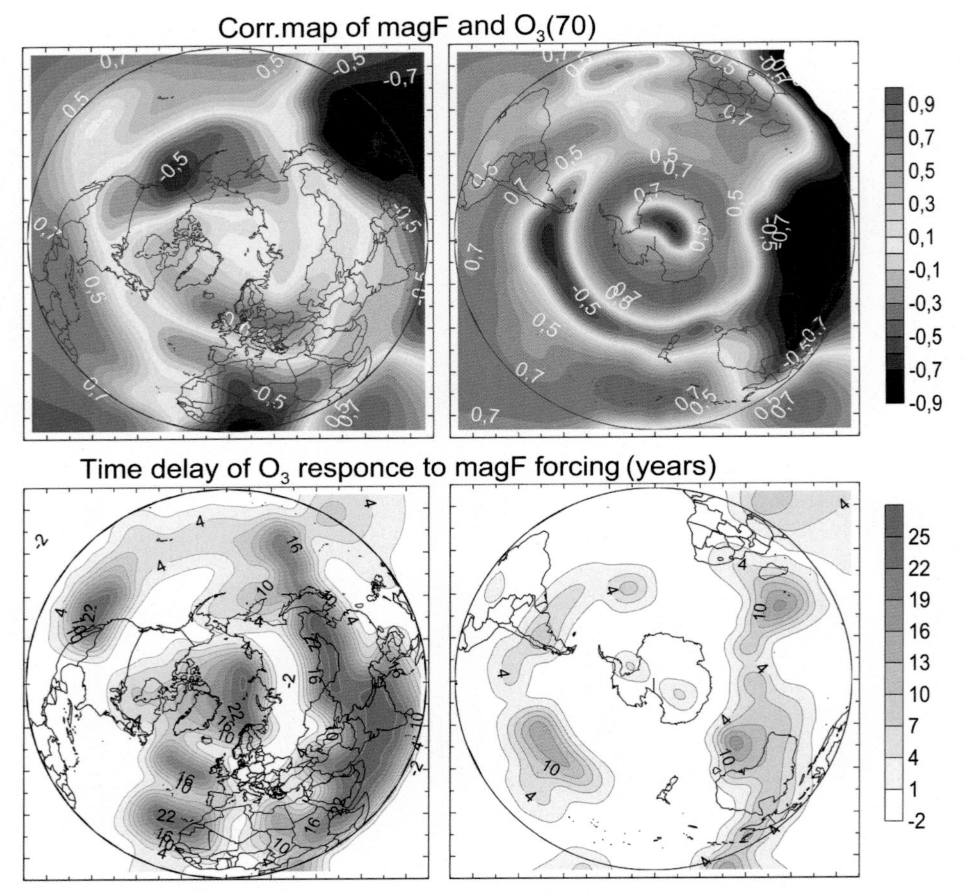

FIG. 4.8 Spatial distribution of covariance between geomagnetic field intensity and winter (Dec–Mar in NH and Jun–Sep in the SH) ozone mixing ratio at 70 hPa during the period 1900–2010. *Data sources: Geomagnetic secular variations are from the IGRF model (https://www.ngdc.noaa.gov/geomag/geomag.shtml); ozone data are from ERA 20th-century reanalysis (https://apps.ecmwf.int/datasets/data/era20c-daily/levtype=sfc/type=an/).*

calculated correlation coefficient (for more details, see Kenny, 1979; Kilifarska, 2015). This procedure allows direct comparison of the strength of the connection between both variables.

Time lags of ozone response to the changes of geomagnetic field strength are shown in the bottom panels of Fig. 4.8. It is clear that in regions with the strongest relation, the ozone responds to geomagnetic forcing without or with a minimal delay.

4.3.2 Evidence for existing relations between galactic cosmic rays and some climate variables

Direct geomagnetic influence on the nonmagnetic atmosphere is infeasible, so the supposed mediator of their relation should explain not only the time delay, but also the spatial heterogeneity of the atmospheric response. One of the most explored candidates for the role of mediator is galactic cosmic rays (e.g. Gallet et al., 2005; Usoskin et al., 2004; Dergachev et al., 2006, 2007; Courtillot et al., 2007; Kirkby, 2007; Miyahara et al., 2008; De Santis et al., 2012; Dorman, 2012; Kilifarska et al., 2015; Kilifarska, 2017). Cosmic ray intensity, and the depth of their penetration in Earth's atmosphere, is strongly modulated by the geomagnetic field, which determines the effect they have on the atmospheric chemical composition and energy balance. One of atmospheric compounds that is especially vulnerable to precipitating energetic particles is the atmospheric ozone. The near tropopause ozone is of special interest due to its ability to influence the planetary energy balance (Hansen et al., 1997; de F. Forster and Shine, 1997; Gauss et al., 2006; Randel et al., 2007; Aghedo et al., 2011).

Comparison of temporal variability of GCR and winter ozone at 70 hPa is shown in Fig. 4.9, which presents the correlation maps (created from statistically significant correlation coefficients at 2σ level) between both variables. In order to compare the strength of connectivity between GCRs and ozone in regions with different time lags, their correlation coefficients have been preliminary weighed by the value of GCR autocorrelation function, corresponding to the particular time lag of ozone response (for more details, see Kenny, 1979; Kilifarska, 2015). This procedure weakens the delayed correlation coefficients substantially, thus reducing the uncertainty of comparing correlations with different time lags. The centennial climatology of geomagnetic field intensity and its longitudinal gradient is shown as a background shading in Fig. 4.9. It is worth noting that the highest in-phase correlation between GCRs and ozone coincides fairly well with the strongest Northern Hemisphere geomagnetic field

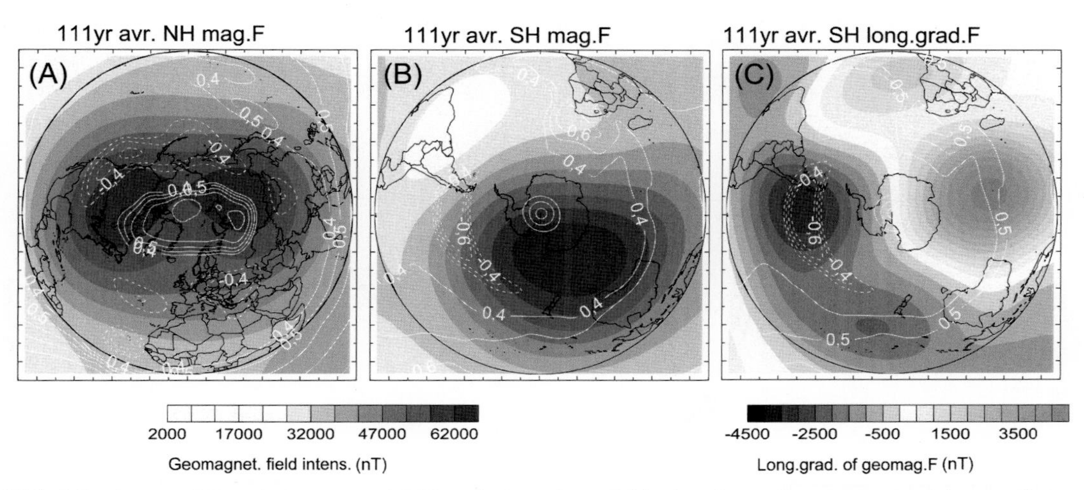

FIG. 4.9 *(contours)* Correlation maps of GCR and winter O_3 at 70 hPa (*continuous lines* present positive correlation, *dashed lines* negative); *(background shading)* spatial distribution of geomagnetic field intensity (A and B panels) and its longitudinal gradient (panel C). Note that in the Northern Hemisphere the strongest in-phase covariance between GCRs and ozone is found in regions with the strongest geomagnetic field. In SH, however, the impact of GCRs on the lower stratospheric O_3 corresponds better to the longitudinal gradient of the geomagnetic field, instead to its intensity. *Data source: O_3 data are taken from ERA 20th-century reanalysis; centennially reconstructed GCR are from Usoskin, I.G., et al., 2008. Cosmic Ray Intensity Reconstruction. IGBP PAGES/World Data Center for Paleoclimatology, Data Contribution Series # 2008-013, NOAA/NCDC Paleoclimatology Program, Boulder, CO; geomagnetic field data are from the IGRF model (https://www.ngdc.noaa.gov/geomag/geomag.shtml).*

intensity in polar regions (Fig. 4.9A). In addition to the positive correlation, a moderate negative one is visible in latitudinal band 45–60°N, suggesting that centennial decrease of GCR intensity is accompanied by an increase of the lower stratospheric ozone at these latitudes (see Fig. 4.10A and B).

This peculiarity in the centennial evolution of the lower stratospheric ozone is clearly visible in Fig. 4.10, presenting zonally averaged O_3 mixing ratio at 70 hPa, and at different latitudes. Unlike the polar and tropical latitudes, where a persistent decrease of the ozone mixing ratio is found (Fig. 4.10A and C), at mid-latitudes—notably between 45°N and 60°N latitude—the lower stratospheric O_3 has progressively increased with time (see Fig. 4.10B). This tendency is in contrast to the temporal evolution of GCR and is visible as a negative correlation in the correlation map provided in Fig. 4.9A.

The middle panel in Fig. 4.9 shows that there is no correspondence between GCR-ozone correlation and the strength of the Southern Hemisphere geomagnetic field. Consequently,

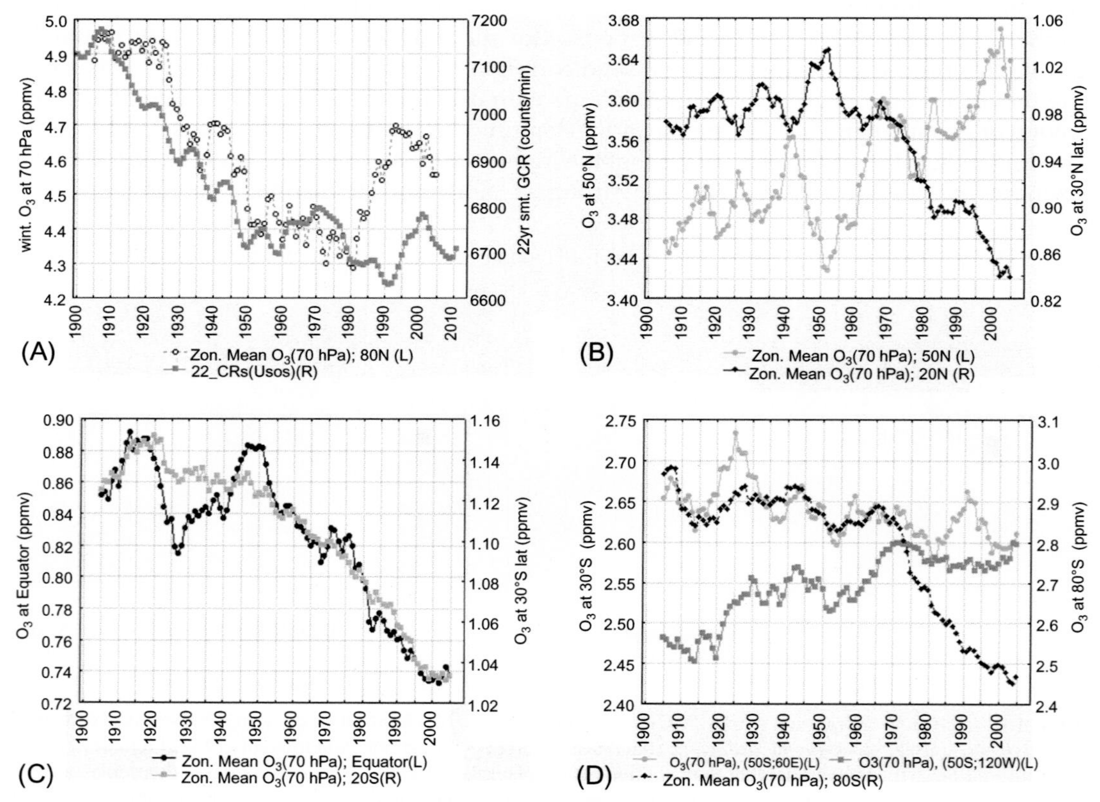

FIG. 4.10 Time series of zonally averaged winter (Dec–Mar in NH and Jun–Sep in the SH) ozone volume mixing ratio at 70 hPa, covering the period 1900–2010 at different latitudes; (A) O_3 at 80°N latitude *(open black circles)* and 22-year running mean of GCR annual values *(full grey squares)*; (B) O_3 at 50°N *(pale grey dots)* and 20°N *(black diamonds)* latitudes; (C) O_3 at the Equator *(black dots)* and 20°S lat. *(pale grey squares)*; (D) O_3 at 50°S lat. and at two longitudes *(grey symbols)*, and 80°S lat. *(black dots). Data source: ERA 20th-century reanalysis (https://apps.ecmwf.int/datasets/data/era20c-daily/levtype=sfc/type=an/).*

the simplistic expectation for strongest GCR impact on the polar lower stratospheric ozone is not confirmed in the Southern Hemisphere. On the other side, the map of longitudinal variations of geomagnetic field intensity shows good correspondence to the GCR-ozone correlation map (Fig. 4.9C). For example, the strong antiphase temporal synchronization of GCR and ozone variations is placed in the region of a negative longitudinal magnetic gradient (near the southernmost edge of Latin America), while the in-phase GCR-ozone covariance corresponds to the region of a positive geomagnetic gradient (near South Africa). This implies that the tendencies of ozone evolution in the two regions are opposite, which can also be seen in Fig. 4.10D. Although the centennial changes are weak, the tendencies are exactly opposite, decreasing near South Africa and increasing near southern Latin America. Fig. 4.10D shows in addition that the Southern Hemisphere polar ozone has remained almost unaffected by GCRs during the better part of the past century. Ever since 1970, however, it has decreased abruptly. Figs 4.9 and 4.10 illustrate that despite the overall decrease of the lower stratospheric O_3 density, during the 20th century (e.g. IPCC, 2007, 2013), there are regions in extra-tropics where ozone enhancement has been detected.

Fig. 4.11 compares the connectivity of lower stratospheric ozone with geomagnetic field intensity and GCRs (shown in Figs 4.8 and 4.9 correspondingly), during the period 1900–2010. The hemispherical asymmetry in the spatial distribution of ozone correlations with geomagnetic field intensity and GCRs is the first one to note in Fig. 4.11. This asymmetry could be reasonably attributed to the heterogeneous geomagnetic field, modulating the depth of penetration of GCRs in the atmosphere. The latter, as will be described in Chapter 6, determines the type of chemical reactions activated in the lower stratosphere and the upper troposphere, having a different impact on the near tropopause ozone. Moreover, in Chapter 5 it will be shown that the longitudinal gradient of geomagnetic field could act as a magnetic lensing of charged particles in some regions of the world.

The other interesting feature, noticeable in Fig. 4.11, is the coincidence of the antiphase covariance between GCRs and ozone on one side, and between the geomagnetic field and ozone on the other. In contrast, the positive GCR-ozone correlation tends to coincide with the in-phase covariance between the geomagnetic field and ozone. Fig. 4.11 illustrates that geomagnetic influence on the GCRs is unevenly distributed over the globe (see also Chapter 5), and that the pattern of their relation is projected on the lower stratospheric ozone density, which could further impact the surface temperature (see Chapter 6).

However, examination of the temporal covariance between GCRs and the near surface air temperature reveals that their connectivity is quite weak in the Northern Hemisphere, at middle to high latitudes (with the exception of northern Canada, the Aleutian Islands, and East Mediterranian Sea); see Fig. 4.12A. On the other hand, there is a relatively strong antiphase covariation between GCRs and surface temperature in the Southern Hemisphere (Fig. 4.12B), which forms a kind of belt, roughly coinciding with the Antarctic convergent zone (extended approximately between the latitudes 48°S and 60°S). Coincidentally or not, this result brings an association with the fact that the strongest weakening of geomagnetic field intensity (during the 20th century) has been detected in the Southern Hemisphere. Moreover, Fig. 4.5 illustrates that the strongest rise of the near-surface temperature is found in the South Atlantic–Southern Ocean (refer to Fig. 4.5B), thus raising the question of causality. The average time delay of surface temperature response to GCR forcing is 7 years in the Northern Hemisphere and 9 years in the Southern Hemisphere (Fig. 4.12C and D).

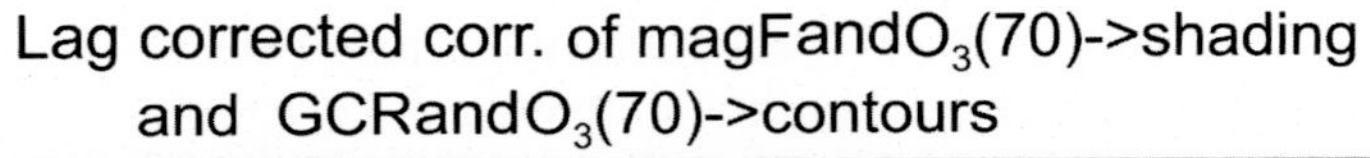

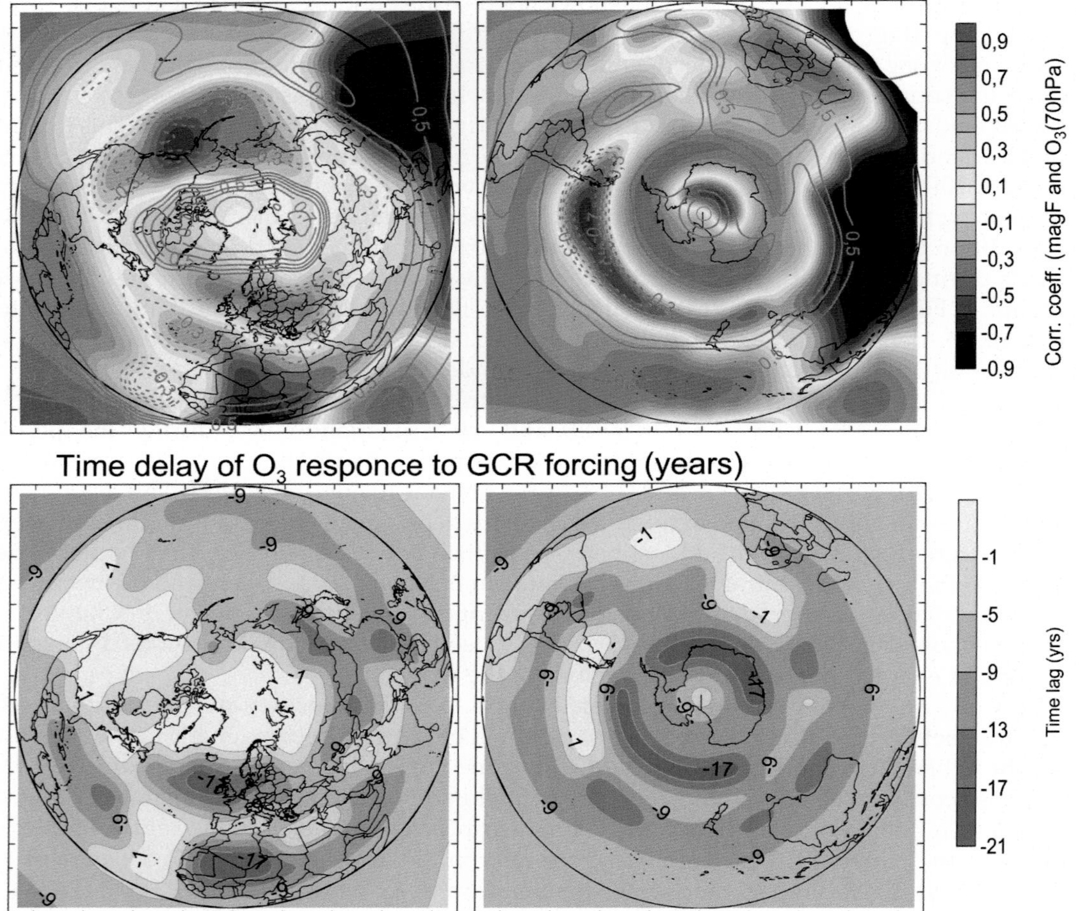

FIG. 4.11 Lag-corrected correlation maps of winter (Dec–Mar in NH and Jun–Sep in the SH) O_3 at 70 hPa with geomagnetic field *(shading)* and with GCR *(contours)* calculated for the period 1900–2010. The bottom panels illustrate the time lag of temperature response to GCR impact. Note the coincidence of regions with stronger ozone correlations with geomagnetic field and GCR, as well as their hemispherical asymmetry. *Data source: ERA 20th-century reanalysis for ozone at 70 hPa (https://apps.ecmwf.int/datasets/data/era20c-daily/levtype=sfc/type=an/); centennial reconstruction of GCR are provided by Usoskin, I.G., et al., 2008. Cosmic Ray Intensity Reconstruction. IGBP PAGES/World Data Center for Paleoclimatology, Data Contribution Series # 2008-013, NOAA/NCDC Paleoclimatology Program, Boulder, CO; geomagnetic field is from the IGRF model (https://www.ngdc.noaa.gov/geomag/geomag.shtml).*

The surface temperature correlation maps (lag corrected) with geomagnetic field and GCR intensities (i.e. Figs 4.6 and 4.12) are combined and presented in Fig. 4.13. Note that similary to the influence of these factors on the lower stratospheric ozone, there is a tendency for overlapping of regions with stronger temperature anticorrelation with both the geomagnetic field and GCRs. Thus Fig. 4.13 is a good illustration that climate response to geomagnetic and

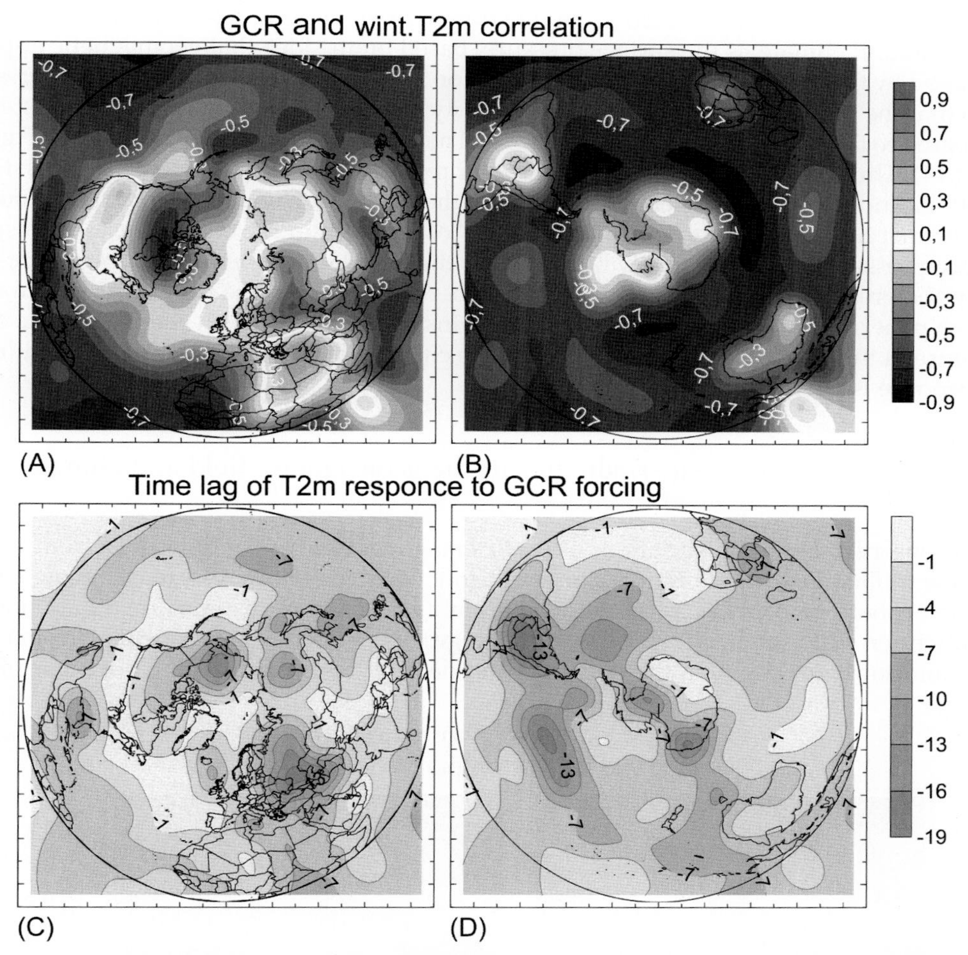

FIG. 4.12 Lag-corrected correlation maps of galactic cosmic rays and winter (Dec–Mar in NH and Jun–Sep in the SH) near-surface air temperature calculated for the period 1900–2010. Note the strong negative correlation in the Southern Hemisphere and its coincidence with the region of strongest temperature rise during the 20th century, illustrated in Fig. 4.5B. The bottom panels show the time lag of temperature response to GCR impact. *Data source: ERA 20th-century reanalysis for air temperature at 2 m above the surface; centennial reconstruction of GCR are provided by Usoskin, I.G., et al., 2008. Cosmic Ray Intensity Reconstruction. IGBP PAGES/World Data Center for Paleoclimatology, Data Contribution Series # 2008-013, NOAA/NCDC Paleoclimatology Program, Boulder, CO.*

respectively GCR forcing is heterogeneously distributed in space. Moreover, the sign of connectivity varies in different regions over the world. Consequently, the expectations for homogeneous climate response to the geomagnetic field variations, visible in many studies comparing geomagnetic and climate variations, appears to be very simplistic (at least on a decadal timescale). In addition, the geomagnetic influence on the surface temperature is probably mediated by the GCRs, whose intensity and depth of penetration in Earth's atmosphere are controlled by the geomagnetic field.

Another climate variable, used as an illustration of an existing relation between geomagnetic field and climate, is the surface pressure. Based on the idea that geomagnetic impact is mediated by GCRs, Fig. 4.14 shows the correlation map between GCRs and surface pressure. Note that the spatial patterns of their connectivity are more consistent in both hemispheres, compared to the relationship between GCRs and near-surface temperature (compare Figs 4.12 and 4.14). Thus, the positive correlation in polar regions and a ring of negative correlation at 40–50° latitudes (especially well pronounced in the Southern Hemisphere) could be found in both hemispheres. The more sustainable 'signal' of GCRs, found in the Southern Hemisphere surface pressure, should probably be related to the long-lasting memory of the ocean, occupying the greater part of the Southern Hemisphere. Analysis of the time lags (shown in the bottom panel of Fig. 4.14) reveals a tendency for instantaneous (or with very short time delay) response of the mid-latitude surface pressure to changes of GCR intensity. At high latitudes, its response is delayed by ~7 years.

4.3.3 Multidecadal quasiperiodicities in the geomagnetic field and climate variations

The observed temporal variations of Earth's magnetic field cover timescales from seconds to a few million years and originate in two distinct source regions: external (having solar or magnetosphere-ionospheric origin) and internal (generated at the core–mantle boundary). The attention in this book is focused on the changes of the internal component of geomagnetic field, and more precisely on the decadal–multidecadal scale of variability, because of their

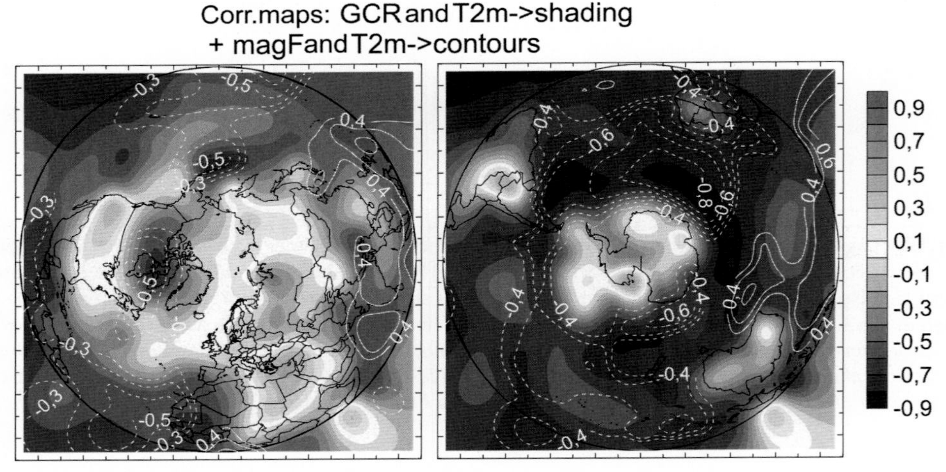

FIG. 4.13 Comparison of winter air temperature correlation maps with *(shading)* galactic cosmic rays (GCRs) and *(contours)* geomagnetic field intensity; *continuous lines* denote positive correlation, *dashed lines* denote the negative ones. Note the coincidence of the strongest temperature anticorrelations with GCRs and geomagnetic field intensities. *Data source: ERA 20th-century reanalysis for air temperature at 2 m above the surface (https://apps.ecmwf.int/datasets/data/era20c-daily/levtype=sfc/type=an/), geomagnetic field intensity is from the IGRF model (https://www.ngdc.noaa.gov/geomag/geomag.shtml) and centennial reconstruction of GCR are from Usoskin, I.G., et al., 2008. Cosmic Ray Intensity Reconstruction. IGBP PAGES/World Data Center for Paleoclimatology, Data Contribution Series # 2008-013, NOAA/NCDC Paleoclimatology Program, Boulder, CO.*

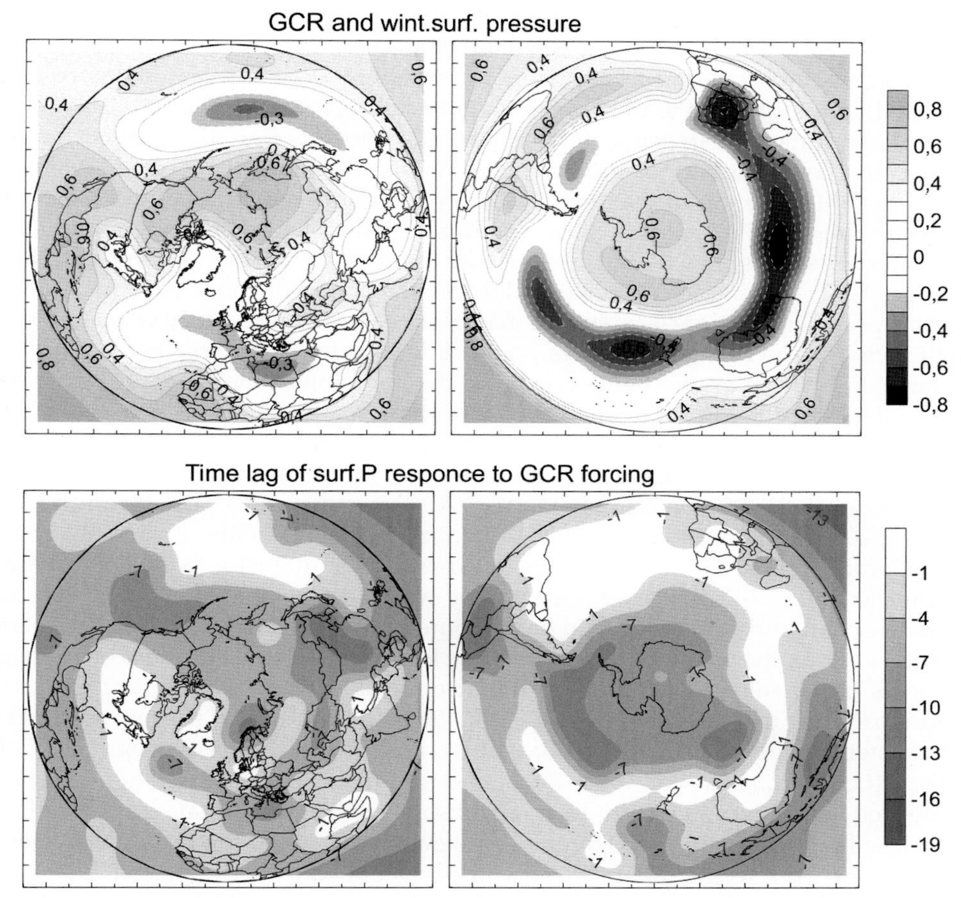

FIG. 4.14 Correlation map of galactic cosmic rays and winter (Dec–Mar in NH and Jun–Sep in the SH) surface pressure derived for the period 1900–2010. Note the consistency of pressure response to galactic cosmic rays' impact in both hemispheres: *in-phase* covariance at high latitudes and antiphase one at mid-latitudes, with a stronger 'signal' in the Southern Hemisphere. *Data source: ERA 20th-century reanalysis for the surface pressure and centennial reconstruction of GCR provided by Usoskin, I.G., et al., 2008. Cosmic Ray Intensity Reconstruction. IGBP PAGES/World Data Center for Paleoclimatology, Data Contribution Series # 2008-013, NOAA/NCDC Paleoclimatology Program, Boulder, CO.*

comparability with climate variations having similar periodicities (Bakhmutov, 2006a). These variations are thought to result from the magnetic induction and diffusion in the core and the mantle.

There exists a bundle of periods, which are supposed to be inherent features of the geomagnetic secular variations. In geomagnetism, the term 'secular variations' describes the long-term variations of Earth's magnetic field. Unlike the main field, which is dominantly dipolar, the secular variations are clearly nondipolar, which is well illustrated by their different signs and magnitude over the globe. Moreover, a tendency of isoporic foci (i.e. areas of maximum secular variation) to drift westward with an average velocity of ~0.3 degrees per year is well-established (Vestine and Kahle, 1968; Bloxham et al., 1989).

One prominent long period in geomagnetic variability is the 23-year (or more precisely, 22.9) period. It is found in the data of geomagnetic observatories (Alldredge, 1977) as well as in secular variation coefficients, derived by the spherical harmonics analysis of the first-time derivatives of observatory annual means of vector intensity (Langel et al., 1986). Despite the closeness of this period to the double solar cycle period, the origin of the 23-year period seems to be internal. This conclusion is supported by the fact that it does not appear in all analysed observatories, and is not in-phase at those observatories at which it is present (Alldredge, 1977). Among the most studied is also the period of about 60 years. The true nature of this periodicity is not clear, since it matches the period of about 60 years found in geomagnetic activity and sunspot numbers, suggesting its external origin. On the other hand, it could have an internal origin, being related to the variation of the westward drift and torsional oscillations of the Earth's fluid outer core (Braginsky, 1970).

The existence of quasi bi-decadal and 60-year periodicities is found also in different climate parameters: surface temperature, atmospheric circulation, rainfall, etc. (Almeida et al., 2004; Miyahara et al., 2008; Souza Echer et al., 2012; Barlyaeva, 2013; Scafetta and West, 2005; Sinha et al., 2005; Yadava and Ramesh, 2007; etc.). These periodicities are usually attributed to the heliomagnetic modulation of galactic cosmic rays' intensity, because at these time-scales the variability of solar irradiance is very small (e.g. Beer et al., 2000). If the 22-year periodicity of climatic variables could be attributed to the average period of the solar magnetic field reversal, the origin of the 60-year one is unknown. Some authors have suggested that planets in the solar system could influence the solar variation through gravitational spin-orbit coupling mechanisms and tides, which could be projected furthermore on Earth's climate system (Landscheidt, 1988; Charvátová and Střeštík, 2004; Mackey, 2007; Hung, 2007; Scafetta, 2010). However, if the quasi bi-decadal and 60-year periodicities of climate are attributed to this extra-terrestrial forcing, this means that the climate response should be more or less longitudinally homogenous. However, analysis of the near surface temperature deviations from its stationary mean value (calculated over the period 1900–2010) reveals that this is not the case (see Fig. 4.15A). It is easily visible, from the Hovmoller diagram presented in Fig. 4.15, that multidecadal periods in surface temperature variability exist, but they have a well-pronounced regional character.

A similar spatial–time distribution has been found in the stationary anomalies of the ozone at 70 hPa, shown in Fig. 4.16A. This similarity is an indication that the 60-year periodicity found in the climate parameters is more likely to be related to the similar one obtained in geomagnetic secular variations, with their regional specificity. The regional character of the secular variations is well-known, and is supposedly related to a torsional wave exerted from the outer to the inner core (Braginsky, 1970; Buffett, 1996). According to the current understandings, this wave is excited by the redistribution of Earth's angular momentum between the mantle, fluid, and solid cores.

Other authors (e.g. Raspopov et al., 2007; Veretenenko and Ogurtsov, 2012) explain the multidecadal and centennial periodicities of climate variability through corresponding changes in atmospheric circulation regimes, having periodicities of ~60-year, ~200-year, etc. However, the atmospheric circulation itself is determined by the temperature and pressure gradients. Consequently, the explanation of regional character of temperature response to external forcings by the changing regime of atmospheric circulation looks like an attempt to explain the reason by the consequence.

In resume: Comparison of the spatial–temporal variation of geomagnetic field intensity and some climatic variables (e.g. surface temperature and pressure and lower stratospheric ozone) during the period 1900–2010 reveals the existence of various degrees of synchronization between them (see Section 4.3.1). According to many authors, galactic cosmic rays (GCRs) are one of the most probable mediators of geomagnetic influence on climate. Statistical analyses reveal that climate variables covariate not only with geomagnetic, but also with GCR intensity (see Section 4.3.2). This is a hint that geomagnetic influence is mediated by GCRs, whose intensity and depth of penetration in Earth's atmosphere are controlled by the geomagnetic field. Interestingly, the strength and the phase of geomagnetic–climate connectivity are heterogeneously distributed over the globe. The latter is probably related to the geomagnetic lensing of GCRs in some regions of the world.

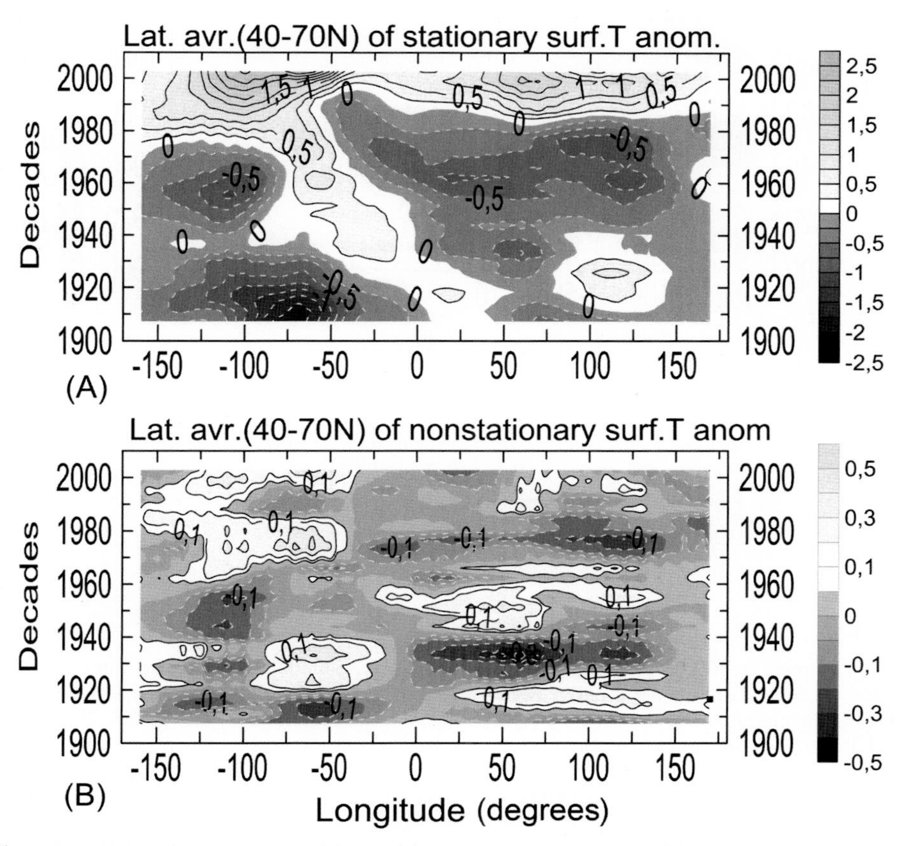

FIG. 4.15 (A) Multidecadal periodicity of the surface temperature deviations from their stationary mean over the period 1900–2010; (B) temperature deviations from dynamically varying decadal mean—note the existence of quasi bi-decadal periodicity. *Data source is ERA-20C reanalysis (https://apps.ecmwf.int/datasets/data/era20c-daily/levtype=sfc/type=an/). From Kilifarska, N., 2017. Mechanism of Relation Between Cosmic Rays, Geomagnetic Field and Earth's Climate, (DSc thesis), Sofia, Bulgaria.*

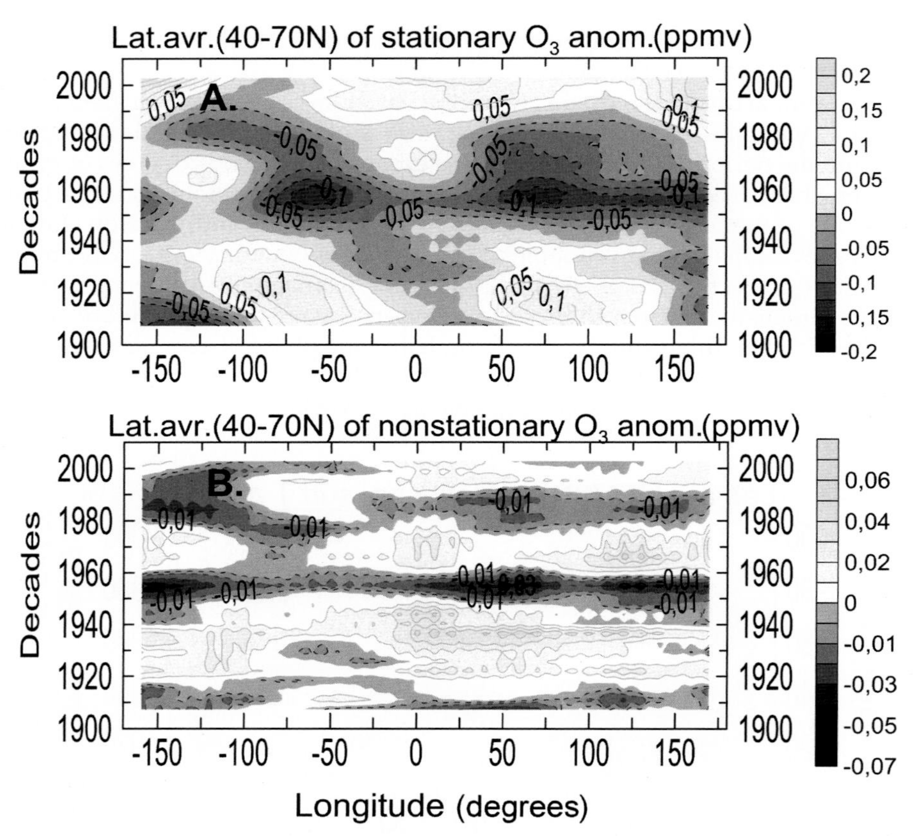

FIG. 4.16 (A) Multidecadal periodicity of the deviations of O_3 at 70 hPa from their stationary mean over the period 1900–2010; (B) nonstationary deviations O_3 at 70 hPa from its dynamically varying mean—note the existence of quasi bi-decadal periodicity. *Data source: ERA 20th century.* From *Kilifarska, N., 2017. Mechanism of Relation Between Cosmic Rays, Geomagnetic Field and Earth's Climate, (DSc thesis), Sofia, Bulgaria.*

References

Aghedo, A.M., Bowman, K.W., Worden, H.M., Kulawik, S.S., Shindell, D.T., Lamarque, J.F., Faluvegi, G., Parrington, M., Jones, D.B.A., Rast, S., 2011. The vertical distribution of ozone instantaneous radiative forcing from satellite and chemistry climate models. J. Geophys. Res. Atmos. 116. https://doi.org/10.1029/2010JD014243.

Alldredge, L.R., 1977. Geomagnetic variations with periods from 13 to 30 years. J. Geomagn. Geoelectr. 29, 123–135. https://doi.org/10.5636/jgg.29.123.

Almeida, A., Gusev, A., Mello, M.G.S., Martin, I.M., Pugacheva, G., Pankov, V.M., Spjeldvik, W.N., Schuch, N.J., 2004. Rainfall cycles with bidecadal periods in the Brazilian region. Geofis. Int. 43 (2), 271–279.

Bakhmutov, V.G., 2006a. Paleo-Secular Variations of Geomagnetic Field. Naukova Dumka, Kiev (in Russian).

Bakhmutov, V.G., 2006b. The connection between geomagnetic secular variation and long-range development of climate changes for the last 13,000 years: the data from NNE Europe. Quat. Int. 149, 4–11.

Bakhmutov, V.G., Martazinova, V.F., Ivanova, E.K., Melnyk, G.V., 2011. Changes of the main magnetic field and the climate in the XX-th century. Rep. Natl. Acad. Sci. Ukr. 7, 90–94.

Bakhmutov, V.G., Martazinova, V.F., Kilifarska, N.A., Melnyk, G.V., Ivanova, E.K., 2014. Climate changes and their relations with geomagnetic field. Part 1: spatio-temporal structure of the Earth's magnetic field and climate of the XX century. Geofiz. Zh. 36 (1), 81–104 (in Russian).

Barlyaeva, T.V., 2013. External forcing on air–surface temperature: geographical distribution of sensitive climate zones. J. Atmos. Sol. Terr. Phys. 94, 81–92. https://doi.org/10.1016/j.jastp.2012.12.014.

Beer, J., Mende, W., Stellmacher, R., 2000. The role of the sun in climate forcing. Quat. Sci. Rev. 19, 403–415. https://doi.org/10.1016/S0277-3791(99)00072-4.

Beltrami, H., Smerdon, J.E., Matharoo, G.S., Nickerson, N., 2011. Impact of maximum borehole depths on inverted temperature histories in borehole paleoclimatology. Clim. Past 7, 745–756. https://doi.org/10.5194/cp-7-745-2011.

Blackett, P.M.S., 1952. A negative experiment relating to magnetism and the Earth's rotation. Philos. Trans. R. Soc. Lond. Ser. A Math. Phys. Sci. 245, 309–370. https://doi.org/10.1098/rsta.1952.0024.

Bloxham, J., Gubbins, D., Jackson, A., 1989. Geomagnetic Secular Variation. Royal Society.

Braginsky, S.I., 1970. Torsional magnetohydrodynamic vibrations in the Earth's core and variations in day length. Geomagn. Aeron. 10, 1–10.

Bucha, V., 2002. Long-term trends in geomagnetic and climatic variability. Phys. Chem. Earth A/B/C 27, 427–431. https://doi.org/10.1016/S1474-7065(02)00022-0.

Bucha, V., Bucha, V., 1998. Geomagnetic forcing of changes in climate and in the atmospheric circulation. J. Atmos. Sol. Terr. Phys. 60, 145–169. https://doi.org/10.1016/S1364-6826(97)00119-3.

Bucha, V., Zikmunda, O., 1976. Variations of the geomagnetic field, the climate and weather. Stud. Geophys. Geod. 20, 149–167. https://doi.org/10.1007/BF01626048.

Bucha, V., Taylor, R.E., Berger, R., Haury, E.W., 1970. Geomagnetic intensity: changes during the past 3000 years in the Western Hemisphere. Science 168, 111–114. https://doi.org/10.1126/science.168.3927.111.

Buffett, B.A., 1996. A mechanism for decade fluctuations in the length of day. Geophys. Res. Lett. 23, 3803–3806. https://doi.org/10.1029/96GL03571.

Butchvarova, V., Kovacheva, M., 1993. European changes in the paleotemperature and Bulgarian archaeomagnetic data. Bulg. Geophys. J. 19, 19–23.

Cai, S., Tauxe, L., Paterson, G.A., Deng, C., Pan, Y., Qin, H., Zhu, R., 2017. Recent advances in Chinese archeomagnetism. Front. Earth Sci. 5. https://doi.org/10.3389/feart.2017.00092.

Campuzano, S.A., De Santis, A., Pavón-Carrasco, F.J., Osete, M.L., Qamili, E., 2018. New perspectives in the study of the Earth's magnetic field and climate connection: the use of transfer entropy. PLoS One 13, e0207270. https://doi.org/10.1371/journal.pone.0207270.

Channell, J.E.T., Hodell, D.A., McManus, J., Lehman, B., 1998. Orbital modulation of the Earth's magnetic field intensity. Nature 394, 464. https://doi.org/10.1038/28833.

Channell, J.E.T., Xuan, C., Hodell, D.A., 2009. Stacking paleointensity and oxygen isotope data for the last 1.5 Myr (PISO-1500). Earth Planet. Sci. Lett. 283, 14–23. https://doi.org/10.1016/j.epsl.2009.03.012.

Charvátová, I., Střeštík, J., 2004. Periodicities between 6 and 16 years in surface air temperature in possible relation to solar inertial motion. J. Atmos. Sol. Terr. Phys. 66, 219–227. https://doi.org/10.1016/j.jastp.2003.10.003.

Cnossen, I., Liu, H., Lu, H., 2016. The whole atmosphere response to changes in the Earth's magnetic field from 1900 to 2000: an example of 'top-down' vertical coupling. J. Geophys. Res. Atmos. 121, 7781–7800. https://doi.org/10.1002/2016JD024890.

Courtillot, V., Gallet, Y., Le Mouël, J.-L., Fluteau, F., Genevey, A., 2007. Are there connections between the Earth's magnetic field and climate? Earth Planet. Sci. Lett. 253, 328–339. https://doi.org/10.1016/j.epsl.2006.10.032.

de F. Forster, P.M., Shine, K.P., 1997. Radiative forcing and temperature trends from stratospheric ozone changes. J. Geophys. Res. Atmos. 102, 10841–10855. https://doi.org/10.1029/96JD03510.

De Santis, A., Qamili, E., Spada, G., Gasperini, P., 2012. Geomagnetic South Atlantic Anomaly and global sea level rise: a direct connection? J. Atmos. Sol. Terr. Phys. 74, 129–135. https://doi.org/10.1016/j.jastp.2011.10.015.

Dergachev, V.A., Dmitriev, P.B., Raspopov, O.M., Jungner, H., 2006. Cosmic ray flux variations, modulated by the solar and earth's magnetic fields, and climate changes. 1. Time interval from the present to 10–12 ka ago (the Holocene Epoch). Geomagn. Aeron. 46, 118–128. https://doi.org/10.1134/S0016793206010130.

Dergachev, V.A., Dmitriev, P.B., Raspopov, O.M., Jungner, H., 2007. Cosmic ray flux variations, modulated by the solar and terrestrial magnetic fields, and climate changes. Part 2: the time interval from ~10000 to ~100000 years ago. Geomagn. Aeron. 47, 109–117. https://doi.org/10.1134/S0016793207010173.

Dobrica, V., Demetrescu, C., Boroneant, C., Maris, G., 2009. Solar and geomagnetic activity effects on climate at regional and global scales: case study—Romania. J. Atmos. Sol. Terr. Phys. 71, 1727–1735. https://doi.org/10.1016/j.jastp.2008.03.022.

Dorman, L.I., 2012. Cosmic rays and space weather: effects on global climate change. Ann. Geophys. 30, 9–19. https://doi.org/10.5194/angeo-30-9-2012.

Driscoll, P.E., Wilson, C., 2018. Paleomagnetic biases inferred from numerical dynamos and the search for geodynamo evolution. Front. Earth Sci. 6. https://doi.org/10.3389/feart.2018.00113.

Elsasser, W., Ney, E.P., Winckler, J.R., 1956. Cosmic-ray intensity and geomagnetism. Nature 178, 1226. https://doi.org/10.1038/1781226a0.

Frank, M., 2000. Comparison of cosmogenic radionuclide production and geomagnetic field intensity over the last 200 000 years. Philos. Trans. R. Soc. Lond. *A* 358, 1089–1107

Freon, A., McCracken, K.G., 1962. A note on the vertical cutoff rigidities of cosmic rays in the geomagnetic field. J. Geophys. Res. (1896–1977) 67, 888–890. https://doi.org/10.1029/JZ067i002p00888.

Gallet, Y., Genevey, A., Fluteau, F., 2005. Does Earth's magnetic field secular variation control centennial climate change? Earth Planet. Sci. Lett. 236, 339–347. https://doi.org/10.1016/j.epsl.2005.04.045.

Gauss, M., Myhre, G., Isaksen, I.S.A., Grewe, V., Pitari, G., Wild, O., Collins, W.J., Dentener, F.J., Ellingsen, K., Gohar, L.K., Hauglustaine, D.A., Iachetti, D., Lamarque, F., Mancini, E., Mickley, L.J., Prather, M.J., Pyle, J.A., Sanderson, M.G., Shine, K.P., Stevenson, D.S., Sudo, K., Szopa, S., Zeng, G., 2006. Radiative forcing since preindustrial times due to ozone change in the troposphere and the lower stratosphere. Atmos. Chem. Phys. 6, 575–599. https://doi.org/10.5194/acp-6-575-2006.

Guyodo, Y., Gaillot, P., Channell, J.E.T., 2000. Wavelet analysis of relative geomagnetic paleointensity at ODP Site 983. Earth Planet. Sci. Lett. 184, 109–123. https://doi.org/10.1016/S0012-821X(00)00313-7.

Halko, G., Tsilika, K., 2014. Nonlinear time series analysis of annual temperatures concerning the global Earth climate. http://mpra.ub.uni-muenchen.de/59140/ MPRA Paper No. 59140.

Hansen, J., Sato, M., Ruedy, R., 1997. Radiative forcing and climate response. J. Geophys. Res. Atmos. 102, 6831–6864. https://doi.org/10.1029/96JD03436.

Harrison, C.G.A., Funnell, B.M., 1964. Relationship of palæomagnetic reversals and micropalæontology in two late cænozoic cores from the Pacific Ocean. Nature 204, 566. https://doi.org/10.1038/204566a0.

Hung, C.-C., 2007. Apparent relations between solar activity and solar tides caused by the planets. NASA Technical Rep. TM-2007-214817. https://ntrs.nasa.gov/archive/nasa/casi.ntrs.nasa.gov/20070025111.pdf.

IPCC, 2007. In: Solomon, S. et al., (Eds.), (Intergovernmental Panel on Climate Change), Climate Change 2007: The Physical Science Basis. Cambridge University Press, New York, p. 996.

IPCC, 2013. Climate Change 2013: The Physical Science Basis. Cambridge University Press, Cambridge and New York, NY.

Kenny, D.A., 1979. Correlation and Causality. John Wiley & Sons Inc., New York.

Kent, D.V., 1982. Apparent correlation of palaeomagnetic intensity and climatic records in deep-sea sediments. Nature 299, 538. https://doi.org/10.1038/299538a0.

Kent, D.V., Opdyke, N.D., 1977. Palaeomagnetic field intensity variation recorded in a Brunhes epoch deep-sea sediment core. Nature 266, 156. https://doi.org/10.1038/266156a0.

Kerton, A.K., 2009. Climate change and the Earth's magnetic poles, a possible connection. Energy Environ. 20, 75–83. https://doi.org/10.1260/095830509787689286.

Kilifarska, N.A., 2012a. Statistical methods for analysis of climatic time series and factors controlling their variability. In: Proceedings of the 2nd Workshop of the EU FP7 Project BlackSeaHazNet, September 2011, 2233-3681 pp. 111–124 Tbilisi, Georgia.

Kilifarska, N.A., 2012b. Climate sensitivity to the lower stratospheric ozone variations. J. Atmos. Sol. Terr. Phys. 90–91, 9–14.

Kilifarska, N.A., 2015. Bi-decadal solar influence on climate, mediated by near tropopause ozone. J. Atmos. Sol. Terr. Phys. 136, 216–230. https://doi.org/10.1016/j.jastp.2015.08.005i.

Kilifarska, N.A., 2017. Hemispherical asymmetry of the lower stratospheric O3 response to galactic cosmic rays forcing. ACS Earth Space Chem. 1, 80–88. https://doi.org/10.1021/acsearthspacechem.6b00009.

Kilifarska, N.A., Bakhmutov, V.G., Melnik, G.V., 2015. Geomagnetic field and climate: causal relations with some atmospheric variables. Izv. Phys. Solid Earth 51, 768–785. https://doi.org/10.1134/S1069351315050067.

King, J.W., 1974. Weather and the Earth's magnetic field. Nature 247, 131. https://doi.org/10.1038/247131a0.

King, T., 1996. Quantifying nonlinearity and geometry in time series of climate. Quat. Sci. Rev. 15, 247–266. https://doi.org/10.1016/0277-3791(95)00060-7.

Kirkby, J., 2007. Cosmic rays and climate. Surv. Geophys. 28, 333–375. https://doi.org/10.1007/s10712-008-9030-6.

Kitaba, I., Hyodo, M., Katoh, S., Dettman, D.L., Sato, H., 2013. Midlatitude cooling caused by geomagnetic field minimum during polarity reversal. PNAS 110, 1215–1220. https://doi.org/10.1073/pnas.1213389110.

Knudsen, M.F., Riisager, P., 2009. Is there a link between Earth's magnetic field and low-latitude precipitation? Geology 37, 71–74. https://doi.org/10.1130/G25238A.1.

Landscheidt, T., 1988. Solar rotation, impulses of the torque in the Sun's motion, and climatic variation. Clim. Chang. 12, 265–295. https://doi.org/10.1007/BF00139433.

Langel, R.A., Kerridge, D.J., Arraclough, D.R., Malin, S.R.C., 1986. Geomagnetic temporal change. J. Geomagn. Geoelectr. 38, 573–597. https://doi.org/10.5636/jgg.38.573.

Li, Y., Lu, H., Jarvis, M.J., Clilverd, M.A., Bates, B., 2011. Nonlinear and nonstationary influences of geomagnetic activity on the winter North Atlantic Oscillation. J. Geophys. Res. Atmos. 116. https://doi.org/10.1029/2011JD015822.

Lu, H., Clilverd, M.A., Seppälä, A., Hood, L.L., 2008. Geomagnetic perturbations on stratospheric circulation in late winter and spring. J. Geophys. Res. Atmos. 113. https://doi.org/10.1029/2007JD008915.

Mackey, R.P., 2007. Rhodes Fairbridge and the idea that the solar system regulates the Earth's climate. J. Coast. Res. SI 50, 955–968 (Proceedings of the 9th International Coastal Symposium, Gold Coast, Australia). ISSN 0749.0208.

Mikšovský, J., Raidl, A., 2006. Testing for nonlinearity in European climatic time series by the method of surrogate data. Theor. Appl. Climatol. 83, 21–33. https://doi.org/10.1007/s00704-005-0130-7.

Miyahara, H., Yokoyama, Y., Masuda, K., 2008. Possible link between multi-decadal climate cycles and periodic reversals of solar magnetic field polarity. Earth Planet. Sci. Lett. 272, 290–295. https://doi.org/10.1016/j.epsl.2008.04.050.

Muxworthy, A.R., 2017. Considerations for latitudinal time-averaged-field palaeointensity analysis of the last five million years. Front. Earth Sci. 5. https://doi.org/10.3389/feart.2017.00079.

Nurgaliev, D.K., 1991. Solar activity, geomagnetic variations, and climate changes. Geomagn. Aeron. 31, 14–18.

Okeke, F.N., Audu, M.O., 2017. Influence of solar and geomagnetic activity on climate change in Nigeria. IJPS 12, 184–193. https://doi.org/10.5897/IJPS2017.4655.

Palamara, D.R., Bryant, E.A., 2004. Geomagnetic activity forcing of the Northern Annular Mode via the stratosphere. Ann. Geophys. 22, 725–731. https://doi.org/10.5194/angeo-22-725-2004.

Paluš, M., Novotná, D., 2009. Phase-coherent oscillatory modes in solar and geomagnetic activity and climate variability. J. Atmos. Sol. Terr. Phys. 71, 923–930. https://doi.org/10.1016/j.jastp.2009.03.012.

Pavón-Carrasco, F.J., Gómez-Paccard, M., Campuzano, S.A., González-Rouco, J.F., Osete, M.L., 2018. Multi-centennial fluctuations of radionuclide production rates are modulated by the Earth's magnetic field. Sci. Rep. 8, 9820. https://doi.org/10.1038/s41598-018-28115-4.

Randel, W.J., Wu, F., Forster, P., 2007. The extratropical tropopause inversion layer: global observations with GPS data, and a radiative forcing mechanism. J. Atmos. Sci. 64, 4489–4496. https://doi.org/10.1175/2007JAS2412.1.

Raspopov, O.M., Dergachev, V.A., Kuzmin, A.V., Kozyreva, O.V., Ogurtsov, M.G., Kolström, T., Lopatin, E., 2007. Regional tropospheric responses to long-term solar activity variations. Adv. Space Res. 40, 1167–1172. https://doi.org/10.1016/j.asr.2007.01.081.

Roberts, A.P., Winklhofer, M., Liang, W.-T., Horng, C.-S., 2003. Testing the hypothesis of orbital (eccentricity) influence on Earth's magnetic field. Earth Planet. Sci. Lett. 216, 187–192. https://doi.org/10.1016/S0012-821X(03)00480-1.

Scafetta, N., 2010. Empirical evidence for a celestial origin of the climate oscillations and its implications. J. Atmos. Sol. Terr. Phys. 72, 951–970. https://doi.org/10.1016/j.jastp.2010.04.015.

Scafetta, N., West, B.J., 2005. Estimated solar contribution to the global surface warming using the ACRIM TSI satellite composite. Geophys. Res. Lett. 32. https://doi.org/10.1029/2005GL023849.

Seppälä, A., Randall, C.E., Clilverd, M.A., Rozanov, E., Rodger, C.J., 2009. Geomagnetic activity and polar surface air temperature variability. J. Geophys. Res. Space Physics. 114. https://doi.org/10.1029/2008JA014029.

Seppälä, A., Lu, H., Clilverd, M.A., Rodger, C.J., 2013. Geomagnetic activity signatures in wintertime stratosphere wind, temperature, and wave response. J. Geophys. Res. Atmos. 118, 2169–2183. https://doi.org/10.1002/jgrd.50236.

Shea, M.A., Smart, D.F., McCracken, K.G., 1965. A study of vertical cutoff rigidities using sixth degree simulations of the geomagnetic field. J. Geophys. Res. (1896–1977) 70, 4117–4130. https://doi.org/10.1029/JZ070i017p04117.

Sinha, A., Cannariato, K.G., Stott, L.D., Li, H.-C., You, C.-F., Cheng, H., Edwards, R.L., Singh, I.B., 2005. Variability of Southwest Indian summer monsoon precipitation during the Bølling-Ållerød. Geology 33, 813–816. https://doi.org/10.1130/G21498.1.

Souza Echer, M.P., Echer, E., Rigozo, N.R., Brum, C.G.M., Nordemann, D.J.R., Gonzalez, W.D., 2012. On the relationship between global, hemispheric and latitudinal averaged air surface temperature (GISS time series) and solar activity. J. Atmos. Sol. Terr. Phys. 74, 87–93. https://doi.org/10.1016/j.jastp.2011.10.002.

Sternberg, R.S., Damon, P.E., 1979. Re-evaluation of possible historical relationship between magnetic intensity and climate. Nature 278, 36. https://doi.org/10.1038/278036a0.

Usoskin, I.G., Gladysheva, O.G., Kovaltsov, G.A., 2004. Cosmic ray-induced ionization in the atmosphere: spatial and temporal changes. J. Atmos. Sol. Terr. Phys. 66, 1791–1796. https://doi.org/10.1016/j.jastp.2004.07.037.

Valet, J., Meynadier, L., 1993. Geomagnetic field intensity and reversals during the past four million years. Nature 366, 234. https://doi.org/10.1038/366234a0.

Veretenenko, S., Ogurtsov, M., 2012. Regional and temporal variability of solar activity and galactic cosmic ray effects on the lower atmosphere circulation. Adv. Space Res. 49, 770–783. https://doi.org/10.1016/j.asr.2011.11.020.

Vestine, E.H., Kahle, A.B., 1968. The westward drift and geomagnetic secular change. Geophys. J. R. Astron. Soc. 15, 29–37. https://doi.org/10.1111/j.1365-246X.1968.tb05743.x.

Vieira, L.E.A., da Silva, L.A., 2006. Geomagnetic modulation of clouds effects in the Southern Hemisphere Magnetic Anomaly through lower atmosphere cosmic ray effects. Geophys. Res. Lett. 33. https://doi.org/10.1029/2006GL026389.

Vieira, L.E.A., da Silva, L.A., Guarnieri, F.L., 2008. Are changes of the geomagnetic field intensity related to changes of the tropical Pacific sea-level pressure during the last 50 years? J. Geophys. Res. Space Phys. 113. https://doi.org/10.1029/2008JA013052.

Winkler, H., Sinnhuber, M., Notholt, J., Kallenrode, M.-B., Steinhilber, F., Vogt, J., Zieger, B., Glassmeier, K.-H., Stadelmann, A., 2008. Modeling impacts of geomagnetic field variations on middle atmospheric ozone responses to solar proton events on long timescales. J. Geophys. Res. Atmos. 113. https://doi.org/10.1029/2007JD008574.

Wollin, G., Ericson, D.B., Ryan, W.B.F., Foster, J.H., 1971. Magnetism of the earth and climatic changes. Earth Planet. Sci. Lett. 12, 175–183. https://doi.org/10.1016/0012-821X(71)90075-6.

Worm, H.-U., 1997. A link between geomagnetic reversals and events and glaciations. Earth Planet. Sci. Lett. 147, 55–67. https://doi.org/10.1016/S0012-821X(97)00008-3.

Wu, Z., Huang, N.E., Long, S.R., Peng, C.-K., 2007. On the trend, detrending, and variability of nonlinear and nonstationary time series. Proc. Natl. Acad. Sci. U. S. A. 104, 14889–14894. https://doi.org/10.1073/pnas.0701020104.

Xuan, C., Channell, J.E.T., 2008. Testing the relationship between timing of geomagnetic reversals/excursions and phase of orbital cycles using circular statistics and Monte Carlo simulations. Earth Planet. Sci. Lett. 268, 245–254. https://doi.org/10.1016/j.epsl.2007.12.021.

Yadava, M.G., Ramesh, R., 2007. Significant longer-term periodicities in the proxy record of the Indian monsoon rainfall. New Astron. 12, 544–555. https://doi.org/10.1016/j.newast.2007.04.001.

Yamazaki, T., 1999. Relative paleointensity of the geomagnetic field during Brunhes Chron recorded in North Pacific deep-sea sediment cores: orbital influence? Earth Planet. Sci. Lett. 169, 23–35. https://doi.org/10.1016/S0012-821X(99)00064-3.

Yamazaki, T., Oda, H., 2002. Orbital influence on Earth's magnetic field: 100,000-year periodicity in inclination. Science 295, 2435–2438. https://doi.org/10.1126/science.1068541.

CHAPTER

5

Galactic cosmic rays and solar particles in Earth's atmosphere

OUTLINE

5.1 Origin, compositional structure, and variability of energetic particles entering Earth's atmosphere

5.1.1 Galactic cosmic rays

The Earth's atmosphere is continuously bombarded by energetic particles from several tens of KeV to 3.10^{20} eV per particle. Discovered by Victor Hess in the early 20th century (1911–13), the existence of extra-terrestrial radiation reaching Earth's surface was confirmed

https://doi.org/10.1016/B978-0-12-819346-4.00005-X

© 2020 Elsevier Inc. All rights reserved.

in 1925 by Robert Millikan. Millikan gave this radiation the name 'cosmic rays', because he erroneously thought that the radiation penetrating deep under the ground water was gamma rays, i.e. energetic photons. Later on, Jacob Clay (1927) revealed the latitudinal dependence of cosmic rays' (CRs') intensity, which indicates that they are not photons, but most probably charged particles, which are affected by the geomagnetic field through the Lorentz force. Now it is well established that CRs are charged energetic particles: mainly protons, electrons, and nuclei, but also containing traces of antimatter (positrons) (Berezinskii et al., 1990; Ginzburg and Syrovatskii, 1964).

Depending on the energy and region of their generation and acceleration, CRs could have solar or galactic/extragalactic origin. Observationally, it is well-established that particles with energies less than 100 MeV are originated by the Sun. In contrast, particles with energies higher than several hundred MeV and lower than 10^{17}–10^{18} eV are usually considered to be of galactic origin. Their spatial distribution, determined by gamma-ray observations, possesses a well-defined radial gradient with higher intensity in the inner galaxy than in its outer disc (Drury, 2012). This is a conclusive argument that the particles are produced in the galaxy rather than diffusively penetrating from outside.

Given that the averaged magnetic field of the Milky Way galaxy is $\sim$3 μG, the Larmor radius (see Eq. 5.2) of protons with energy 10^{18} eV will be $\sim$360 pc[a] (Hörandel, 2008). This radius is comparable to the thickness of the galactic disc, which indicates that particles with higher energies will not be magnetically confined to the galaxy. For this reason, they are considered of extragalactic origin. Being injected by unknown sources outside the solar system, galactic cosmic rays (GCRs) diffuse through the interstellar medium and its turbulent interstellar magnetic field. Analysis of their diffusive propagation provides some constraints on the energy output of the sources of these highly energetic particles.

Taking into account that CRs' lifetime in the galaxy disc (before they escape to the intergalactic medium) is no more than 10^7 years, it was shown that the power needed to maintain the observed cosmic ray energy density of $\sim$1 eV/cm^3 (Neronov et al., 2012) must be $\sim 10^{41}$ erg/s $= 10^{34}$ W (Berezinskii et al., 1990; Blasi and Amato, 2011; Ginzburg and Syrovatskii, 1964; Strong et al., 2010). This estimation raises an important question: what energy sources in the galaxy are powerful enough to produce this output beam power?

Since the early 1930s, the main candidate that is able to provide such a power is the supernovae explosion, if we assume that at least 10% of its output kinetic energy is transferred to the GCRs (e.g. Blasi and Amato, 2011). The usual mechanism for particle acceleration is related to the Fermi I process, known also as diffusive shock acceleration. Briefly, the idea consists of the following: the charged particles, while spiralling along the galactic magnetic field lines, could be reflected by magnetic irregularities with a stronger magnetic field (for a detailed explanation, see Section 5.2). If this is a moving magnetic inhomogeneity, the reflected charged particle could gain speed when moving against the magnetic cloud (because its new velocity, after reflection, is the sum of its initial value and that of the moving magnetic irregularity). If the particle and magnetic cloud move in the same direction, the reflection will reduce the particle's speed by the velocity of the magnetic inhomogeneity.

[a]The *parsec* is a unit of length used to measure large distances to astronomical objects outside the solar system. One parsec is equal to about 3.26 light years or 31×10^{12} km.

The explosion of super nova star drives an expanding, fast-moving, shock wave of gas and dust (called ejecta) in the surrounding interstellar medium, which is observed as a supernova remnant. Magnetic irregularities, being in close proximity to the front of the shock wave, attain the shock velocity. An energetic particle, travelling through the shock wave, encounters the moving gradient magnetic field, which could reflect it back (increasing its initial speed with that of the shock wave velocity). The energy gained by the moving particle each time it diffuse across the shock is proportional to the velocity of the shock divided by the speed of light ($\Delta E/E \approx V_s/c$). This first-order relation determines the name of the acceleration, also known as 'first-order Fermi acceleration'.

If the shock is nonrelativistic, the particles make many shock crossings before finally being advected downstream. As a result, the whole energy spectrum of particles, diffusing through the shock, is formed from their smallest (initial energies) up to the maximal ones gained during the finite age and size of the shock. This spectrum is described by a power law: $N(p) \propto p^{-\gamma_p}$, with a universal slope, close to what is implied from CR observations (refer to Fig. 5.2). Here $N(p)$ is the number of particles per unit momentum interval and $\gamma_p = 3R_T/(R_T - 1)$. $R_T = u_1/u_2$ is the compression ratio of the shock, where u_1 and u_2 are the fluid velocities upstream and downstream of the shock, respectively. For strong shocks, $R_T \approx 4$ and hence $\gamma_p = 4$. For such a shock compression, the simple theory predicts an universal energy spectrum at relativistic energies of the form $N(E) \sim E^{-2}$, close to what is inferred from observations.

Consequently, while supernova explosions may be the ultimate source of energy, the acceleration process itself takes place at later times in the supernova remnants, which are formed around the explosion site. The efficiency of the first-order Fermi acceleration depends on the irreversibility of the process, which is guaranteed by the random scattering of particles by magnetic inhomogeneities within the shock wave. However, the acceleration remains effective for only a relatively short period. After some time (~few hundred years) the mass of the interstellar medium (swept up by the shock) becomes comparable to the mass of the ejecta, and starts to slow down the movement of the shock.

The first-order Fermi acceleration explains qualitatively fairly well the process of particles' acceleration by the supernovae remnants shock wave. However, estimations of the maximum achievable energy of particles (gained during their multiple crossing of the supernovae shock) end up with values 10^4–10^5 GeV, which are 30–100 times lower than the upper energy limit for particles originating in the galaxy (Amato, 2014). This result has brought scientists to the idea that the effective magnetic field strength of the shock may be substantially larger than the standard interstellar magnetic field (Bell, 2004). This means that the assumption for quasistationarity of the shock at intermediate length scales is not fully correct.

In order to describe more satisfactory the acceleration process, a new paradigm has been determined: Nonlinear Diffusive Shock Acceleration with Magnetic Field Amplification (e.g. Amato, 2014, 2016; Blasi, 2008). This new understanding takes into account the nonlinear evolution of the shock, as well as the feedback of accelerated particles over the shock wave. Without going into detail, we only mention here that particles, accelerated to velocities much larger than the shock speed, can stream ahead of the shock, creating magnetic instabilities there, which could amplify the interstellar magnetic field. This theoretical assumption has been confirmed by the high-resolution X-ray observations of several supernovae remnants, showing that the magnetic field at the shock location is ~100 times stronger than the typical interstellar magnetic field (Warren et al., 2005).

Regarding the chemical composition of CRs, these consist of nuclei of all elements in the periodic table. About 89% of them are hydrogen (protons), 10% helium, and about 1% heavier elements. Although overall similarities in compositional structure of solar and galactic cosmic rays, some differences have also been found. Thus, compared to the solar upper atmosphere and solar wind composition, GCRs are characterized by a general overabundance (by a factor of about 30) of iron and other heavy elements, relative to the hydrogen H and helium He densities (Ellison et al., 1997; Drury, 2012).

The ion composition of the solar upper atmosphere is determined by the energy of ionization of its constituents, i.e. the most abundant elements are more easily ionized. For a long time this was believed to be valid for GCRs as well. Therefore, the discovery that heavier ions in GCRs are much more than in solar wind was unexpected. Meyer and co-workers (1997) revealed, furthermore, that acceleration of the greater part of heavy elements depends on the condensation temperature T_c (determining their ability to condense into solid compounds) instead of on the energy of ionization. These authors found that the ratio of elements easily condensed at high temperature T_c, to the other heavy elements is enhanced by a factor of ~5 in GCRs, compared to that in solar corona and solar wind. This difference is increased by a factor of ~30 when the ratio of easily condensed elements to hydrogen is adjusted.

Moreover, numerous observations in UV, IR, and optical range of the spectrum reveal that the easily condensed elements (e.g. Mg, Al, Si, Ca, Fe, Ni, Sr, Zr) are largely locked into solid dust grains in most of the interstellar medium and supernovae ejecta. These weakly charged grains can behave like ions with extremely high rigidity, and thus be very efficiently accelerated. Taking into account all these facts, Ellison and co-workers (1997) have shown that the lighter gas components and dust grains (holding condensed heavier elements) are accelerated simultaneously, but with different mechanisms, by the same supernovae remnant blast wave. The gas ions are accelerated directly to cosmic ray energies, with well-pronounced dependence on the atomic mass to charge ratio ($\boldsymbol{A/Q}$), i.e. heavier atoms are accelerated to higher energies. The acceleration of the dust grains, however, is a two-stage process consisting of: (i) acceleration of entire grains to ~100 KeV/nucleon, followed by (ii) sputtering of accelerated grains and further acceleration of sputtered ions to GeV and TeV cosmic ray energies (Meyer et al., 1997; Ellison et al., 1997).

Another intriguing problem is the significantly lower density of electrons in GCRs, despite diffusive shock acceleration being capable of accelerating both protons and electrons. The low electron density is usually attributed to their strong radiative losses (e.g. by synchrotron radiation, inverse Compton scattering on thermal photons accompanied by γ-ray emission, leakage to metagalactic space, etc.), which can limit their maximum attainable energy. As a result, only the most energetic electrons are able to overcome the barrier of the heliomagnetic field, while the greater part is deflected back in the galaxy. Indeed, the measurements show that the lower energy end of the observed GCR spectrum is cut off due to interaction with solar wind. This is a reason to believe that in interstellar space, the cosmic ray spectrum extends far below what we can observe on Earth (i.e. above several GeV).

5.1.1.1 Long-term variability

Temporal variations of galactic particles penetrating in the solar system are related mainly to variability of the solar magnetic field. The most prominent is the 11-year modulation, characterized by a stronger deflection of GCRs by solar wind in periods of high solar activity (Jokipii, 1971). Satellites' measurements have shown in addition that periodical reversal of heliomagnetic field (from north to south and vice versa) with period of ~22 years modulates the compositional structure of galactic charged particles entering the heliosphere. For example, even solar cycles are characterized by a higher density of the positively charged particles compared to the odd cycles. Oppositely, the amount of negatively charged particles (e.g. electrons) is higher in odd solar cycles compared to even ones (Heber et al., 2009; Potgieter, 2014). During the even cycles (having maxima in the 1970s, 1990s, 2010s, etc.), the solar magnetic field is directed outwards from the solar North Pole. In odd cycles (with maxima in the 1960s, 1980s, 2000s, etc.), the solar magnetic field direction is towards the solar North Pole.

This bi-decadal modulation of charged particles entering the heliosphere is supposedly due to the curvature and the gradient of the heliomagnetic field (Jokipii, 1989). According to present understanding, the solar wind (blowing radially outwards from the Sun) carries with itself the solar magnetic field, which fills the heliosphere. However, closer to the Sun, the magnetic field lines rotate with the speed of the star, but much more slowly at far distances. This differential rotation (see Fig. 5.1A) of the heliomagnetic field produces its spiral structure. In solar minima, the heliomagnetic field is close to the field of a magnetic dipole (Jokipii, 1989) and the magnetic field in both hemispheres is separated by a thin layer known as a current sheet, across which the field changes its direction (Fig. 5.1B). With enhancement of solar activity, the structure of the heliosphere becomes more and more disturbed and its

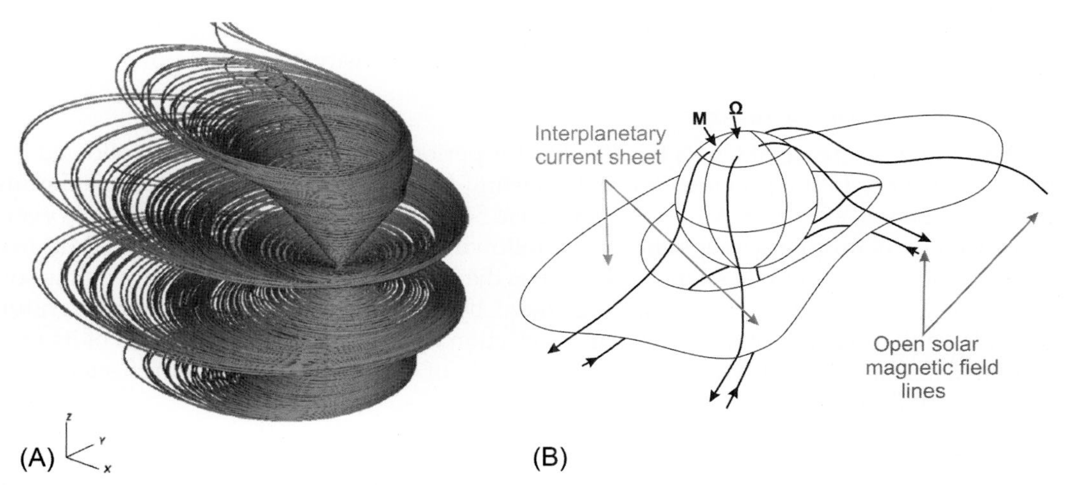

FIG. 5.1 Heliomagnetic field and equatorial current sheet. (A) 3D view of heliospheric magnetic field and (B) Equatorial current sheet in solar minimum. *(A) From Senanayake, U.K., Florinski, V., 2013. Is the acceleration of anomalous cosmic rays affected by the geometry of the termination shock? Astrophys. J. 778, 122. (B) Courtesy to Smith, E.J., Tsurutani, B.T., 1978. Observations of the Interplanetary Sector Structure up to Heliographic Latitudes of 16^0, Pioneer 11. J. Geophys. Res. 83 (A2), 717–7214.*

topology is changed substantially. Thus a nearly planar (in periods of quiet Sun) current sheet becomes more and more wavy with increasing solar activity.

To understand the mechanism of compositional modulation of GCRs by these topological changes of the heliomagnetic field, let us examine the case when the solar magnetic field is directed outwards from its North Pole (assumed to be in its positive direction). The numerical experiments with the transport equation show that in this case, the positively charged particles (mostly protons) drift inward from the heliosphere (Jokipii, 1989). Moreover, Jokipii noticed that nearly all of the particles found in the inner solar system (for this configuration of heliomagnetic field) came from a very small area near the North Pole. In contrast, when the north solar magnetic field is directed inward to the Sun's North Pole, the positively charged particles are deflected by the magnetic gradient-curvature drift out of the solar system. The other possibility for them to enter the heliosphere is through the neutral equatorial current sheet. This pathway, however, takes much more time due to the wavy structure of the heliospheric current sheet, which slows down their speed and reduces their intensity. This effect is detected by satellite instruments as a delayed appearance of positively charged particles compared to the electrons and other negatively charged particles in periods of negative polarity of solar magnetic field (Ferreira et al., 2003; Heber et al., 2009; Potgieter, 2014). As explained above, this configuration favours the inward diffusion of electrons and other negatively charged particles from the solar North Pole, a phenomenon detected by satellite instruments as an earlier appearance and enhancement of electrons concentration.

The longer timescales of galactic cosmic ray variability (e.g. their centennial variations) should probably be attributed to the secular variations of the geomagnetic field, which additionally modulate the intensity of particles flux entering Earth's magnetosphere. Some authors, analysing the sign of correlation between solar activity and Earth's surface temperature, have reported existing secular variations, which they related to the changing Earth's exposure to the more active solar hemisphere (Georgieva and Kirov, 2005; Georgieva et al., 2007). Whether such a periodicity affects GCRs' intensity is still unknown.

5.1.1.2 Short-term variations

The sudden decrease of GCRs by 3%–20% for periods of minutes to hours, followed by a gradual recovery to the previous intensity (lasting for hours or days), is called the Forbush decrease (named after the American physicist Scott Forbush). These short-lasting events are caused by heliospheric magnetic shock, following the coronal mass ejection (CME) from the Sun. The main reason for such a reduction is the sweeping of galactic particles by the solar wind plasma. Forbush decreases are registered by particle detectors and their magnitude depends on the strength of the coronal mass ejection (CME), and on the direction of its propagation. The rate of the CME is higher in periods of active Sun, which is reflected in the frequency of Forbush events as well.

5.1.1.3 Energy spectrum

Experimental cosmic ray research reveals that within the energy range 10^{11}–10^{20} eV, the energetic particles spectrum has a power-law like behaviour (i.e. proportional to $E^{-\gamma}$, with $\gamma \approx 2.7$). There are two changes of the curve slope: at energy range 2–5.10^{15} eV, known as a knee, and at 2–8.10^{18} eV, called an ankle (see Fig. 5.2). At the knee the spectrum steepens to $\gamma \approx 3.0$, while at the ankle it is flattened again by roughly the same change of the spectral index, i.e. back to $\gamma \approx 2.7$. It is currently accepted that cosmic rays with energies up to the

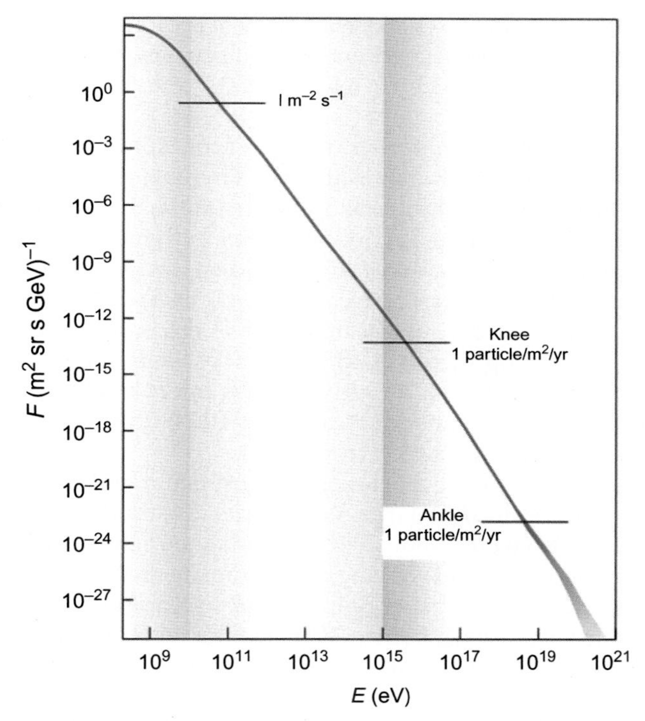

FIG. 5.2 Observed energy spectrum of primary cosmic rays in Earth's atmosphere. *Based on Wikipedia, https://en.wikipedia.org/wiki/Cosmic_ray#/media/File:Cosmic_ray_flux_versus_particle_energy.svg.*

ankle are of galactic origin, while those with higher energies originate most probably from outside the Milky Way (Abraham et al., 2007), i.e. the transition of cosmic rays of galactic to extragalactic origin is in the energy range from 10^{16} eV to a few 10^{18} eV.

There are two possible explanations for changing steepness of the cosmic rays energy spectrum. The astrophysical one relates it to the different processes of production, acceleration, and propagation of CR, which may affect not only their energy, but also their composition. According to the other point of view, the knee and the ankle are atmospheric effects, related to the particles interaction with atmospheric nuclei. Thus the formation of the knee is attributed to the formation of muons and neutrinos, which receive a part of the primary CRs' energy, but remain undetected by the surface detectors areas, measuring the air showers of secondary charged particles (Petrukhin, 2002). This 'missing' energy leads to the steepening of the energy slope of the primary CRs. Oppositely, the ankle appearance is supposedly due to the interaction of the muons with the atmospheric nuclei, and newly generated particles give rise to the intensity of measured extensive atmospheric showers.

5.1.2 Solar energetic particles

Massive bursts of ionized gas and magnetic field from the solar corona, known as coronal mass ejections (CMEs), are sources of energetic charged particles propagating in the heliosphere. The ejected plasma fluxes consist mainly of protons, with a variable admixture of He^{++} and heavier nuclei, and associated energetic electrons. The protons and other nuclei have kinetic energies between several tens of KeV and several hundred MeV (Reid et al., 1976).

Energetic particles of solar eruption travel to Earth, spiralling along the lines of the interplanetary magnetic field. Their arrival at Earth's magnetopause is dependent upon several factors such as the location of the eruption centre on the Sun, the energy of the particles, and the turbulence of the interplanetary medium. For extremely energetic events, in which particles are accelerated to energies exceeding 500 MeV, the X-ray and radio emissions can reach the Earth within a few minutes (Shea and Smart, 1995). If the solar eruption is located within about 30–40 degrees of the central meridian of the Sun (as viewed from Earth), the interplanetary shock preceding the fast CME will most likely intersect Earth's orbit within 1 or 2 days (Shea and Smart, 1995). The arrival of the interplanetary shock and subsequent coronal mass material, with its embedded magnetic field, typically manifest themselves as geomagnetic disturbance and increased particle flux.

The temporal variations of solar particles are determined mainly by the 11-year solar cycle, due to the higher frequency of the solar proton events in periods of increased solar activity. The semiannual cycle is also well-established in geomagnetic storm activity (and respectively particles fluxes reaching the ground) having maximums near the equinoxes. This periodicity is attributed to the increased storm energy input during equinoxes compared to that in solstices (Clúa de Gonzalez et al., 2001; Currie, 1966; Hakkinen et al., 2003; Patowary et al., 2013; Russell and McPherron, 1973). There are several hypotheses attempting to explain this effect: (i) the 'axial' hypothesis relates the seasonal variations to the changes of Earth's heliographic latitude during the year; (ii) the 'equinoctial' hypothesis relates them to the changes of the direction of solar wind flow relative to Earth's magnetic axis; and (iii) the 'Russell–McPherron' hypothesis attributes this effect to the inclination between the ecliptic (i.e. a trajectory of Earth's rotation around the Sun) and heliospheric equatorial current sheet (centred on the solar magnetic equator). The changing position of Earth relative to the equatorial current sheet, i.e. above or below it, determines the different orientation of the interplanetary magnetic field (IMF) near Earth, i.e. towards or outwards from the Sun (Prölss, 2004). The azimuthal warps and radial deformation of the equatorial current sheet forces additional variations in a direction perpendicular to the solar magnetic equator, which is more effective in generation of geomagnetic disturbances (reconnection between the IMF and geomagnetic fields occurs when they are antiparallel to each other).

5.2 Charged particles movement in heterogeneous magnetic field

The flux of cosmic rays (with galactic or solar origin) arriving at the outer boundary of Earth's magnetosphere are approximately homogeneous. While moving towards the Earth, these particles start to experience the action of the Lorentz's force, which in a nonrelativistic case and without an electric field could be written in the following form:

$$m\frac{d\mathbf{v}}{dt} = q(\mathbf{v} \times \mathbf{B}); \quad \mathbf{v} = \frac{d\mathbf{r}}{dt} \tag{5.1}$$

where **B(r,*t*)** is the external magnetic field (a function of the spatial dimensions and time), **r** and **v** are particle radius vector and velocity, respectively, ***m*** is particle mass, and ***q*** is its charge. Particles that are coming along the open magnetic field lines are either repelled back in space by the geomagnetic field or lost in the atmosphere due to the multiple collisions with the atmospheric molecules. Particles approaching the closed magnetic field lines, whose energy is insufficient to overcome them, become trapped or quasitrapped in the geomagnetic field. They are impelled by the Lorentz force to move along the magnetic field lines on spiral trajectories (the result of a combined circular and a field align motion), continuously bouncing between the Northern and the Southern Hemispheres. The radius of the particles circular trajectory, known as a Larmor radius (Eq. 5.2), is determined by: (i) the projection of particle velocity on a direction perpendicular to ***B***; (ii) the particle charge and mass; and (iii) the magnetic field strength, i.e.:

$$r_B = \frac{m \cdot v_{\perp 0}}{q \cdot B} \tag{5.2}$$

Formula (5.2) clearly shows that the electrons and protons rotate in opposite directions (by convention the positive, clockwise, is the angular velocity of electrons when ***B*** is away from the observer). The field align velocity of a charged particle is determined by the projection of its initial speed along the magnetic field line, which in the case of a constant magnetic field remains unchanged (e.g. Moisan and Pelletier, 2012).

In a gradient magnetic field, however, the situation is substantially changed. Thus, supposing that nonuniformity is along the magnetic field lines, and choosing the z-axis to be parallel to the magnetic field, the solution of the equation of motion (5.1), for the guiding centre of a particle (i.e. the centre of its circular motion), has the following form (Moisan and Pelletier, 2012):

$$v_{II}(t) = v_{II0} - \frac{v_{\perp}^2}{2 \cdot B_z}\left(\frac{\partial B_z}{\partial z}\right) \cdot t \tag{5.3}$$

where $\boldsymbol{v_{II}}$ denotes particle velocity parallel to the magnetic field line, with $\boldsymbol{v_{II0}}$ its initial value, and $\boldsymbol{v_{\perp}}$ is the velocity perpendicular to the magnetic field line. Eq. (5.3) shows that when a particle enters a region of increasing field strength, its parallel to the magnetic field velocity is retarded and at some level, when it becomes zero, the particle is reflected backward. The region where charged particles are subject to a reflection is called the magnetic mirror. In this case the constant of motion (i.e. the first adiabatic invariant) is already not the particle's velocity (as in the case of a constant and linear magnetic field), but its orbital magnetic moment ($\boldsymbol{\mu}$). The physical meaning of $\boldsymbol{\mu}$ is the magnetic field, created by the particle cyclotron motion around the magnetic line, which opposes the direction of the main field ***B***, defined as:

$$\mu = \frac{1}{2}\frac{m \cdot v_{\perp 0}}{B} = \frac{E_{kin}}{B}, \tag{5.4}$$

Here $\boldsymbol{E_{kin}}$ is the particle kinetic energy in the plane perpendicular to ***B***. In the absence of electric field, the total kinetic energy of the particle is a constant, because the inserted magnetic force is directed perpendicular to its velocity (see Eq. 5.1). This means that no mechanical work (i.e. no displacement in the direction of the force) is done. Consequently, the invariance of $\boldsymbol{\mu}$ implies synchronous changes in ***B*** and $\boldsymbol{E_{kin}}$. Since the total kinetic energy is conserved, the strengthening of ***B*** will lead to an increase of the velocity perpendicular to ***B***, i.e. ($\boldsymbol{v_{\perp}}$) and to a

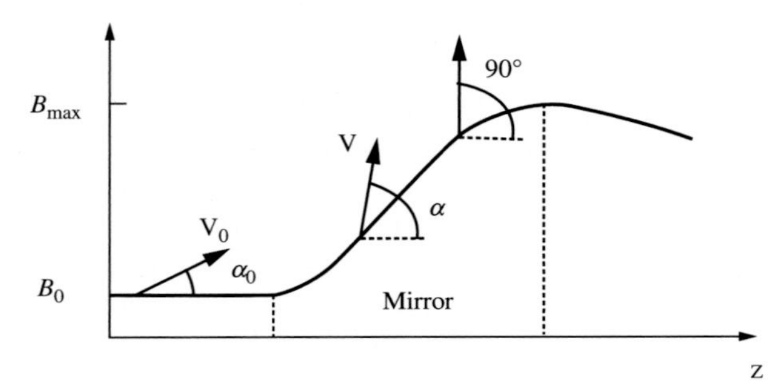

FIG. 5.3 Orientation of the particle's velocity vector, with respect to the equatorial magnetic field $\boldsymbol{B_0}$, and changing particles pitch angle $\boldsymbol{\alpha}$ (from $\boldsymbol{\alpha_0}$ at equator, to 90 degrees at magnetic mirror point). *Reproduced with permission from Moisan, M., Pelletier, J., 2012. Individual motion of a charged particle in electric and magnetic fields. In: Physics of Collisional Plasmas. Springer, Dordrecht.*

decrease of particle velocity parallel to the magnetic field line $\boldsymbol{v_{II}}$ (see Eq. 5.3). As a result, the radius $\boldsymbol{r_B}$ and the angle between particle helical trajectory and magnetic field line, known as pitch angle ($\boldsymbol{\alpha}$), are not a constant (Moisan and Pelletier, 2012); see Fig. 5.3 and Eq. (5.5):

$$\alpha = 2\pi\left(\frac{v_{II}}{v_{\perp}}\right)\cdot r_B, \text{ where } r_B = \frac{m}{qB}v_{\perp}; \tag{5.5}$$

The nonuniformity of Earth's magnetic field in the direction perpendicular to $\boldsymbol{B}$ (and it is easier to prove that the gradient along magnetic field lines is always accompanied by a gradient in radial direction, i.e. perpendicular to the field lines; Moisan and Pelletier, 2012), together with magnetic field curvature $\boldsymbol{\rho}$, induce a drift of particle guiding centre perpendicular to both the geomagnetic field and its gradient, described by the following formula:

$$v_{drift} = -\frac{\boldsymbol{\rho}\times\mathbf{B}\cdot m}{q\cdot B^2\cdot\rho^2}\left(\frac{1}{2}\cdot v_{\perp}^2 + v_{II}^2\right) \tag{5.6}$$

where the first term in the brackets corresponds to the magnetic gradient perpendicular to the magnetic field lines, while the second term relates to the curvature of the magnetic field lines. Formula (5.6) shows that particles' drift across the magnetic field lines depends on their charge $\boldsymbol{q}$, and consequently leads to charge separation, which generates electric field $\boldsymbol{E}$ along the drift direction. The combined effect of $\boldsymbol{E}$ and $\boldsymbol{B}$ fields induces an $\boldsymbol{E \times B/B^2}$ drift of particles, which displaces positive ions and negative electron in the same direction, perpendicular simultaneously to $\boldsymbol{B}$ and to $\boldsymbol{E}$. These charged particles are then 'lost' in the ambient atmosphere, where they release their energy by producing showers of secondary particles.

5.3 Energy losses of particles moving through the atmosphere

The different radiations behave differently in their passage through matter. Heavy particles, for example, lose little of their energy in a single collision, and are practically never deflected, except by a rare encounter with a nucleus. Thus the heavy particles lose their energy primarily through Coulomb scattering on the outer electron shells, delivering small amounts of energy to the orbital electrons and leaving behind excited molecules or electron-ion pairs (Wilson et al., 1991). Light particles (i.e. electrons or positrons) can lose a large fraction of their energy in a single collision with an electron. In fact, they can lose all of it, creating a knock-on electron of about the same energy. They are much more easily deflected, so their paths are not as straight as those of heavy particles.

The problem for estimation of energy losses of high-speed particles, traversing the matter, arises soon after their discovery. The energy loss of ions (as proposed by N. Bohr) could be divided into two components: nuclear stopping (i.e. energy loss due to the ion interaction with the nuclei in the medium) and electronic stopping (energy loss by the traversing ion due to its inelastic collisions, i.e. excitations of the bound electrons in the medium).

The relatively low energetic particles contribute to the energy loss mainly through the ionization of atoms and molecules filling the medium, while the very energetic ones (with energies greater than 1 GeV) also contribute with nuclear interactions with atomic nuclei of the atmosphere. After Bohr, Bethe showed than less than 0.1% of the energy loss is due to the interactions with target nuclei (ignoring nuclear reactions) and excluded this component from his analyses. Finally, Bloch evaluated the differences between the classical (Bohr) and quantum-mechanical (Bethe) approaches, for particles with velocities much larger than the target electrons. So today the Bohr-Bethe-Bloch formula, describing the energy loss of energetic particles moving through matter, has the following form (Leung, 1989; Ziegler, 1999):

$$-\left[\frac{dE}{dx}\right] = \frac{4\pi n z^2}{m_e v^2} \cdot \left(\frac{e^2}{4\pi\varepsilon_0}\right)^2 \cdot \left[\ln\left(\frac{2m_e v^2}{I(1-\beta^2)}\right) - \beta^2\right], \tag{5.7}$$

where z are charges of incident particles, e is the electron charge, n is the electron density of the target, I is the mean excitation energy of the target atom (molecule), m_e is the mass of the electron, v and c are the speeds of incident particle and speed of light, respectively, and $\beta = v/c$.

Formula (5.7) shows that energy lost by a particle in the atmosphere depends inversely on the square of its speed, $1/v^2$. This indicates that the faster the particle is, the less energy it transfers to the molecules of the medium. Consequently, in regions with a higher geomagnetic gradient, where according to Eq. (5.3) particles' velocity slows down, the energy lost on ionization of atmospheric molecules must increase. That is why dE/dx increases towards the end of particles' ionization range, reaching a maximum (known as the Bragg peak) shortly before the energy drops to zero.

The energy loss dE/dx is used to calculate the amount of secondary electrons produced by energetic particles in the atmosphere. Most existing models are based on the assumption for a dipole geomagnetic field, which determines the minimal energy ensuring particles access to a given point in the magnetosphere, called vertical magnetic rigidity R (e.g. Blinov and Kropotina, 2009; Usoskin et al., 2009; Mishev, 2013; Velinov et al., 2013; Mishev and Velinov, 2015, 2018). The calculation of the latter is based on Störmer's theory (Störmer, 1930). The number of ion pairs produced in one gram of the ambient air per second at a given atmospheric depth h is usually calculated by the following formula:

$$Q(h, \lambda_m) = \frac{1}{35} \sum_i \int_{E_{ci}}^{\infty} \int_{\Omega} D_i(E) \frac{\Delta E(h, E)}{\Delta h} \cdot \rho(h) \cdot dE \cdot d\Omega \tag{5.8}$$

where ΔE is the energy deposited in an atmospheric layer Δh and h is the residual atmospheric depth—the mass of the atmosphere above a given altitude in g cm^{-2} (recall that atmospheric depth h could be converted to atmospheric pressure by the formula: $\mathbf{1[g\ cm^{-2}] = 0.980665\ hPa}$). $D_i(E)$ in Eq. (5.8) is a differential cosmic ray spectrum for a given i-th component of primary CR (e.g. protons p, helium (α-particles), light nuclei with atomic numbers $\mathbf{Z}$ between 3 and 5, medium nuclei with $6 \leq \mathbf{Z} \leq 9$, heavy nuclei with $\mathbf{Z} \geq 10$, and very heavy nuclei with $\mathbf{Z} \geq 20$, having the units $[\mathrm{cm}^2\ \mathrm{sr\ GeV\ s}]^{-1}$); ρ is the atmospheric density in g cm^{-3}; λ_m is the geomagnetic latitude; and E_{ci}, is the initial kinetic energy of a particle of i-th type, corresponding to the local vertical geomagnetic cut-off rigidity R. The parameter R determines the minimum momentum per unit charge, which the given particle must possess to overcome the shielding effect of the geomagnetic field; Ω is a solid angle defined by the directions of arriving particles.

Formula (5.8) illustrates that regardless of the complexity of the code, simulating hadronic and leptonic cascades in the atmosphere, the calculated ionization suffers from the assumption of freely propagating charged particles along the open geomagnetic field lines, which excludes the impact of trapped and quasitrapped particles in Earth's radiation belts. Measurements of the Alfa Magnetic Spectrometer on the low orbiting satellites reveal an existence of under cut-off protons, electrons, and positrons at ionospheric levels with energies up to several GeV, near the maximum of the ionization layer (Fiandrini et al., 2004). By presumption, such particles are not included in calculation of atmospheric ionization by Eq. (5.8), despite their certain contribution to the atmospheric ion density.

Moreover, even at close distance to Earth (less than 1 Earth radius), the Larmor radii of protons are large enough ($\sim 10^3$–10^4 km), so the geomagnetic field cannot be treated as uniform during their gyration along the magnetic field line. Consequently, particles' trajectories can be very complex, even when assuming a simple dipole field, to say nothing about the complications resulting from the heterogeneous nature of the geomagnetic field.

All this means that theory based on the assumption for a dipole magnetic field topology (Störmer, 1930) and invariance of particles' velocity along the geomagnetic field line, which is equivalent to the assumption for constancy of magnetic field intensity, is obsolete nowadays. Störmer's theory does not take into account the impact of trapped and quasitrapped highly energetic protons (with hundreds MeV to GeV energies) from the inner Van Allen radiation belt, or the variations of the interplanetary magnetic field direction, leading to a continuous dynamical reconfiguration of magnetospheric topology and particle trajectories (Lemaire, 2003). Therefore, if the assumption for a dipolar character of geomagnetic field is still applicable to some spacecraft purposes (at levels where deviations from the magnetic dipole are relatively small), in the lower atmosphere, where particles are more sensitive to geomagnetic irregularities, the use of the vertical cut-off rigidity is quite a rough assumption. Thus the accurate calculation of geomagnetic rigidity is of special importance for climatic studies and should be calculated by the use of a trajectory-tracing technique, in a higher-order geomagnetic field model (Smart and Shea, 2009).

5.4 Hemispherical asymmetry in charged particles' confinement by the geomagnetic field

The number of particles with very high energy entering Earth's atmosphere is relatively small and their effect is negligible (Jackman et al., 2016; Semeniuk et al., 2011), because the Bohr-Bethe-Bloch formula (5.7) shows that energy released in the atmosphere is inversely proportional to the particles' speed (Leung, 1989; Ziegler, 1999). The less energetic particles are attracted by the geomagnetic field and are compelled by the Lorentz force to move around and along the magnetic field lines on spiral trajectories.

The confinement of any particle in the gradient magnetic field $\boldsymbol{B}$ depends on the ratio between the maximum field strength B_{max} in the polar regions (where the backward reflection of trapped particles occurs) and the equatorial magnetic field strength $\boldsymbol{B_0}$ (see Fig. 5.4), i.e.:

$$\mathbf{sin}\,(\alpha) = \mathbf{sin}\,(\alpha_0) \cdot \sqrt{\frac{B_{max}}{B_0}} \tag{5.9}$$

where the angle $\boldsymbol{\alpha_0}$ between velocity vector of an arriving particle and the equatorial magnetic field is known as an equatorial pitch angle, and $\boldsymbol{\alpha}$ is the continuously changing (with particle motion along the magnetic field line) pitch angle (see illustrations in Figs 5.3 and 5.4 from Moisan and Pelletier, 2012).

Any particle is assumed trapped by the magnetic field, when the angle $\boldsymbol{\alpha}$ becomes greater than $\frac{\pi}{2}$, because at this point—known as the magnetic mirror—the particle reverses its direction of movement, remaining confined by the magnetic field line. Obviously, if a charged particle approaches Earth's magnetosphere at a very small angle, the left-hand side of Eq. (5.9) could not exceed the value of $\frac{\pi}{2}$ (certainly true for $\boldsymbol{\alpha_0} = 0$). Entering the region of higher magnetic gradient, such a particle could not be reflected back and will therefore be 'lost' in the ambient atmosphere. The minimum value of the equatorial pitch angle $\boldsymbol{\alpha_{0m}}$ for which the maximum magnetic field is still able to reflect particles is called the 'loss cone'. If a particle arrives at an angle lower than the solid angle defined by $\boldsymbol{\alpha_{0m}}$, it will be lost in the ambient atmosphere on its motion along the magnetic field line. Note that the efficiency of the magnetic mirror to reflect charged particles does not depend on the particles' speed, or on their charge and mass.

The stronger geomagnetic field of the polar Southern Hemisphere ensures a smaller loss cone for energetic particles than those in the Northern Hemisphere. Thus, assuming an isotropic flux of the arriving particles, almost every third particle will be trapped by the geomagnetic field of the Southern Hemisphere, while in the Northern Hemisphere less than one-quarter of all arriving particles are trapped, due to its larger loss cone. Consequently, some

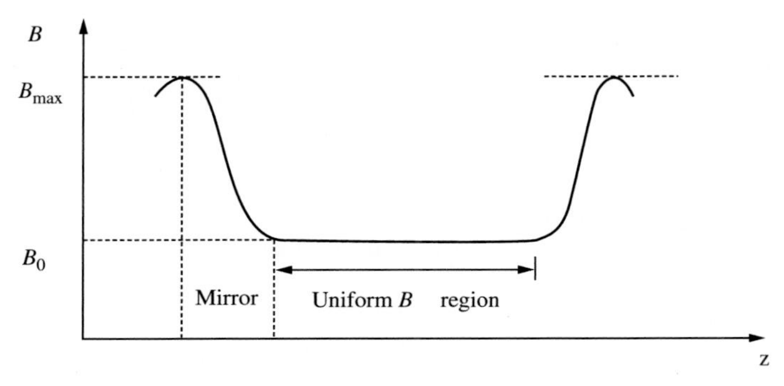

FIG. 5.4 An exemplar configuration of a confining magnetic field, keeping charged particles trapped between both regions with increased field (known as magnetic mirrors). Such a magnetic field configuration is known as a 'minimum B'. *Reproduced with permission from Moisan, M., Pelletier, J., 2012. Individual motion of a charged particle in electric and magnetic fields. In: Physics of Collisional Plasmas. Springer, Dordrecht (License agreement with Springer is available).*

particles trapped in the Southern Hemisphere could not be held by the weaker geomagnetic field in the Northern Hemisphere. The expected result is more particles precipitating in the Northern Hemisphere. Similarly, the magnetic field maximum in the mirror point over the South Atlantic magnetic anomaly is not symmetrical to its counterpart in the Northern Hemisphere. Being weaker, it allows particle penetration in the denser atmosphere in the region of the South Atlantic anomaly, where they are 'lost' due to the numerous interactions with neutral atoms and molecules (Fiandrini et al., 2004; Dachev et al., 1992, 2015).

The life span of the magnetically trapped charged particles is 2–3 years at lower altitudes (Filz and Holeman, 1965; Lazutin, 2010) and more than 10 years at higher altitudes (Selesnick et al., 2007), because of the smaller probability for collisions with atmospheric constituents at a distance of several Earth's radii.

Energetic particles penetrating deeper in the atmosphere create showers of secondary particles, produced from their interaction with atmospheric molecules: the deeper the penetration, the wider the showers. In the lower stratosphere, the number of secondary products dramatically increases, becoming maximal at a certain level. This level is known as the Regener–Pfotzer maximum. Beneath it, the concentration of secondary ions and electrons decreases again. It is well-documented that the altitude of the Regener–Pfotzer maximum is higher in polar regions and gradually decreases towards the tropics (Rosen and Hofmann, 1981; Stozhkov et al., 1996; Sarkar et al., 2017). This altitude dependence should possibly be attributed to the trapped radiation from the inner radiation belt, affecting mostly the tropics and extra-tropics. Satellite measurements show that protons in the inner radiation belt could be accelerated to several hundred MeV or even GeV energies (Fiandrini et al., 2004), so they could reach the lower atmospheric levels, before being expelled from the magnetic trap, by the drift-induced electric field (see Section 5.2).

5.5 Regional focusing of cosmic rays in the lower atmosphere

When entering Earth's atmosphere, the energetic particles collide with the atmospheric molecules, generating secondary ions and electrons, as well as different products of nuclear reactions. The efficiency of ion-molecular and nuclear reactions depends on the atmospheric characteristics such as density, temperature, humidity, etc., not only near the ground but also within the layer 50–100 hPa (Duperier, 1941, 1949; Trefall, 1957; Mendonça et al., 2013). Some particles, such as muons and neutrons, are highly penetrating and travel long distances without an interaction with the atmospheric molecules. They are the main component of the cosmic rays' secondaries, which are detected by sea-level neutron monitors, and even underground. The CR flux measured near the ground depends also on geomagnetic rigidity and the altitude of the observational point. The influence of all these factors on the particles' flux reaching Earth's surface is analysed in this section.

5.5.1 Analysis of neutron monitors' annual means

The temporal and spatial variability of GCR during 2009 has been analysed in 33 neutron monitors (NMs), with freely available data. The data used (courtesy of IZMIRAN: http://cr0.izmiran.ru/common/links.htm and NMDB portal: http://www01.nmdb.eu) are corrected for pressure variability and detectors' efficiency. The chosen 2009—an year of minimal solar activity and maximal GCR intensity—ensures easier detection of meteorological influences, as well as the effect of geomagnetic field spatial irregularities on the NMs measurements.

According to current understanding, the dominant part of the particles reaching Earth's surface arrives along the open magnetic field lines. This means that NMs with lower geomagnetic rigidity (i.e. magnetic field efficiency for particles deflection) receive a higher radiation dose than those situated at lower geomagnetic latitudes (with higher magnetic rigidity). The intensity of the measured particle fluxes depends also on the altitude of the monitoring station, because only particles with very high energies are able to reach sea level. To clarify how these controversial factors affect the spatial distribution of the particles' intensity near the surface, two maps have been combined: the map of NMs' elevation above sea level and the map of NMs' annual mean counting rates. The result is presented in Fig. 5.5A and shows that the dominant factor, determining the measured particles' intensity at the Earth's surface, is the altitude

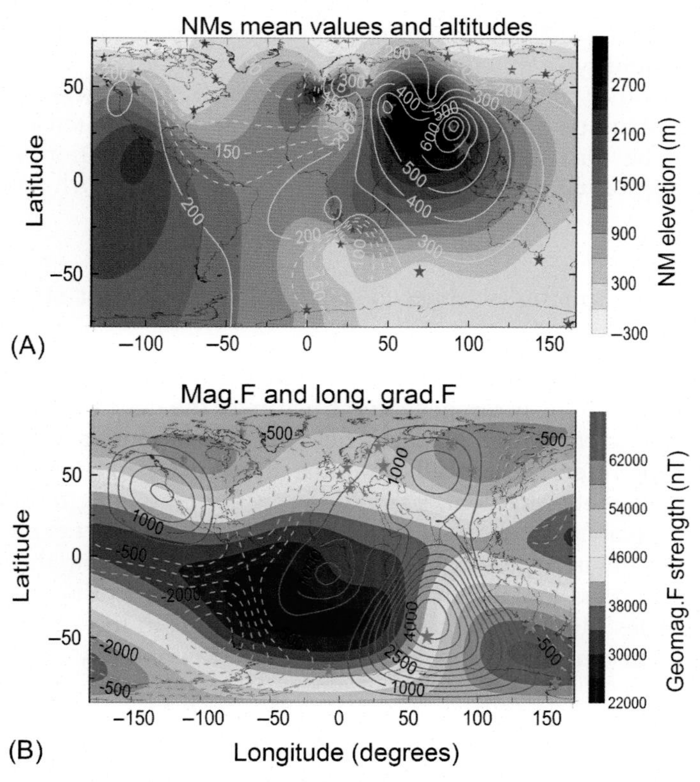

FIG. 5.5 (A) Maps of neutron monitors (NMs) elevation above the sea level (*coloured shading*). Overdrawn contours, and different size stars, present the spatial distribution of annual mean counting rates for 2009, based on the individual values calculated for each NM. The counting rates lower than 200 are given by *dashed contours*. Note that larger stars denote higher cosmic ray intensity. (B) Maps of geomagnetic field intensity (*coloured shading*) and longitudinal gradient of geomagnetic field (*contours*). The annual mean cosmic radiation received by NMs with altitudes less than 500 m (during 2009) is indicated by the different size stars. *Based on Kilifarska, N., Bojilova, R., 2019. Geomagnetic focusing of the cosmic ray's neutron component—evidence and mechanism. C. R. Acad. Bulg. Sci. 72 (3), 365–374.*

of the measurements point. It is clear that the highest particle fluxes are observed in the monitors with higher elevation above sea level, e.g. at Tibetan Plato (refer to Table 5.1).

On the other hand, a close examination of the annual particles fluxes—detected by monitors with similar elevations above sea level, but with different magnetic rigidity (***R***)—reveals some unexpected results. Thus, the annual mean impulses measured in Thule, Inuvik, and Nain (all of them with $\boldsymbol{R} = 0.3$) are almost half the size of that measured in Moscow ($\boldsymbol{R} = 2.43$); see Table 5.1. Similarly, Table 5.1 shows that particles' intensity in Fort Smith ($\boldsymbol{R} = 0.3$) is smaller than that measured in a station with seven times higher magnetic rigidity, Magadan ($\boldsymbol{R} = 2.1$), despite their similar elevations. However, the stations Tixie Bay ($\boldsymbol{R} = 0.48$) and Norilsk ($\boldsymbol{R} = 0.63$), both being at sea level, having similar latitudes and magnetic rigidities, count substantially different particle fluxes (with Norilsk being much higher than Tixie Bay).

Another example of deviation from the 'elevation-rigidity' rule gives the comparison of the African stations: Tsumeb ($\boldsymbol{R} = 9.15$) and Potchefsroom ($\boldsymbol{R} = 6.98$). Table 5.1 shows that particles' flux intensity, measured in the station with lower elevation and with substantially higher magnetic rigidity, is almost six times higher. The situation with stations Kerguelen ($\boldsymbol{R} = 1.14$) and Sinae ($\boldsymbol{R} = 0.73$) is similar; the annual mean particles' flux in an elevated station, with a lower rigidity, is much weaker.

These and many other 'anomalies', found in the intensity of particles reaching the Earth's surface, suggest that another factor(s) might exist that affects the intensity of particle distribution over the globe. One such factor could be the trapped radiation in Earth's radiation belts. The following section describes the possible mechanism of such an influence.

5.5.2 Geomagnetic lensing of charged particles in the lower stratosphere–upper troposphere

The geomagnetic field is a vector sum of a magnetic dipolar field (the dominant part), a nondipole field related to the heterogeneous structure of the deep Earth's interior, magnetic properties of the crustal rocks, and magnetic field of external sources. The resultant vector at the planetary surface differs significantly from the dipole magnetic field and is accompanied by a nonuniform magnetic gradient, particularly in the cross-longitudinal direction. This means that when advancing towards Earth's surface, particles start experiencing magnetic irregularities, especially in the lower part of their spiral motion along the geomagnetic field lines. The additional cross-longitudinal magnetic gradient affects the speed of trapped and quasitrapped particles' drift (simultaneously perpendicular to the geomagnetic field lines, the radius of their curvature, and the gradient vector), described by Eq. (5.10):

$$\mathbf{v}_{drift} = \frac{m}{q \cdot B^2}\left(v_\perp^2 \cdot \frac{\mathbf{B} \times \nabla \mathbf{B}}{2B} + v_{II}^2 \cdot \frac{\boldsymbol{\rho} \times B}{\rho^2}\right) \tag{5.10}$$

where $\boldsymbol{B}$ is the magnetic vector, $\boldsymbol{\rho}$ is the radius of curvature of the geomagnetic field lines, v_{II} and $v_\perp$ are projections of particles velocities parallel and perpendicular to geomagnetic field line, respectively, and $\boldsymbol{q}$ and $\boldsymbol{m}$ are particles' charge and mass, respectively. The first term in the brackets corresponds to the magnetic gradient perpendicular to the field lines, while the second term relates to their curvature. Close to the point of the magnetic mirror, the field aligned particles' velocity is approaching zero, so the drift velocity is determined mainly by the cross-latitudinal and cross-longitudinal magnetic gradients (the first term in Eq. 5.9).

TABLE 5.1 List of analysed neutron monitors with their geographic coordinates, geomagnetic rigidity, elevation above sea level, and annual mean value of the measured cosmic ray flux.

	NM name	NM code	Lat. (degrees)	Long. (degrees)	Rigidity (GV)	Alt (m)	Ann. mean
1	Alma-Ata	AANM	43.04N	76.94E	6.69	3340	167.4
2	Apatity	APTY	67.57N	33.39E	0.65	181	185.5
3	Aragats	ARNM	40.47N	47.44E	7.1	3200	670.1
4	Athens	ATHN	37.97N	23.78E	8.53	260	57.3
5	Calgary	CALG	51.08N	114.13W	1.08	1123	349.2
6	Dourbes	DRBS	50.1N	4.59E	3.18	225	115.3
7	Fort Smith	FRSM	60.02N	111.93W	0.3	180	137.3
8	Hermanus	HRMS	34.43S	19.23E	4.58	26	75.9
9	Inuvik	INUVIK	68.36N	133.72W	0.3	21	122.5
10	Irkutsk	IRKS	52.47N	104.03E	3.64	435	130.2
11	Jungfraujoch	JUNG	46.55N	7.98E	4.49	3475	168.1
12	Kerguelen	KERG	49.35S	70.25E	1.14	33	236.7
13	Kingston	KGSN	42.98S	147.29E	1.88	65	219.3
14	Kiel	KIEL	54.34N	10.12E	2.36	54	180.4
15	Lomnicky Stit	LMKS	49.2N	20.22E	3.84	2634	472.0
16	Mc Murdo	MCMU	77.9S	166.6E	0.3	48	174.8
17	Magadan	MGDN	60.04N	151.05E	2.1	220	148.0
18	Moscow	MOSC	55.47N	37.32E	2.43	200	241.5
19	Mexico	MXCO	19.8N	99.18W	8.28	2274	232.7
20	Nain	NAIN	56.55N	61.68W	0.3	46	136.0
21	Nor-Amberd	NANM	40.37N	44.25E	7.1	2000	496.2
22	Newark	NEWK	39.68N	75.75W	2.4	50	102.7
23	Norilsk	NRLK	69.26N	88.05E	0.63	0	180.8
24	Oulu	OULU	65.05N	25.47E	0.81	15	113.4
25	Princess Sirindhorn	PSNM	18.59N	98.49E	16.8	2565	620.3
26	Potchefstroom	PTFM	26.68S	27.09E	6.98	1351	59.4
27	Rome	ROME	41.86N	12.47E	6.27	0	158.2
28	Sanae	SNAE	70.17S	2.35W	0.73	856	178.7
29	Thule	THULE	76.5N	68.7W	0.3	26	130.7
30	Tibet	TIBT	30.11N	90.56E	14.1	4300	3166.7
31	Tsumeb	TSMB	19.2S	17.58E	9.15	1240	338.9
32	Tixie Bay	TXBT	71.59N	128.78E	0.48	0	104.1
33	Yakutsk	YKTS	61.99N	129.7E	1.65	105	107.8

From Kilifarska, N., Bojilova, R., 2019. Geomagnetic focusing of the cosmic ray's neutron component—evidence and mechanism. C. R. Acad. Bulg. Sci. 72 (3), 365–374.

Under the influence of the bi-directional magnetic gradient (in the x–y plane, with x directed to the east, while y is to the north), the protons, entering Earth's atmosphere from the west (at the lowest part of their circular trajectories around geomagnetic field lines), are shifted south-westwards when entering regions with a positive cross-longitudinal gradient, and south-eastwards in regions with a negative gradient (see Eq. 5.9). Consequently, the overall westward drift (forced by the magnetic curvature and cross-latitudinal gradient) is reduced by the eastward component, exerted by the cross-longitudinal magnetic gradient in regions with negative longitudinal magnetic gradient such as the East American–Atlantic region, Eastern Asia–Western Pacific, and the South Pacific Ocean. Furthermore, the drift-aligned electric field—expelling the confined particles outside the magnetic trap (due to the $(\boldsymbol{E} \times \boldsymbol{B})/\boldsymbol{B}^2$ electric drift; see Section 5.2)—is reduced significantly in these regions. As a result, only a few particles have a 'chance' to be lost in the atmosphere in these regions, and the ground-based neutron monitors should measure low counting rates.

Oppositely, in regions with positive cross-longitudinal gradients (i.e. North-Western America, Eastern Europe-Western Asia, the central part of the South Atlantic Ocean, and the South Indian Ocean) the southward component, induced by the cross-longitudinal magnetic gradient, changes slightly the direction but not the amplitude of the westward drift, impelled by the magnetic curvature and latitudinal gradient. Consequently, in these regions the drift-induced charge separation, and related electric field, will intensively expel the charged particles outside the magnetic trap. Furthermore, these particles interact with the atmospheric molecules, creating secondary electrons, ions, and nuclear products, giving rise to the ionization of the lower atmosphere, and to the radiation measured by the ground-based neutron monitors. Moreover, the effect of the cross-longitudinal magnetic heterogeneity should be stronger in regions with a steeply decreasing negative gradient and in the rising part of the positive cross-longitudinal gradient, i.e. in regions of geomagnetic field strengthening.

The validity of these theoretical considerations is presented in Fig. 5.5B, which compares the maps of the cross-longitudinal geomagnetic gradient (contours) with annual mean values of NMs counting rates (differently sized stars). In order to eliminate the altitude dependence of received radiation, shown are only NMs with altitudes less than 500 m. Although the map of the near-surface particles' intensity is quite rough (due to the relatively small number of NMs and their irregular distribution globally), Fig. 5.5B illustrates fairly well the fact that the lowest counting rates are detected in regions with longitudinaly decreasing geomagnetic field. This effect could be a reasonable explanation for the higher particle intensity encountered in Moscow compared to Inuvik, Tule, Nain, and other neighbouring stations (Kiel, Oulu, Apatite). Similarly, the dose measured in Norilsk and Irkutsk (situated in a region with a positive longitudinal geomagnetic field gradient) is higher than that detected in Tixie Bay, Yakutsk, and Magadan, which are placed in a region with a decreasing (along the path of the arrival protons) geomagnetic field.

5.5.3 Spatial distribution of neutron monitors' seasonal variations

In addition to the annual mean values, we have also analysed the seasonal variation of all examined neutron monitors, which vary quite a lot between the individual stations. More

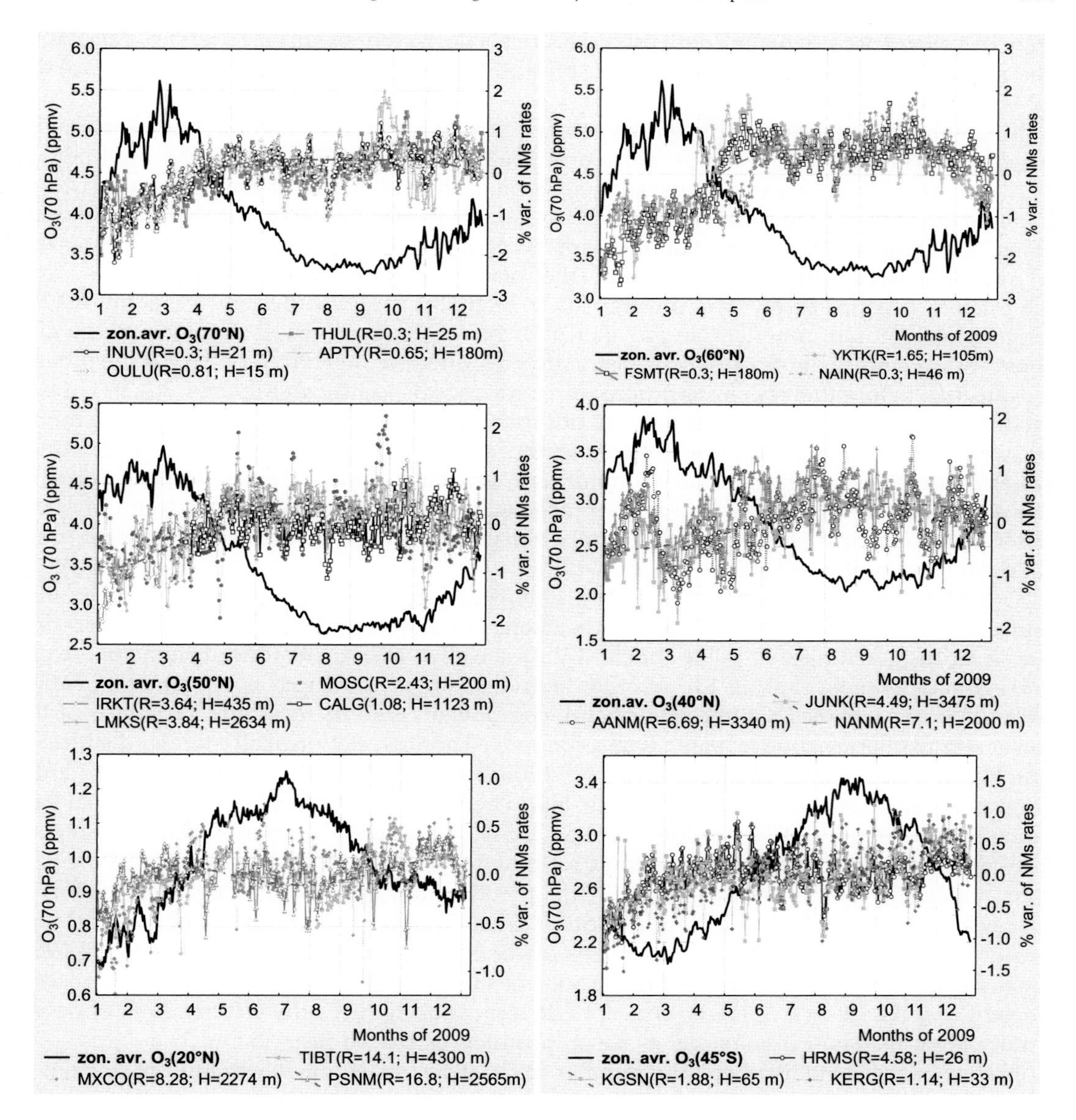

FIG. 5.6 Seasonal variation of NMs counting rates *(different grey-black symbols)* and ozone mixing ratio at 70 hPa *(black contour)*, shown for different geographic latitudes. *Based on Kilifarska, N., Bojilova, R., 2019. Geomagnetic focusing of the cosmic ray's neutron component—evidence and mechanism. C. R. Acad. Bulg. Sci. 72 (3), 365–374.*

detailed analysis reveals, however, that the shape of the seasonal variability is mostly confined to the geographic latitude (see Fig. 5.6). This implies that the seasonal variations of neutron monitors are more probably related to meteorological rather than geomagnetic effects.

Here it is worth noting that the data used are already corrected for surface pressure variability and monitor efficiency, which means that some other atmospheric effect should influence the NMs' measurements.

Meanwhile, in the mid-20th century, some authors found out the dependence of NMs' seasonal variations on the temperature and pressure between 50 and 100 hPa (Duperier, 1941; Trefall, 1955). As well as this is the altitude of the maximum of **π**-mesons production (detected by NMs), Duperier (1949) has attributed the relation between meteorological variables and ground based measurement of GCR intensity to two competing processes: (i) mesons' decay into muons, being the main atmospheric source of muons, and (ii) nuclear capture of mesons through their interaction with other nuclei, acting as an stratospheric sink of muons. The prevalence of any of these processes depends on the mesons' energy and the atmospheric density, determining the mean free path of atmospheric mesons ($\boldsymbol{d_\pi}$) before their decay to muons and neutrinos (Cecchini and Spurio, 2012). Thus, if the free path of atmospheric nuclei is greater than $\boldsymbol{d_\pi}$, the mesons' decay dominates, ending up with production of muons. Otherwise the atmospheric mesons interact with atmospheric nuclei, producing tertiary, quaternary, etc. subatomic particles. Even if some of the newly appearing products are other mesons, which furthermore decay to muons, the latter have much less energy and their probability of reaching the Earth's surface is severely reduced (Cecchini and Spurio, 2012).

Our comparative analysis of the NMs' counts and temperature at 70 hPa reveals their antiphase covariance at latitudes where the ozone at the same level controls the temperature (i.e. in latitudinal range: 40–10°N and 30–40°S), as well as over the Southern Hemisphere polar region (Kilifarska et al., 2018). At other latitudes, however, there is no systematic relation between both variables. On the other hand, comparison with the seasonal variations of the lower stratospheric ozone reveals a well-pronounced anticorrelation at all examined latitudes (see Fig. 5.6). The winter reduction of NMs' counting rates, when the ozone density at 70 hPa is raised, is clear in both hemispheres (Fig. 5.6). The maximal amplitude of this seasonal variation is observed at 60°N latitude equal to 2.75%. It gradually decreases poleward (being 2.3% at 70°N) and towards the equator, dropping to 1% at 20°N. The weakening of the seasonal variability of the ground-level CR flux could be attributed to the higher elevation of ozone layer at tropical and subtropical latitudes. On the other hand, the level of maximal ionization in the lower stratosphere decreases toward the equator (Rosen and Hofmann, 1981; Stozhkov et al., 1996; Sarkar et al., 2017). Thus, the increased distance between the peak ozone density and the maxim of lower atmospheric ionization reduces the effectiveness of ozone influence on the layer of maximal **π**-meson production, placed near 15 km.

In the Southern Hemisphere, the number of CR detectors is much less, but the calculated amplitude of seasonal variation at 45°S latitude is equal to that found at 40–50°N ones (i.e. 1.5%). The higher amplitude of seasonal variability, found in the NM rates at the surface of the Ice Cube in Antarctica, i.e. 5% (Tilav et al., 2009), should probably be attributed to the higher elevation of the detector, which is placed 2835 m above sea level.

These results suggest that the lower stratosphere influences the NMs measurements not by the virtue of its temperature (as currently believed), but rather thanks to its chemical composition, and more specifically O_3. In fact the lower stratospheric temperature itself is influenced by the ozone density (outside the polar regions), due to the ozone absorption of solar radiation. The suggestion is supported by the fact that the atmospheric **π**-mesons (usually called pions) are produced generally in altitudinal range 10–20 km, which coincides fairly well with the ozone layer (Tilav et al., 2009) at middle and high latitudes. This coincidence could be attributed to the higher interaction cross-section of the nucleons (*N*) in atmospheric showers with the bigger O_3 molecule (σ_N^{air}) and the reduced mean free path of the former in the atmosphere (see Eq. 5.11).

$$\lambda_N = \frac{1}{n \cdot \sigma_N^{air}} \tag{5.11}$$

where *n* is the number density of the air. Thus in a region with an increased O_3 density, the mean free path of energetic nucleons would become smaller than the mean pions' free path before their decay (Grieder, 2010). In this case, the pions' decay is suppressed due to their more frequent interactions with the atmospheric molecules. The tertiary, quaternary, etc. produced pions, due to such interactions, are already less energetic and their probability of reaching the surface and influencing the neutron monitors' measurements is highly reduced (Cecchini and Spurio, 2012).

This reasoning could explain the seasonal antiphase covariance between the lower stratospheric O_3 and most of the analysed neutron monitors. However, in some of them (i.e. Newark, Rome, Athens, Tsumeb, and Potchefstroom) we have found an in phase covariance with ozone at 70 hPa (see Fig. 5.7). Reference to Table 5.1 and Fig 5.5B reveals that all these 'exceptional' NMs are situated in regions with a near-zero or very small longitudinal geomagnetic gradient. So with a relatively small error, we can use the dipole estimations of the height of the Regener–Pfotzer maximum (Sarkar et al., 2017) in these stations. Fig. 5.8 shows that for most of the stations in question, the Regener–Pfotzer maximum is placed beneath 70 hPa (i.e. closer to the earth surface). On the other hand, the peak of the ozone layer increases towards the Equator, being elevated at subtropical latitudes up to 30 hPa. This means that ozone influence on the layer with higher pion density progressively decreases when approaching the Equator. Consequently, the synchronous in phase variability of the O_3 and NMs' counting rates at subtropical latitudes is more likely to be influenced by the atmospheric density seasonal variations, i.e. by the winter compression and summer expansion of the troposphere. Thus the summer uplifting of the denser, and poor of O_3 troposphere, will reduce the ozone density at the level of Regener–Pfotzer maximum of ionization. Despite the ozone depletion, however, the increased atmospheric density at this level will reduce the nucleons' mean free path, due to the more frequent interaction with atmospheric molecules (see Eq. 5.10). This means that most of the lower stratospheric pions will not succeed in decaying to muons and NMs will encounter a decrease of the muonic component of the ground-level CR flux. A weak echo of this effect is visible also at some neutron monitors, shown in Fig. 5.6. Oppositely, the winter compression of the troposphere, followed by a reduction of atmospheric density in the lower stratosphere, will favour the production of muons from the π-mesons' decay. In this context, the hemispherical asymmetry visible in Fig. 5.7 should be attributed to the phase shift of seasons in both hemispheres.

In resume: Analysis of the neutron monitors' measurements in a year with maximal density of GCRs at the end of the 23rd and beginning of the 24th solar cycles (i.e. 2009) reveals the existence of some irregularities in the measured CR flux, which are not consistent with the elevation of the observational point and its geomagnetic rigidity. We show that these

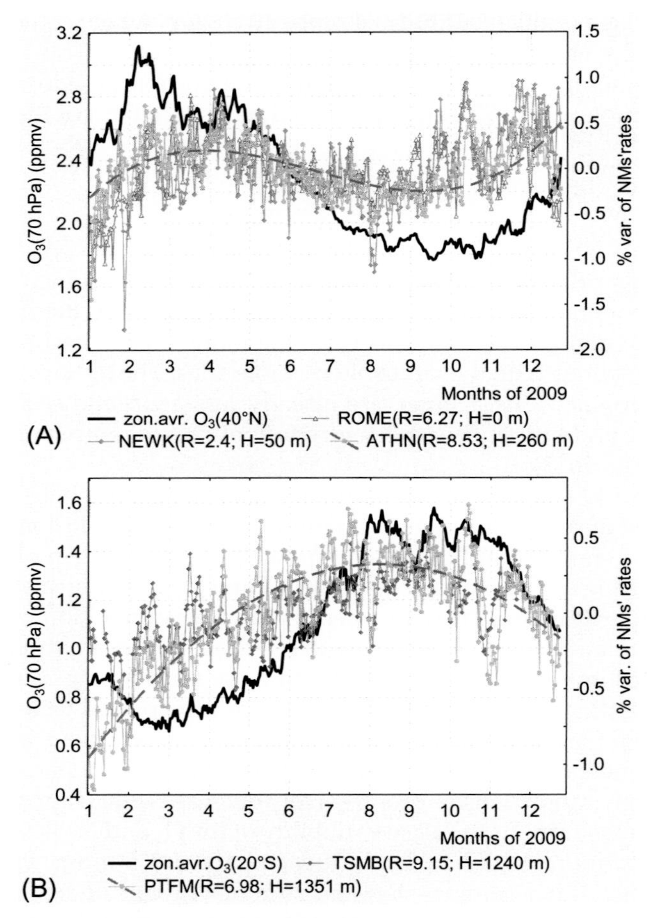

FIG. 5.7 Seasonal variations of neutron monitors counting rates *(different grey symbols)*, situated in regions with near-zero cross-longitudinal magnetic gradients (A) in the Northern Hemisphere and (B) in the Southern Hemisphere. *Dashed grey curves* designate the polynomial fit to data measurements; *continuous black curves* illustrate the seasonal variations of ozone at 70 hPa. *From Kilifarska, N., Bojilova, R., 2019. Geomagnetic focusing of the cosmic ray's neutron component—evidence and mechanism. C. R. Acad. Bulg. Sci. 72 (3), 365–374.*

irregularities could be attributed to the geomagnetic lensing of the precipitating particles (on their trajectories closest to the Earth's surface) in regions with positive geomagnetic gradient, in the direction of eastward-moving protons. Comparative investigation of the seasonal variability of radiation dose (measured in 33 neutron monitors), and the lower stratospheric O_3, implies that the latter modulate the muonic component encountered by the ground-based NMs. For example, enhancement of the heavier ozone molecule density increases the probability for nuclear capture of π-mesons, thus reducing the production of muons. In contrast, depletion of the lower stratospheric O_3 density increases the mean free path of atmospheric nuclei, raising the probability for their decay to muons. As a result, an increase of the counting rates of ground-based NMs is detected.

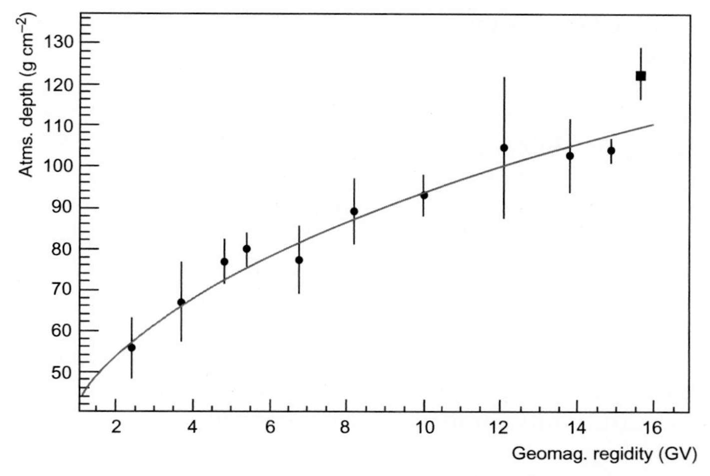

FIG. 5.8 Average altitude of the Regener–Pfotzer maximum in units g cm^{-2} (i.e. atmospheric depth reached by the incoming cosmic rays), depending on the vertical geomagnetic rigidity. Conversion of atmospheric depth in pressure levels could be achieved by the formula: 1[g cm^{-2}] = 0.980665 hPa. *Courtesy of Sarkar, R., Chakrabarti, S.K., Pal, P.S., Bhowmick, D., Bhattacharya, A., 2017. Measurement of secondary cosmic ray intensity at Regener-Pfotzer height using low-cost weather balloons and its correlation with solar activity. Adv. Space Res. 60, 991–998. https://doi.org/10.1016/j.asr.2017.05.014 (copyrights to Elsevier).*

5.6 Cosmogenic isotopes in Earth's atmosphere

Cosmic ray particles with $E > 1$ GeV are able to interact directly with the nucleus of atmospheric atoms. In these inelastic collisions, a large part of the energy of the hitting protons is carried away by protons, neutrons, and alpha particles expelled from the target nucleus. Furthermore, these secondary particles develop hadronic cascades in the atmosphere producing a bulk of new particles like kaons, pions, muons, etc. Some of the energy of cosmic ray protons is transferred to the target nuclei. The excited nucleus can break up into fragments, a process called *spallation*. The large variety of cosmogenic radionuclides has been created by spallation, and all of them have masses smaller than that of the target nucleus. Moreover, the exact structure of the target nucleus is only of minor importance (Beer et al., 2012). For example, beryllium isotope ^{10}Be can be produced by four different reactions:

$$^{14}\mathrm{N} + \mathrm{n} \rightarrow 2\mathrm{n} + 3\mathrm{p} + {}^{10}\mathrm{Be}$$

$$^{14}\mathrm{N} + \mathrm{p} \rightarrow \mathrm{n} + 4\mathrm{p} + {}^{10}\mathrm{Be}$$

$$^{16}\mathrm{O} + \mathrm{n} \rightarrow 3\mathrm{n} + 4\mathrm{p} + {}^{10}\mathrm{Be}$$

$$^{16}\mathrm{N} + \mathrm{p} \rightarrow 2\mathrm{n} + 5\mathrm{p} + {}^{10}\mathrm{Be}$$

where **p** and **n** designate the proton and neutron, respectively. The thresholds energy for **^{10}Be** production from nitrogen is 15 MeV for neutrons, and 33 MeV for protons, while for production from oxygen, it is about 30 MeV for both nucleons. The efficiency of radionuclides production is determined mainly by the hardness of the differential energy spectra of protons and neutrons. Analyses by Beer et al. (2012) reveal that the neutron – nitrogen interactions have the largest impact in ^{10}**Be** production.

The neutrons are created mainly by hadronic cascades, as well as GCRs entering the atmosphere usually do not contain neutrons (except those in the nuclei of helium and heavier elements). The higher efficiency of neutrons in the process of spallation, compared to the energetic protons, is due to the fact that they do not need to overcome the Coulomb barrier of the target atom, experienced by the protons. The difference between the proton and neutron becomes especially obvious at low energies, because the positively charged proton is repelled by the protons in the target nucleus (Coulomb barrier), while the neutron can easily enter the nucleus.

An illustration of the higher impact of neutrons in production of cosmogenic isotopes is the fact that ~99% of the ^{14}**C** isotope is produced by thermal neutrons with a typical energy of 0.025 eV (Beer et al., 2012) through the following reaction:

$$^{14}N + n \rightarrow p + {}^{14}C$$

The production rate ***P*** of a cosmogenic nuclide ***j*** at atmospheric depth ***d*** is given by the following expression (Beer et al., 2012):

$$P_j(D) = \sum_i N_i \sum_k \int_0^\infty \sigma_{ijk}(E_k) \cdot J_k(E_k d) \cdot dE_k \tag{5.12}$$

where the production is summed over interactions with all ***i*** target nuclei (e.g. O, N, or Ar), $\mathbf{J_k(E_k, d)}$ is the intensity of the particle beam of ***k***-component of hadronic cascade, with energy $\boldsymbol{E_k}$, after reaching an atmospheric depth $\boldsymbol{d = \int \rho dx}$, $\boldsymbol{\sigma_{ijk}}$ are the reactions' cross-sections, and $\boldsymbol{N_i}$ is the density of the target atoms in units atoms per gram.

An important part of the production of cosmogenic isotopes (i.e. about 40%–49% of the total production rate) belongs to the alpha particles and the heavier CR components. This has motivated the authors of some recent models to take into account their impact when calculating the isotope production by cosmic rays (McCracken, 2004; Webber et al., 2007; Masarik and Beer, 2009). Some models of atmospheric ionization take into account even the effect of the heavier CR nuclei (e.g. Velinov and Mateev, 2008; Mishev et al., 2009; Mishev and Velinov, 2009, 2014).

5.7 More evidence of charged particles' modulation by the geomagnetic field

In order to estimate the influence of the nondipole part of the geomagnetic field on energetic particles, and how this influence is distributed over the globe, we have calculated the cross-correlation coefficients between GCR counting rates (reconstruction back to the 17th century provided by Usoskin et al., 2008), and the geomagnetic field intensity at the Earth's surface. The IGRF model (International Geomagnetic Reference Field) has been used for geomagnetic field calculation for the period 1900–2013. From statistically significant (at 2σ level) correlations, calculated in each node of our grid (with latitude-longitude increments of 10 degrees) we have drawn the correlation map presented in Fig. 5.9 (gradual shading). The secular

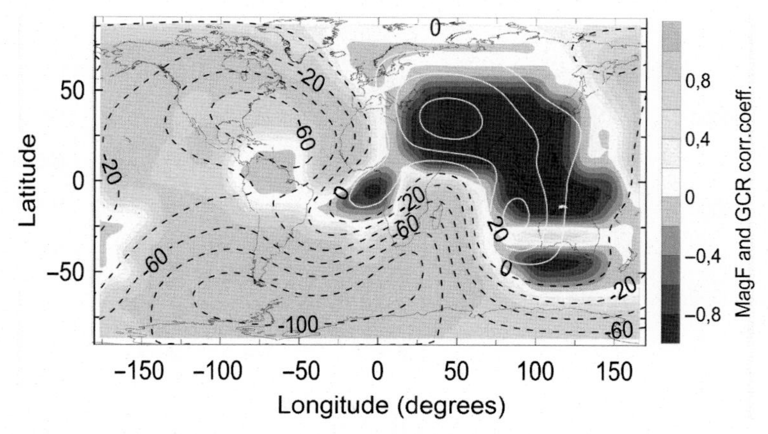

FIG. 5.9 Simultaneous correlation maps of geomagnetic field intensity and galactic cosmic rays *(gradual shading)*, calculated over the period 1900–2013. Overdrawn are contours of geomagnetic field secular variations; *continuous contours* denote positive values, and *dashed contours* negative values. *Data sources: IGRF 2010 model (https://www.ngdc.noaa.gov/geomag/geomag.html) and Usoskin, I.G., et al., 2008. Cosmic Ray Intensity Reconstruction. IGBP PAGES/World Data Center for Paleoclimatology, Data Contribution Series # 2008-013. NOAA/NCDC Paleoclimatology Program, Boulder, CO data for GCRs. From Kilifarska, N., 2017. Mechanism of Relation Between Cosmic Rays, Geomagnetic Field and Earth's Climate, (DSc thesis), Sofia, Bulgaria.*

variations of the geomagnetic field for the period examined are added for comparison in Fig. 5.9.

As might be expected, the spatial distribution of geomagnetic field—GCR correlation resembles fairly well the secular variations of geomagnetic field (contours). The sign of correlations is determined by the long-term trends of the analysed variables. For example, in the North Atlantic and North American regions, the synchronous weakening of GCRs and geomagnetic field intensity (refer to Fig. 5.10) during the period 1900–90 results in a positive correlation between them. In South-Eastern Asia, however, their evolution with time is in antiphase (Fig. 5.10) and correspondingly the correlation is negative (compare Figs 5.9 and 5.10).

Fig 5.10 shows in addition that the trend of GCR intensity changed from negative in the early 1990s to positive. The geomagnetic field, at the same time, changed more smoothly. Consequently, the application of a linear correlation analysis in this case of nonlinearly evolving GCRs is questionable. One way to check the validity of the estimated connectivity between both variables is to split the entire period into two shorter periods with consistent linear trends—e.g. before and after 1960. The recalculated correlation between the geomagnetic field and GCRs, established for the period 1960–2013, is shown in Fig. 5.11. Comparison of Figs 5.9 and 5.11 reveals the change of the sign of correlation in regions with different secular variations. Thus the changed slope of GCR linear trend after 1960 (see Fig. 5.10) specifies the positive correlation coefficients in regions with a geomagnetic field strengthening (continuous contours denoting positive secular variations). In contrast, in regions of geomagnetic field weakening (dashed contours designating negative secular variations), GCRs and the geomagnetic field anticorrelate. Nevertheless, the main pattern and the strength of the relation is more or less similar.

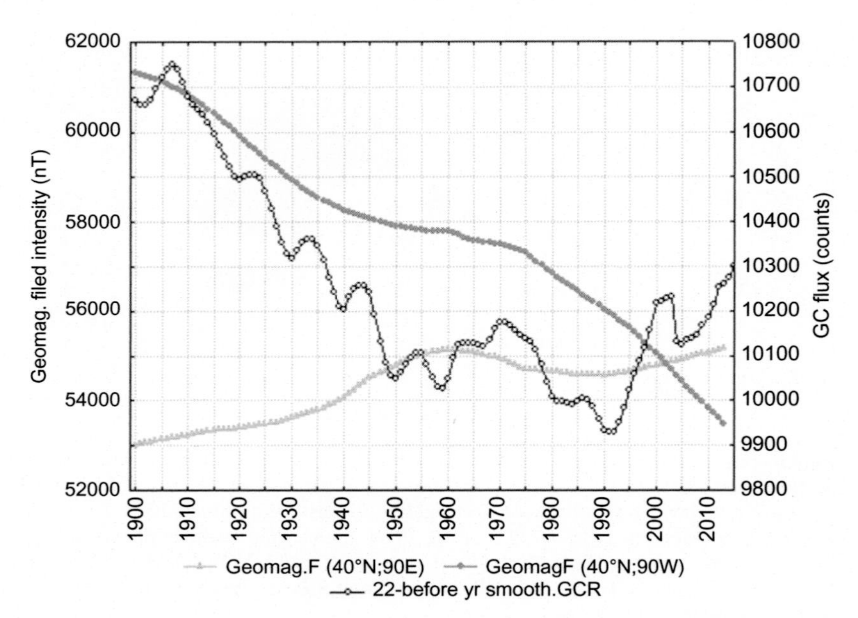

FIG. 5.10 Centennial evolution of geomagnetic field at 40°N latitude and in two longitudinal sectors: 90°W *(dark grey)* and 90°E *(light grey)* longitudes. For comparison, the 22-year running average of the GCRs is shown *(open circles)*. *From Kilifarska, N., Bakhmutov, V., Melnyk, G., 2016. Relations between climate variations and geomagnetic field. Part 3. Northern and Southern Hemispheres. Geophys. J. Geophys. Inst. UAN 38 (3), 52–71 (in Russian).*

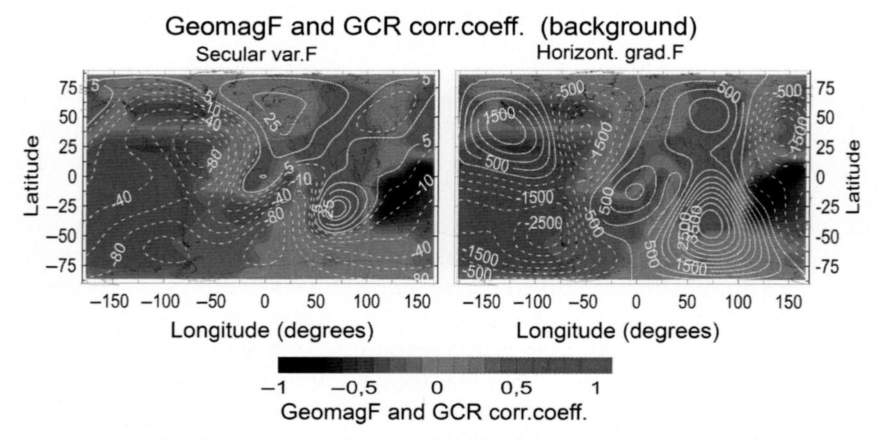

FIG. 5.11 Simultaneous correlation maps of geomagnetic field intensity and galactic cosmic rays *(gradual shading)*, calculated over the period 1960–2013. Overdrawn are contours of geomagnetic field secular variations (*left*) and cross-longitudinal magnetic gradient (*right*); *continuous contours* denote positive values, and *dashed contours* negative values. *Data sources: Annual values of geomagnetic field intensity are from the IGRF 2010 model (https://www.ngdc.noaa.gov/geomag/geomag.html), while centennial reconstruction of GCR intensity is provided by Usoskin, I.G., et al., 2008. Cosmic Ray Intensity Reconstruction. IGBP PAGES/World Data Center for Paleoclimatology, Data Contribution Series # 2008-013. NOAA/NCDC Paleoclimatology Program, Boulder, CO. From Kilifarska, N., Bakhmutov, V., Melnyk, G., 2016. Relations between climate variations and geomagnetic field. Part 3. Northern and Southern Hemispheres. Geophys. J. Geophys. Inst. UAN 38 (3), 52–71 (in Russian).*

Comparison of correlation map with the cross-longitudinal geomagnetic gradient, shown in the right-hand panel of Fig. 5.11 (contours) reveals, in addition, that the stronger in phase connection between the geomagnetic field and GCRs corresponds to regions with a rising positive and steeply decreasing negative cross-longitudinal magnetic gradient (i.e. in regions of geomagnetic strengthening). This result is indirect confirmation that the near-surface irregularities of the geomagnetic field make a substantial contribution to the heterogeneous distribution of secondary ionization in the Regener–Pfotzer ionization maximum. The importance of the spatial-temporal variability of geomagnetic field intensity is also reported by Beer et al. (2012). Based on the investigation of production rate of ^{10}Be radionuclides, the authors conclude that variations of geomagnetic field strength at a fixed latitude lead to nonlinear changes in ^{10}Be production rate.

Besides GCRs (bombarding continuously Earth's atmosphere) the energetic particles generated during eruptions of solar plasma into the interplanetary space are sporadically precipitating in Earth's atmosphere. The relation between these particles (with energies more than 10 MeV) and the geomagnetic field is shown in the left-hand panel of Fig. 5.12. Overdrawn are the geomagnetic secular variations (contours). It is clear that the covariance between the solar protons and geomagnetic field over the globe matches fairly well the spatial distribution of geomagnetic secular variations. This result is reasonably expected, because the greatest part of the lower energetic solar protons is adsorbed at much higher atmospheric levels, which explains their weak sensitivity to the near-surface geomagnetic irregularities. Comparison of Figs 5.11 and 5.12 shows that despite the great similarity in the spatial distribution of geomagnetic control over GCRs and solar protons, some differences are easily noticeable. These differences could also be attributed to the different depth of penetration of both type of particles, specified in turn by their different energies.

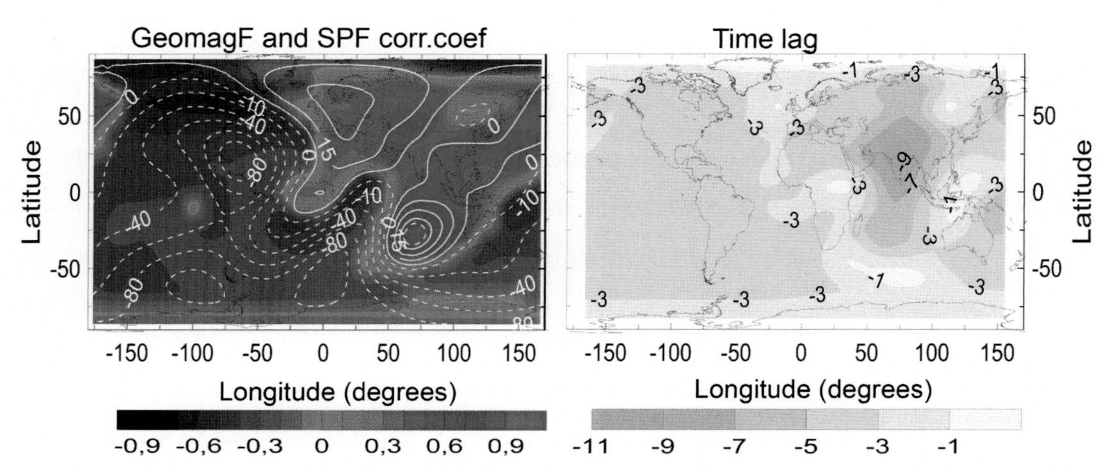

FIG. 5.12 Correlation map *(shading)* of geomagnetic field intensity with solar proton flux with energies greater than 10 MeV, calculated over the period 1970–2013; overdrawn are contours of geomagnetic field secular variation, where *continuous contours* denote positive values, and *dotted contours* negative values. Time lag in years is shown in the *right-hand panel. From Kilifarska, N., 2017. Mechanism of Relation Between Cosmic Rays, Geomagnetic Field and Earth's Climate, (DSc thesis), Sofia, Bulgaria.*

In addition, the right-hand panel of Fig. 5.12 shows that time lag of solar protons' response to geomagnetic action is delayed by 2–4 years on average, unlike the instantaneous correlation between the geomagnetic field and GCRs. This time lag could be related to the protons originated from the inner Van Allen radiation belt. The latter is fed with lower energetic particles (mainly protons) from extra-terrestrial space, during geomagnetic disturbances, holding them for 3–4 years or even more (Lazutin, 2010; Selesnick et al., 2007). Due to the magnetospheric instabilities, substorm activity etc., these particles are gradually lost in the atmosphere, ionizing its neutral components. Protons in the inner radiation belt deserve special attention, because they could be accelerated to energies of more than hundreds MeV (Fiandrini et al., 2004; Lazutin et al., 2007), and are able to penetrate deeper in the lower atmosphere.

In resume: Synchronous analysis of the temporal evolution of GCRs and solar proton fluxes (SPFs) with the geomagnetic field during 1900–2013 shows that: (i) they covariate fairly well in time and space; (ii) the strength of the relations between the geomagnetic field and charged particles is unevenly distributed over the globe; and (iii) some differences in geomagnetic control over the GCRs and solar protons are found. Comparison of correlation maps with geomagnetic cross-longitudinal gradient and secular variations shows that spatial distribution of GCRs reaching the Earth's surface is substantially influenced by the cross-longitudinal gradient of the geomagnetic field. The distribution of SPFs, which are measured at satellite distance from the surface, suits the geomagnetic secular variations better. This difference of geomagnetic influence on particles with different origin should be attributed to the lower sensitivity of solar protons to near-surface geomagnetic irregularities.

References

Abraham, J., Abreu, P., Aglietta, M., Aguirre, C., et al., 2007. Correlation of the highest-energy cosmic rays with nearby extragalactic objects. Science 318, 938, Arxive0711.2256.

Amato, E., 2014. The origin of galactic cosmic rays. Int. J. Mod. Phys. D 23, 1430013. https://doi.org/10.1142/S0218271814300134.

Amato, E., 2016. Particle acceleration in astrophysical sources. In: Proceedings of Frontiers of Fundamental Physics 14—PoS (FFP14). Presented at the Frontiers of Fundamental Physics 14, SISSA Medialab.p. 024. https://doi.org/10.22323/1.224.0024.

Beer, J., McCracken, K., von Steiger, R., 2012. Cosmogenic Radionuclides: Theory and Applications in the Terrestrial and Space Environments. Physics of Earth and Space Environments, Springer-Verlag Berlin Heidelberg.

Bell, A.R., 2004. Turbulent amplification of magnetic field and diffusive shock acceleration of cosmic rays. Mon. Not. R. Astron. Soc. 353, 550–558. https://doi.org/10.1111/j.1365-2966.2004.08097.x.

Berezinskii, V.S., Bulanov, S.V., Ginzburg, V.L., Dogiel, V.A., Ptuskin, V.S., 1990. Ginzburg, V.L. (Ed.), Astrophysics of Cosmic Rays. Physmatgiz, Moscow (in Russian). English translation, North-Holland, Amsterdam.

Blasi, P., 2008. Origin of high energy cosmic rays: a short review. Nucl. Instrum. Methods Phys. Res. Sect. A Accel. Spectrometers Detect. Assoc. Equip.Proceedings of the First International Conference on Astroparticle Physics. 588, 166–170. https://doi.org/10.1016/j.nima.2008.01.033

Blasi, P., Amato, E., 2011. Positrons from pulsar winds. In: Torres, D.F., Rea, N. (Eds.), High-Energy Emission From Pulsars and Their Systems, Astrophysics and Space Science Proceedings. Springer, Berlin Heidelberg, pp. 623–641.

Blinov, A.V., Kropotina, Y.A., 2009. Regions of the maximal cosmic ray energy release in the atmosphere. Geomagn. Aeron. 49, 290–296. https://doi.org/10.1134/S0016793209030025.

Cecchini, S., Spurio, M., 2012. Atmospheric muons: experimental aspects. Geosci. Instrum. Methods Data Syst. 1, 185–196. https://doi.org/10.5194/gi-1-185-2012.

Clay, J., 1927. Penetrating radiation. In: Proceedings of the Section of Sciences, Koninklijke Akademie van Wetenschappen te Amsterdam. vol. 30 (9–10), pp. 1115–1127.

Clúa de Gonzalez, A.L., Silbergleit, V.M., Gonzalez, W.D., Tsurutani, B.T., 2001. Annual variation of geomagnetic activity. J. Atmos. Sol. Terr. Phys. 63, 367–374. https://doi.org/10.1016/S1364-6826(00)00190-5.

Currie, R.G., 1966. The geomagnetic spectrum—40 days to 5.5 years. J. Geophys. Res. 71, 4579–4598. https://doi.org/10.1029/JZ071i019p04579.

Dachev, T.P., Matviichuk, Y.N., Bankov, N.G., Semkova, J.V., Koleva, R.T., Ivanov, Y.J., Tomov, B.T., Petrov, V.M., Shurshakov, V.A., Bengin, V.V., Machmutov, V.S., Panova, N.A., Kostereva, T.A., Temny, V.V., Ponomarev, Y.N., Tykva, R., 1992. 'Mir' radiation dosimetry results during the solar proton events in September–October 1989. Adv. Space Res. 12, 321–324. https://doi.org/10.1016/0273-1177(92)90122-E.

Dachev, T.P., Tomov, B.T., Matviichuk, Y.N., Dimitrov, P.G., Bankov, N.G., Shurshakov, V.V., Ivanova, O.A., Häder, D.-P., Schuster, M.T., Reitz, G., Horneck, G., 2015. 'BION-M' No. 1 spacecraft radiation environment as observed by the RD3-B3 radiometer-dosimeter in April–May 2013. J. Atmos. Sol. Terr. Phys. 123, 82–91. https://doi.org/10.1016/j.jastp.2014.12.011.

Drury, L.O., 2012. Origin of cosmic rays. Astropart. Phys. Cosmic Rays Topical Issue. 39–40, 52–60. https://doi.org/10.1016/j.astropartphys.2012.02.006.

Duperier, A., 1941. The seasonal variations of cosmic-ray intensity and temperature of the atmosphere. Proc. R. Soc. Lond. A 177, 204–216. https://doi.org/10.1098/rspa.1941.0007.

Duperier, A., 1949. The meson intensity at the surface of the earth and the temperature at the production level. Proc. Phys. Soc. A 62, 684. https://doi.org/10.1088/0370-1298/62/11/302.

Ellison, D.C., Drury, L.O., Meyer, J.-P., 1997. Galactic cosmic rays from supernova remnants. II. Shock acceleration of gas and dust. Astrophys. J. 487, 197. https://doi.org/10.1086/304580.

Ferreira, S.E.S., Potgieter, M.S., Heber, B., Fichtner, H., 2003. Charge-sign dependent modulation in the heliosphere over a 22-year cycle. Ann. Geophys. 21, 1359–1366. https://doi.org/10.5194/angeo-21-1359-2003.

Fiandrini, E., Esposito, G., Bertucci, B., Alpat, B., Ambrosi, G., Battiston, R., Burger, W.J., Caraffini, D., Masso, L.D., Dinu, N., Ionica, M., Ionica, R., Pauluzzi, M., Menichelli, M., Zuccon, P., 2004. High-energy protons, electrons, and positrons trapped in Earth's radiation belts. Space Weather 2. https://doi.org/10.1029/2004SW000068.

Filz, R.C., Holeman, E., 1965. Time and altitude dependence of 55-Mev trapped protons, August 1961 to June 1964. J. Geophys. Res. 70, 5807–5822. https://doi.org/10.1029/JZ070i023p05807.

Georgieva, K., Kirov, B., 2005. Secular cycle of the north-south solar asymmetry. arXiv. http://arxiv.org/abs/physics/0509198v1.

Georgieva, K., Kirov, B., Tonev, P., Guineva, V., Atanasov, D., 2007. Long-term variations in the correlation between NAO and solar activity: the importance of north–south solar activity asymmetry for atmospheric circulation. Adv. Space Res. 40, 1152–1166. https://doi.org/10.1016/j.asr.2007.02.091.

The origin of cosmic rays. Ginzburg, V.L., Syrovatskii, S.I. (Eds.), 1964. The Origin of Cosmic Rays. Pergamon, p. VIIIa. https://doi.org/10.1016/B978-0-08-013526-7.50003-7.

Grieder, P.K.F., 2010. Extensive Air Showers: High Energy Phenomena and Astrophysical Aspects—A Tutorial, Reference Manual and Data Book. Springer Science & Business Media.

Hakkinen, L.V.T., Pulkkinen, T.I., Pirjola, R.J., Nevanlinna, H., Tanskanen, E.I., Turner, N.E., 2003. Seasonal and diurnal variation of geomagnetic activity: revised *Dst* versus external drivers. J. Geophys. Res. 108 (A2), 1060. https://doi.org/10.1029/2002JA009428.

Heber, B., Kopp, A., Gieseler, J., Müller-Mellin, R., Fichtner, H., Scherer, K., Potgieter, M.S., Ferreira, S.E.S., 2009. Modulation of galactic cosmic-ray protons and electrons during an unusual solar minimum. Astrophys. J. 699, 1956–1963. https://doi.org/10.1088/0004-637X/699/2/1956.

Hörandel, J.R., 2008. The origin of galactic cosmic rays. Nucl. Instrum. Methods Phys. Res., Sect. A Accel. Spectrometers Detect. Assoc. Equip. Proceedings of the First International Conference on Astroparticle Physics. 588, 181–188. https://doi.org/10.1016/j.nima.2008.01.036.

Jackman, C.H., Marsh, D.R., Kinnison, D.E., Mertens, C.J., Fleming, E.L., 2016. Atmospheric changes caused by galactic cosmic rays over the period 1960–2010. Atmos. Chem. Phys. 16, 5853–5866. https://doi.org/10.5194/acp-16-5853-2016.

Jokipii, J.R., 1971. Propagation of cosmic rays in the solar wind. Rev. Geophys. Space Phys. 9, 27–87.

Jokipii, J.R., 1989. The physics of cosmic-ray modulation. Adv. Space Res. 9, 105–119. https://doi.org/10.1016/0273-1177(89)90317-7.

Kilifarska, N., Bojilova, R., Velichkova, T., 2018. Spatial heterogeneity of cosmic radiation measured at Earth's surface. In: XIV Intern. Conference: Space, Ecology, Safety, Sofia, 7-9 November 2018, pp. 95–100.

Lazutin, L.L., 2010. Belt of protons 1–20 MeV at L = 2. Cosm. Res. 48, 108–112. https://doi.org/10.1134/S0010952510010090.

Lazutin, L.L., Kuznetsov, S.N., Podorol'skii, A.N., 2007. Dynamics of the radiation belt formed by solar protons during magnetic storms. Geomagn. Aeron. 47, 175–184. https://doi.org/10.1134/S0016793207020053.

Lemaire, J.F., 2003. The effect of a southward interplanetary magnetic field on Störmer's allowed regions. Adv. Space Res. Plasma Processes in the Near-Earth Space: Interball and Beyond. 31, 1131–1153. https://doi.org/10.1016/S0273-1177(03)00099-1.

Leung, P.T., 1989. Bethe stopping-power theory for heavy-target atoms. Phys. Rev. A 40, 5417–5419. https://doi.org/10.1103/PhysRevA.40.5417.

Masarik, J., Beer, J., 2009. An updated simulation of particle fluxes and cosmogenic nuclide production in the Earth's atmosphere. J. Geophys. Res.-Atmos. 114 (D11), D11103. https://doi.org/10.1029/2008JD010557.

McCracken, K.G., 2004. Geomagnetic and atmospheric effects upon the cosmogenic Be-10 observed in polar ice. J. Geophys. Res. Space Physics 109 (A4), 4101. https://doi.org/10.1029/2003JA010060.

Mendonça, R.R.S.D., Raulin, J.-P., Echer, E., Makhmutov, V.S., Fernandez, G., 2013. Analysis of atmospheric pressure and temperature effects on cosmic ray measurements. J. Geophys. Res. Space Physics 118, 1403–1409. https://doi.org/10.1029/2012JA018026.

Meyer, J.-P., Drury, L.O., Ellison, D.C., 1997. Galactic cosmic rays from supernova remnants. I. A cosmic-ray composition controlled by volatility and mass-to-charge ratio. Astrophys. J. 487, 182. https://doi.org/10.1086/304599.

Mishev, A., 2013. Short- and medium-term induced ionization in the earth atmosphere by galactic and solar cosmic rays. Int. J. Atmos. Sci. https://doi.org/10.1155/2013/184508.

Mishev, A., Velinov, P.I.Y., 2009. Normalized atmospheric ionization yield functions Y for different cosmic ray nuclei obtained with recent CORSIKA code simulations. Comptes rendus de l'Acad´emie bulgare des Sciences 62 (5), 631–640.

Mishev, A.L., Velinov, P.I.Y., 2014. Influence of hadron and atmospheric models on computation of cosmic ray ionization in the atmosphere—extension to heavy nuclei. J. Atmos. Sol. Terr. Phys. 120, 111–120. https://doi.org/10.1016/j.jastp.2014.09.007.

Mishev, A.L., Velinov, P.I.Y., 2015. Time evolution of ionization effect due to cosmic rays in terrestrial atmosphere during GLE 70. J. Atmos. Sol. Terr. Phys. 129, 78–86. https://doi.org/10.1016/j.jastp.2015.04.016.

Mishev, A.L., Velinov, P.I.Y., 2018. Ion production and ionization effect in the atmosphere during the Bastille day GLE 59 due to high energy SEPs. Adv. Space Res. 61, 316–325. https://doi.org/10.1016/j.asr.2017.10.023.

Mishev, A., Velinov, P., Eroshenko, E., Yanke, V., 2009. The impact of low energy hadron interaction models in CORSIKA code on cosmic ray induced ionization simulation in the Earth atmosphere. In: Proceedings of the 31st ICRC, Łodz. vol. 1–4, p. 2009.

Moisan, M., Pelletier, J., 2012. Individual motion of a charged particle in electric and magnetic fields. In: Physics of Collisional Plasmas. Springer, Dordrecht.

Neronov, A., Semikoz, D.V., Taylor, A.M., 2012. Low-energy break in the spectrum of galactic cosmic rays. Phys. Rev. Lett. 108, 051105. https://doi.org/10.1103/PHYSREVLETT.108.051105.

Patowary, R., Singh, S.B., Bhuyan, K., 2013. A study of seasonal variation of geomagnetic activity. Res. J. Phys. Appl. Sci. 2 (1), 001–011.

Petrukhin, A.A., 2002. Possible explanation of the appearance of the knee and the ankle in cosmic ray energy spectrum. Nucl. Phys. B Proc. Suppl. 110, 484–486. https://doi.org/10.1016/S0920-5632(02)01542-6.

Potgieter, M.S., 2014. The charge-sign dependent effect in the solar modulation of cosmic rays. Adv. Space Res. Cosmic Ray Origins: Viktor Hess Centennial Anniversary. 53, 1415–1425. https://doi.org/10.1016/j.asr.2013.04.015.

Prölss, G., 2004. Physics of the Earth's Space Environment: An Introduction. Springer, Berlin.

Reid, G.C., Isaksen, I.S.A., Holzer, T.E., Crutzen, P.J., 1976. Influence of ancient solar-proton events on the evolution of life. Nature 259, 177–179.

Rosen, J.M., Hofmann, D.J., 1981. Balloon-borne measurements of electrical conductivity, mobility, and the recombination coefficient. J. Geophys. Res. Oceans 86, 7406–7410. https://doi.org/10.1029/JC086iC08p07406.

Russell, C.T., McPherron, R.L., 1973. Semiannual variation of geomagnetic activity. J. Geophys. Res. 78, 92–108. https://doi.org/10.1029/JA078i001p00092.

Sarkar, R., Chakrabarti, S.K., Pal, P.S., Bhowmick, D., Bhattacharya, A., 2017. Measurement of secondary cosmic ray intensity at Regener-Pfotzer height using low-cost weather balloons and its correlation with solar activity. Adv. Space Res. 60, 991–998. https://doi.org/10.1016/j.asr.2017.05.014.

Selesnick, R.S., Looper, M.D., Mewaldt, R.A., 2007. A theoretical model of the inner proton radiation belt. Space Weather 5. https://doi.org/10.1029/2006SW000275.

Semeniuk, K., Fomichev, V.I., McConnell, J.C., Fu, C., Melo, S.M.L., Usoskin, I.G., 2011. Middle atmosphere response to the solar cycle in irradiance and ionizing particle precipitation. Atmos. Chem. Phys. 11, 5045–5077. https://doi.org/10.5194/acp-11-5045-2011.

Shea, M.A., Smart, D.F., 1995. History of solar proton event observations. Nucl. Phys. B Proc. Suppl. Cosmic Rays 94 Solar, Heliospheric, Astrophysical and High Energy. 39, 16–25. https://doi.org/10.1016/0920-5632(95)00003-R.

Smart, D.F., Shea, M.A., 2009. Fifty years of progress in geomagnetic cutoff rigidity determinations. Adv. Space Res. Cosmic Rays From Past to Present. 44, 1107–1123. https://doi.org/10.1016/j.asr.2009.07.005.

Störmer, C., 1930. Twenty-five years' work on the polar aurora. Terr. Magn. Atmos. Electr. 35, 193–208. https://doi.org/10.1029/TE035i004p00193.

Stozhkov, Y.I., Bazilevskaya, G.A., Pokrevsky, P.E., Svirzhevsky, N.S., Martin, I.M., Turtelli, A., 1996. Cosmic rays in the atmosphere: north–south asymmetry. J. Geophys. Res. Space Physics 101, 2523–2528. https://doi.org/10.1029/95JA03317.

Strong, A.W., Porter, T.A., Digel, S.W., Jóhannesson, G., Martin, P., Moskalenko, I.V., Murphy, E.J., Orlando, E., 2010. Global cosmic-ray-related luminosity and energy budget of the milky way. Astrophys. J. 722, L58–L63. https://doi.org/10.1088/2041-8205/722/1/L58.

Tilav, S., Desiati, P., Kuwabara, T., Rocco, D., Rothmaier, F., Simmons, M., Wissing, H., 2009. Atmospheric variations as observed by IceCube. In: Proceedings of the 31st ICRC, ŁODZ. arXiv:1001.0776 [astro-ph.HE].

Trefall, H., 1955. On the positive temperature effect in the cosmic radiation and the p-e decay. Proc. Phys. Soc. A 68, 893–904.

Trefall, H., 1957. On the positive temperature effect in the cosmic radiation and the π—μ decay. Physica 23, 65–72. https://doi.org/10.1016/S0031-8914(57)90472-X.

Usoskin, I.G., et al., 2008. Cosmic Ray Intensity Reconstruction. IGBP PAGES/World Data Center for Paleoclimatology, Data Contribution Series # 2008-013, NOAA/NCDC Paleoclimatology Program, Boulder, CO.

Usoskin, I.G., Desorgher, L., Velinov, P., Storini, M., Flückiger, E.O., Bütikofer, R., Kovaltsov, G.A., 2009. Ionization of the earth's atmosphere by solar and galactic cosmic rays. Acta Geophys. 57, 88–101. https://doi.org/10.2478/s11600-008-0019-9.

Velinov, P.I.Y., Mateev, L., 2008. Analytical approach to cosmic ray ionization by nuclei with charge Z in the middle atmosphere—distribution of galactic CR effects. Adv. Space Res. 42, 1586–1592. https://doi.org/10.1016/j.asr.2007.12.008.

Velinov, P.I.Y., Asenovski, S., Kudela, K., Lastovicka, J., Mateev, L., Mishev, A., Tonev, P., 2013. Impact of cosmic rays and solar energetic particles on the Earth's ionosphere and atmosphere. J. Space Weather Space Clim. 3, A14. https://doi.org/10.1051/swsc/2013036.

Warren, J.S., Hughes, J.P., Badenes, C., Ghavamian, P., McKee, C.F., Moffett, D., Plucinsky, P.P., Rakowski, C., Reynoso, E., Slane, P., 2005. Cosmic-ray acceleration at the forward shock in Tycho's supernova remnant: evidence from Chandra X-ray observations. Astrophys. J. 634, 376. https://doi.org/10.1086/496941.

Webber, W.R., Higbie, P.R., McCracken, K.G., 2007. Production of the cosmogenic isotopes ^{3}H, ^{7}Be, ^{10}Be, and 36Cl in the earth's atmosphere by solar and galactic cosmic rays. J. Geophys. Res. Space Phys. 112(A10). https://doi.org/10.1029/2007JA012499.

Wilson, J.W., Townsend, L.W., Schimmerling, W., Khandelwal, G.S., Khan, F., Nealy, J.E., Cucinotta, F.A., Simonsen, L.C., Shinn, J.L., Norbury, J.W., 1991. Transport methods and interactions for space radiation. Technical Report RP-1257, NASA.

Ziegler, J.F., 1999. The stopping of energetic light ions in elemental matter. J. Appl. Phys. 85, 1249. https://doi.org/10.1063/1.369844.

CHAPTER

6

Energetic particles' impact on the near tropopause ozone and water vapour

OUTLINE

Since the times of Chapman (1930), the scientific community has believed that the single mechanism for ozone production at mesospheric and stratospheric levels is the photolysis of oxygen molecules by solar UV radiation (with wavelengths shorter than 242 nm), and the rapid attraction of the derived oxygen atoms by oxygen molecules (i.e. $O_2 + O + M \rightarrow O_3 + M$; Chapman, 1930; Dütsch, 1971). Having a relatively short lifetime (i.e. ~hours to a day), the upper stratospheric O_3 is nonuniformly mixed throughout the atmosphere and its distribution is controlled by chemical and dynamical processes in the atmosphere (WMO, 2007). In the upper stratosphere, the ozone distribution is determined mainly by the balance between the processes of its production and destruction from catalytic cycles involving

The Hidden Link Between Earth's Magnetic Field and Climate
https://doi.org/10.1016/B978-0-12-819346-4.00006-1

© 2020 Elsevier Inc. All rights reserved.

hydrogen, nitrogen, and halogen radical species. This is the essence of the concept for *photochemical equilibrium* of the upper stratospheric O_3.

In the middle stratosphere, the lifetime of the O_3 molecule substantially increases (~week), which raises the role of atmospheric dynamics in the spatial distribution of ozone density. Up to the ozone maximum, the photochemical equilibrium is still valid. Thus in the middle stratosphere, the stratospheric circulation and photochemical equilibrium compete in determining the ozone concentration (Dütsch, 1971).

In the lower stratosphere, however, the situation is significantly changed. The strong absorption of the solar UV radiation in the Hartley band (200–350 nm), especially in the maximum of the ozone layer, reduces substantially the amount of the short-wave radiation reaching the lower stratospheric levels. In addition, the longer-wave UV radiation, which is able to penetrate in the lower stratosphere, cannot dissociate the molecular oxygen any more. Consequently, the photochemical production of O_3 beneath the maximum of the ozone layer is substantially depressed. Moreover, the absence of atomic oxygen at these levels leads to a significant reduction of the ozone-depleting substances like nitric oxide (NO), because the main source of NO is the oxidation of nitrous oxide ($N_2O + O(^1D) \rightarrow 2NO$; Brasseur and Solomon, 2005).

The other potential source of nitrogen oxides (NO_x) is GCRs (Jackman et al., 1980). However, the dissociative ionization of nitrogen molecule by electron impact (i.e. $N_2 + e^- \rightarrow N^+ + N + e^- + 24.34\,eV$) is very unlikely, by electrons with energies less than 30 eV (Itikawa, 2005). Having in mind that the mean energy of formation of an electron-ion pair in the atmosphere is ~35 eV (Porter et al., 1976), it becomes clear that GCRs as a channel for creation of NO_x molecules are quite ineffective. The primary ionization of N_2 by electron impact ($N_2 + e^- \rightarrow N_2{}^+ + 2e^- + 15.58\,eV$) demands less energy, but created nitrogen cation ($N_2{}^+$) exchanges its charge very rapidly with molecular oxygen ($N_2{}^+ + O_2 \rightarrow N_2 + O_2{}^+ + 3.5\,eV$); see Banks and Kockarts (1973) or Brasseur and Solomon (2005).

Consequently, the photo-dissociative production of O_3 in the lower stratosphere, and its destruction by the NO_x radicals, could be ignored with a save accuracy—at least for latitudes outside the polar regions. At polar latitudes this assumption is not correct, because of the downward transportation of NO_x molecules, from the winter upper stratosphere, mesosphere, or thermosphere, where they are produced by the lower energy particles of solar origin (Funke et al., 2005; Garcia, 1992; Meraner and Schmidt, 2016).

The ozone-destructive chemistry by HO_x (H, OH, HO_2) radical in the lowermost stratosphere is also unimportant, due to the minimum of H_2O vapour density and inability of solar UV radiation to dissociate O_2, H_2O, or H_2 in the lower stratosphere (Banks and Kockarts, 1973; Brasseur and Solomon, 2005). These facts, and the long lifetime of O_3 (in the order of 3–4 years; Brasseur and Solomon, 2005), suggest that spatial distribution of the lower stratospheric ozone is controlled mainly by the stratospheric Brewer-Dobson circulation. However, recent multimodels experiments show that the amount of extra-tropical ozone is determined mainly by local production, and that the impact of the tropical ozone transported by the Brewer–Dobson circulation is no more than 30% (Grewe, 2006). This result immediately raises the question: what is the source of ozone production at these latitudes and altitudes?

The next section makes a reassessment of the efficiency of lower energetic electrons (formed by the galactic cosmic rays in the Regener–Pfotzer maximum) in activation of atmospheric ion-molecular chemistry near the tropopause.

6.1 Production of lower stratospheric ozone by GCRs

In addition to solar UV radiation, the molecular oxygen could be dissociated also by energetic particles. The abundance of low energy secondary electrons near the tropopause, known as the Regener–Pfotzer maximum, created by the highly energetic protons, heavier ions, or neutrons precipitating in Earth's atmosphere. The electron impact ionization, dissociation, or activation of atmospheric constituents at these altitudes is less known, which has been a reasonable motivation for their investigation (Kilifarska, 2013, 2017).

After the experimental confirmation of the real existence of tetraoxygen in laboratory conditions (Cacace et al., 2001), some authors suggested that it could be a source of O_3 in Earth's atmosphere (Cacace et al., 2002; de Petris, 2003). In order to estimate the ionization efficiency of atmospheric molecules by the free electrons in the lower stratosphere, we have selected the following reactions that are energetically possible at these levels:

$$O_2 + e^- \rightarrow O_2^+ + 2e^- + 12.07\,eV \quad (I)$$

$$O_2 + e^- \rightarrow O^+ + O + 2e^- + 18.69\,eV \quad (II)$$

$$N_2 + e^- \rightarrow N_2^+ + e^- + 15.58\,eV \quad (III)$$

$$O_3 + e^- \rightarrow O_3^+ + 2e^- + 12.75\,eV \quad (IV)$$

$$O_3 + e^- \rightarrow O_2^+ + O + 2e^- + 13.125 \quad (V)$$

$$O_3 + e^- \rightarrow O^+ + O_2 + 2e^- + 15.2\,eV \quad (VI)$$

$$O_3 + e^- \rightarrow O_2 + O + e^- + 3.773\,eV \quad (VII)$$

$$O_3^+ + e^- \rightarrow O_2^+ + O + e^- + 0.64\,eV \quad (VIII)$$

$$O_3^+ + e^- \rightarrow O^+ + O_2 + e^- + 2.19\,eV \quad (IX)$$

$$O_3^+ + e^- \rightarrow O + O + O + \sim 0\,eV \quad (X)$$

$$O_2^+ + O_2 + M \rightarrow O_4^+ + M + 3.5\,eV \quad (XI)$$

$$O_2^+ + e^- \rightarrow O + O + 5.12\,eV \quad (XII)$$

$$O_4^+ + O \rightarrow O_2^+ + O_3 + 5.6\,eV \quad (XIII)$$

$$O_4^+ \rightarrow O_3^+ + O + 0.82\,eV \quad (XIV)$$

$$O_4^+ \rightarrow O_2^+ + O_2 + 1.26\,eV \quad (XV)$$

$$N_2^+ + O_2 \rightarrow O_2^+ + N_2 + 3.5\,eV \quad (XVI)$$

$$N_2^+ + O_2 \rightarrow NO + NO^+ + 4.45\,eV \quad (XVII)$$

$$O^+ + O_2 \rightarrow O_2^+ + O + 1.5\,eV \quad (XVIII)$$

The energy on the right-hand side of the chemical reactions is their activation energy (NIST, 2002) in electron-volts (eV). Detailed analysis of the efficiency of ionization of atmospheric compounds and ion-molecular reactions themselves reveals that in the lowermost stratosphere, where the air is dry enough, there are conditions for activation of an *autocatalytic cycle* for O_3 production (Kilifarska, 2013, 2017).

The main source of ozone at these levels appears to be the short live tetraoxygen complex O_4^+, experimentally detected for the first time by Cacace et al. (2001). Produced by reaction (XI), it dissociates rapidly in two different channels. The first one produces O_3^+ and O (reaction XIV), and Cacace with co-authors (2002) have suggested that this channel could be an additional source of atmospheric O_3. The second channel restores the amount of ions in the atmosphere (reaction XV), ensuring in such a way the continuous production of the O_4^+ complex. On the other hand, the ozone cation O_3^+ is weakly bounded and could be easily dissociated by photon absorption, collision, or recombination. Its dissociative recombination to three oxygen atoms (reaction X) demands almost no energy and occurs with a very high probability (Zhaunerchyk et al., 2008). For the prevailing conditions in the lower stratosphere (i.e. ground state ozone ions and lower energetic electrons), reaction (X) occurs in 94% of all cases (Zhaunerchyk et al., 2008).

In total, the dissociation of one tetraoxygen ion (via reaction XIV) is a source of four ozone molecules. The amount of O_4^+ ions is restored by reaction (XI), while O_2^+ ions are continuously produced by GCR and partially recovered by a reaction (XV). Thus reactions (XI), (XIV), (XV), and (X) form an autocatalytic cycle for continuous O_3 production in the lower stratosphere. The maximum efficiency of this ozone-producing cycle should be expected near the level of the highest secondary ionization produced by GCRs (known as the Regener–Pfotzer maximum).

Due to the short lifespan of O_4^+ ions, there are no measurements of its density in the atmosphere. However, its concentration could be calculated using the Saha equation, written for the specific case of reaction (XI) in the form:

$$\frac{[O_2^+][O_2]}{[O_4^+]} = \left[\frac{2\pi\frac{m(O_2^+)m(O_2)}{m(O_4^+)}\cdot kT}{h^2}\right]^{3/2} \cdot \frac{Z(O_2^+)Z(O_2)}{Z(O_4^+)} \tag{6.1}$$

where [.] denotes the number density of reactants or product, $\boldsymbol{m}$(.) are their masses in grams, so the calculated O_4^+ density will be in cm^{-3}, $\boldsymbol{k}$ is the Boltzmann const., $\boldsymbol{T}$ is the temperature of the reaction, $\boldsymbol{h}$ is the Plank constant in eV s, and $\boldsymbol{Z}$(.) is the partition function of reactants or product.

According to our lower stratospheric ion chemistry model (i.e. reactions I–XVIII), O_4^+ is produced only through the reaction (XI) and its partition function is defined as $\mathbf{Z}(O_4^+) = \exp(-\mathbf{\Delta E/kT})$, where $\boldsymbol{\Delta E}$ is the energy of appearance of tetraoxygen. The partition functions of O_2 and O_2^+ include all reactions leading to their disappearance accounted for by our model, i.e.:

$$Z(O_2) = \exp\left(-\frac{3}{2}\cdot\frac{12.07}{35}\right) + \exp\left(-\frac{3}{2}\cdot\frac{18.69}{35}\right) + \exp\left(-\frac{3}{2}\cdot\frac{3.5}{35}\right) + \exp\left(-\frac{3}{2}\cdot\frac{4.45}{35}\right) + \exp\left(-\frac{3}{2}\cdot\frac{1.5}{35}\right)$$

and:

$$Z(O_2^+) = \exp\left(-\frac{3}{2}\cdot\frac{3.5}{35}\right) + \exp\left(-\frac{3}{2}\cdot\frac{5.12}{35}\right).$$

In the above equations, the thermal energy of the reactions $\boldsymbol{kT}$ is expressed through the energy of appearance of electron-ion couple $\boldsymbol{E_e} \sim \mathbf{35}$ eV (Porter et al., 1976), by the formula $\boldsymbol{kT} = \mathbf{2/3}.\boldsymbol{E_e}$. The values in the numerators are the activation energy of each reaction. The thermal energy of reaction (XI) is assumed to be ~0.01%–0.02% of the mean kinetic energy of the impacting electrons energy $\boldsymbol{E_e}$. The right-hand side of Eq. (6.1) is also known as an equilibrium rate of reaction (XI).

The amount of the ionized molecular oxygen could be calculated from the relation $\mathbf{O_2}^+ = \mathbf{0.978}.\boldsymbol{N_e}$, determined by Kilifarska et al. (2013) after estimation of the reactions' efficiency in the lower stratosphere; $\boldsymbol{N_e}$ is the free electron density profile produced by GCRs, which could be calculated from the relation $\boldsymbol{Q} = \boldsymbol{\alpha}_{\mathbf{eff}}.(\boldsymbol{N_e})^2$, where $\boldsymbol{Q}$ is the ion production rate, provided in many publications, while $\boldsymbol{\alpha_{eff}}$ is an effective recombination coefficient (Rosen and Hofmann, 1981).

The branching ratio between reactions (XIV) and (XV), determined by the Saha equation, is found to be 0.505/0.495. Thus the amount of ozone produced through the autocatalytic cycle could be calculated from the O_4^+ density by the following formula:

$$O_3prod = (0.505 \times O_4^+) \times 4 \tag{6.2}$$

where the multiplication by the factor 4 denotes the number of oxygen atoms produced by reactions (XIV) and (X), which are immediately converted to O_3 molecules by a three body reaction: $O_2 + O + M \rightarrow O_3 + M$ (Chapman, 1930).

The ozone profiles shown in Fig. 6.1 are calculated with ionization rate $\boldsymbol{Q}$ taken from Bazilevskaya et al. (2008), for a period of solar minimum, and $\boldsymbol{\alpha_{eff}}$ from the balloon-borne measurements of Rosen and Hofmann (1981). The mean vertical profiles of the main atmospheric constituents (O_2, N_2, and O_3) are taken from the US Standard Atmosphere (1976).

Fig. 6.1 presents a comparison between O_3 profiles at latitudes 70°N, 50°N, and 30°N (taken from ERA Interim reanalysis, at the Greenwich Meridian, for January 2009), and autocatalytically produced ozone by GCRs with energies 1.5, 2.5, and 5 GeV (corresponding to the average cut-off rigidity of the given latitudes; Dorman et al., 1967). This comparison shows that despite the great uncertainty in estimation of the effect of electron-impacted ion-molecular reactions at stratosphere–troposphere levels (based on the simplifying assumption for a given energy of impacting electron and imperfect input information), the O_3 profile in the lower stratosphere/upper troposphere could be significantly affected by the ionization in the Regener–Pfotzer maximum.

If the impact of ion-molecular reactions on the lower stratospheric ozone is really significant, the difference between solar minimum and solar maximum profiles should be encountered, due to the modulation of GCRs by solar activity. Such a comparison is given in Fig. 6.2, which presents the real ozone profiles in the maximum (2001) and minimum (2009) of solar activity, at various longitudes and latitudes. It is obvious that the winter O_3 data (December–March), taken from ERA Interim reanalysis behaves quite unexpectedly. From the perspective of the photochemical ozone production, one should expect a higher ozone density during solar maximum. Surprisingly, Fig. 6.2 shows that at polar and middle latitudes, the ozone density in the peak of the O_3 layer is higher in solar minimum than in solar maximum.

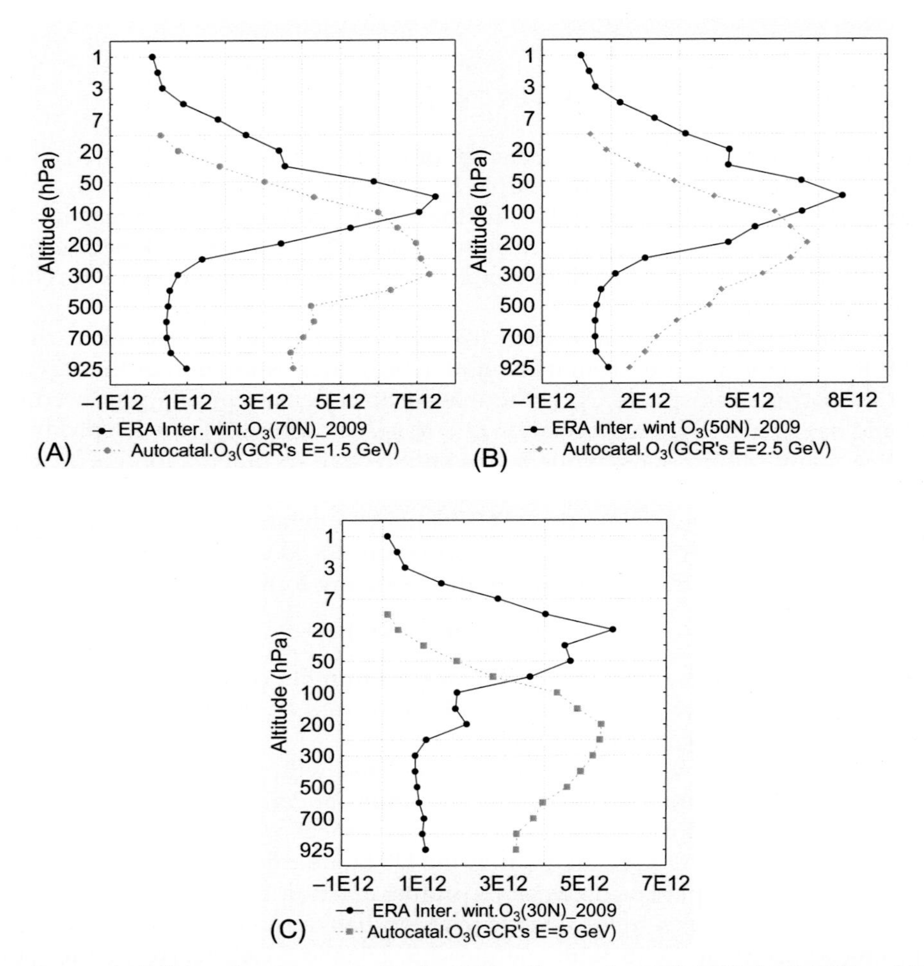

FIG. 6.1 Autocatalytically produced ozone profiles by GCRs with energies 1.5 GeV (A), 2.5 GeV (B), and 5 GeV (C), compared with ERA Interim O_3 profile for January 2009, derived at 70°N latitude (A), 50°N latitude (B), and 30°N latitude (C) at the Greenwich Meridian.

The effect is strongest in the 120°E longitudinal sector (Fig. 6.2A and C) and weaker near the 120°W longitude (Fig. 6.2B and D). At tropical latitudes, differences between high and low solar activities are detected only at the tropospheric levels, in the Western Hemisphere (Fig. 6.2F).

This longitudinal dependence of O_3 response to GCR forcing could hardly be attributed to the equatorward meandering of geomagnetic rigidity isolines, allowing easier access of energetic particles at mid-latitudes of American continent. Rather, this effect should be explained in the light of geomagnetic lensing of charged particles confined in Earth's radiation belts. This problem has been partially discussed in Chapter 5 (Sections 5.5.2 and 5.5.3), in relation to the spatial inhomogeneity of the worldwide neutron monitors' (NMs') measurements and their seasonal variability. Ground-based measurements of cosmic rays provide

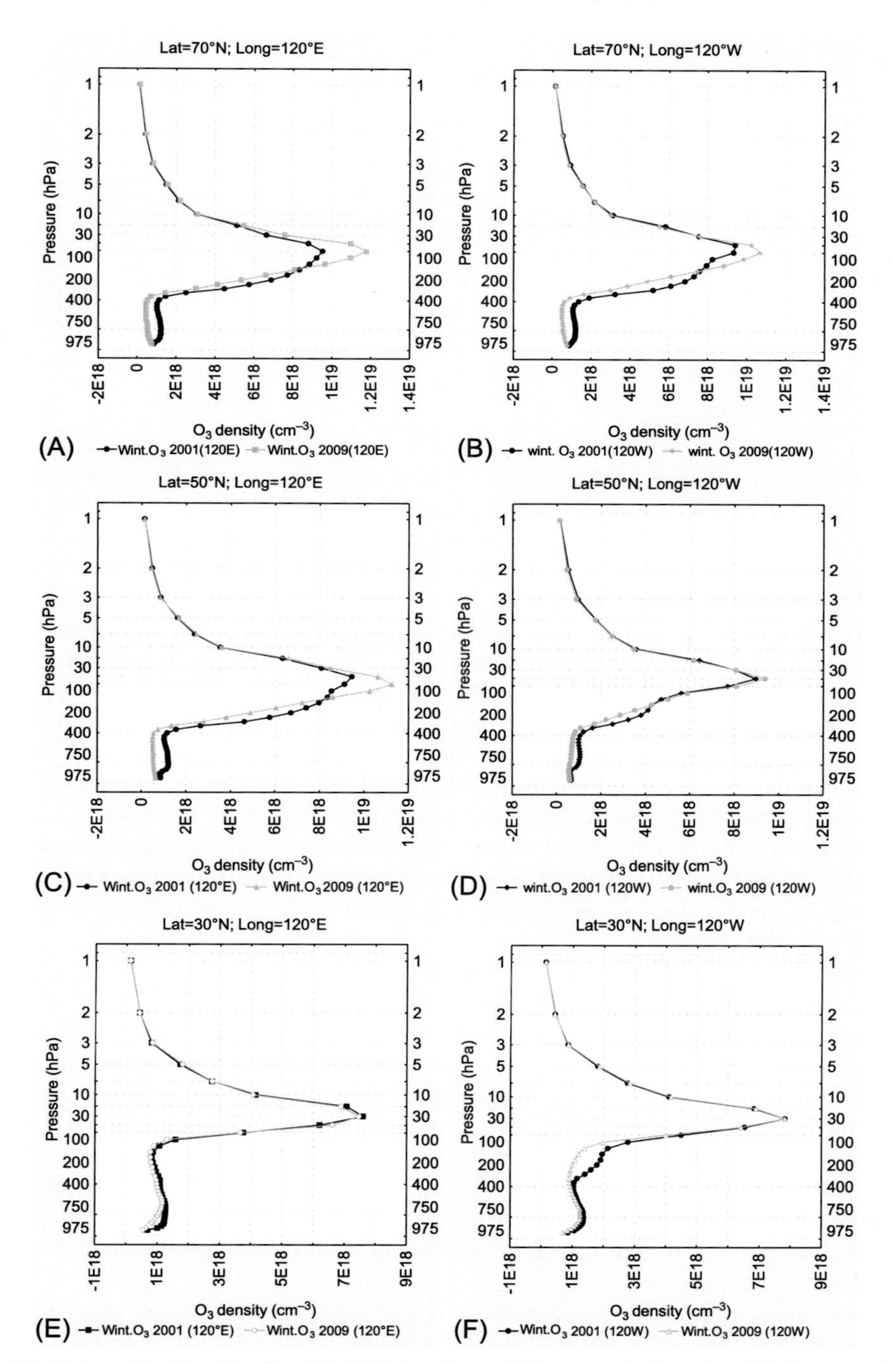

FIG. 6.2 Vertical profiles of winter (Dec–Mar) O_3 density for 2001, year of high activity *(black contours)*, and 2009, year of low solar activity *(grey contours)*, at 120°E and 120°W longitudes, and three latitudes 70°N (A, B), 50°N (C, D), and 30°N (E, F) latitudes. *Data source: ERA Interim reanalysis. Based on Kilifarska, N., 2017. Mechanism of Relation Between Cosmic Rays, Geomagnetic Field and Earth's Climate, (DSc thesis), Sofia, Bulgaria.*

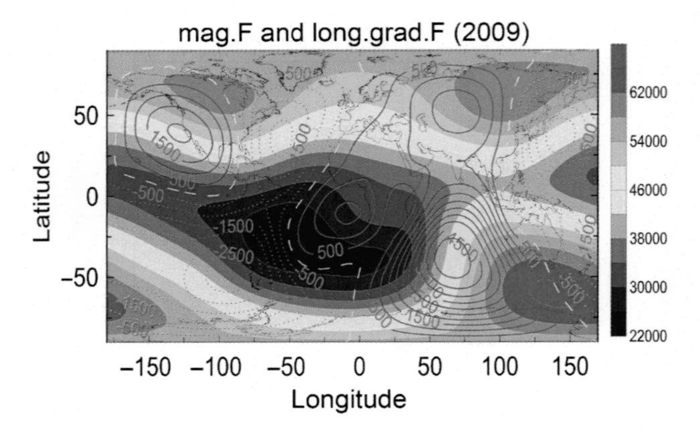

FIG. 6.3 Geomagnetic field *(coloured shading)* and its longitudinal variations calculated by the IGRF 2010 model (https://www.ngdc.noaa.gov/geomag/geomag.shtml) for 2009.

evidence that the cross-longitudinal gradient of the geomagnetic field act as a lens focusing cosmic ray showers in regions with geomagnetic field strengthening along the trajectory of eastward-propagating protons. In this context, the map of geomagnetic field intensity and its longitudinal gradient, shown in Fig. 6.3, could help us to understand the longitudinal variability of the atmospheric ozone profile.

Comparison of Figs 6.2 and 6.3 reveals that the longitudinal sector with a maximal difference between ozone profiles in high and low solar activity (at 120°E) is found in a region of geomagnetic weakening in Eastern Asia. The difference is much smaller in the western hemisphere (at 120°W) and progressively decreases towards the equator. Having in mind the variety of motions of charged particles in heterogeneous geomagnetic field (see Chapter 5), it is reasonable to assume that the geomagnetic lensing of GCRs is imprinted on the O_3 density of winter lower stratosphere. An additional test for this conjecture is shown in Fig. 6.4A–C, which compares ozone profiles at 50°N latitude in regions with geomagnetic strengthening and weakening. Fig. 6.3 shows that there are two regions with a stronger and two regions with a weaker field in the Northern Hemisphere, against the single maximum and single minimum in the Southern Hemisphere.

The juxtaposition of ozone profiles in regions with geomagnetic strengthening and weakening reveals that peak O_3 density is higher, and the ozone layer is thicker in the regions with a decreasing geomagnetic field (grey curves) in the eastern (Fig. 6.4A and B) and western hemispheres (Fig. 6.4C and D). The effect is much clearer in solar minimum than in solar maximum conditions. In the Southern Hemisphere, however, the difference between ozone profiles in regions with decreasing and increasing geomagnetic field is very small (see Fig. 6.4E and F), and with opposite sign. Thus, unlike the Northern Hemisphere, a slight gradual rise from western to eastern longitudes (i.e. in the direction of geomagnetic strengthening) is well visible. Note also that the height of the ozone peak layer in the Southern Hemisphere is substantially higher in solar maximum conditions unlike the opposite tendency found in the Northern Hemisphere.

The problem for longitudinal and hemispherical asymmetries in the ozone spatial distribution was noticed long ago, but still is not explained. An attempt of reducing the uncertainties, by introducing an analogy with the spatial distributions of geomagnetic field (see Figs 6.2 and 6.4), seems to be mined, however, by the opposite sing of their relation in both hemispheres. For example, the depth of the Southern Hemispheric ozone layer slightly

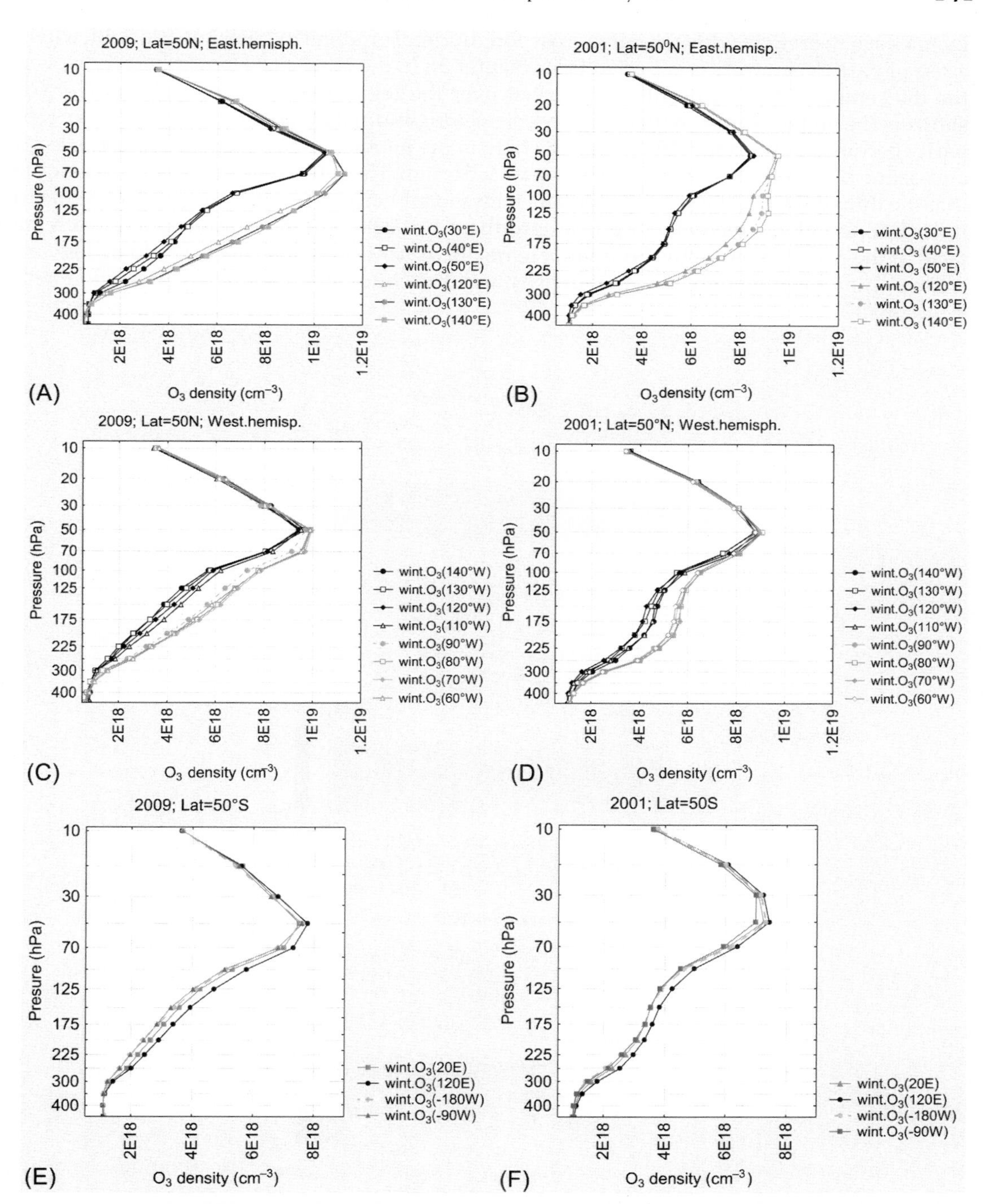

FIG. 6.4 Longitudinal variations in O_3 profiles at 50°N latitude, detected in two sectors with rising—i.e. positive longitudinal gredient (140°W–110°W; 30°E–50°E) *(grey curves)*, and decreasing geomagnetic field intensity—i.e. negative longitudinal gredient (90°W–50°W; 120°E–140°E) *(black curves)*. Comparison is given for two levels of solar activity: low (2009), panels A, C, E, and high (2001), panels B, D, and F. Panels E and F compare the Southern Hemisphere sectors with geomagnetic strengthening (20°E–120°E) and weakening (110°W–50°W), at 50°S latitude. *Data source: ERA Interim reanalysis.*

increases in the region with a positive cross-longitudinal gradient of geomagnetic field, where a rise of particle showers is expected (see Chapter 5). In the Northern Hemisphere, however, the thickening of the O_3 profile is observed over the regions with supposedly weaker CR showers (i.e. in the regions with a negative cross-longitudinal magnetic gradient). This peculiarity becomes understandable within the light of the interaction between unstable π-mesons and ozone molecules (for details see Chapter 5, Section 5.5.3). It seems reasonable to assume that the thinner O_3 profile over the regions with more intensive cosmic ray showers is due to its collisions with the secondary particles produced by GCRs, having a maximum beneath the peak of the ozone layer. In the Southern Hemisphere, however, this effect should be less important due to significantly lower O_3 density and consequently less effective interaction with π-mesons.

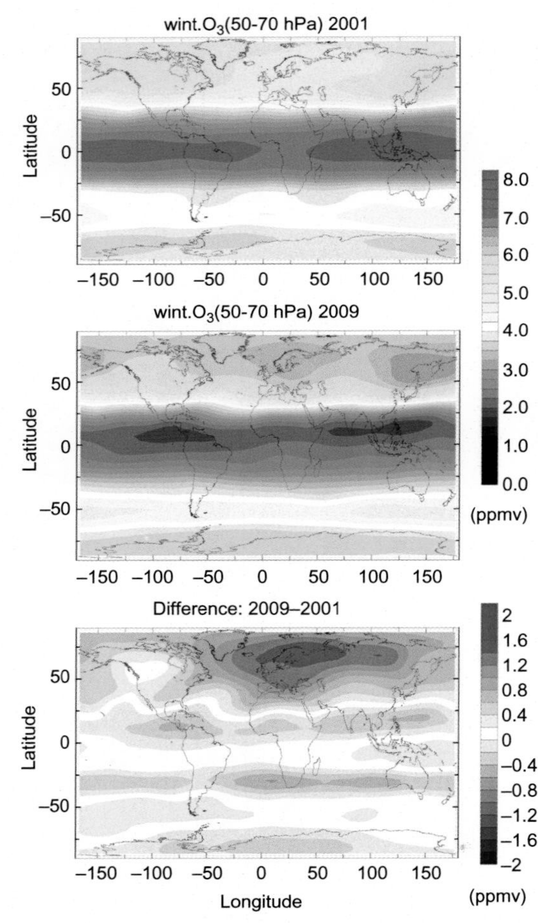

FIG. 6.5 (*top and middle*) Spatial distribution of winter O_3 between 50 and 70 hPa in 2001 (solar maximum), and 2009 (solar minimum); (*bottom*) difference between solar minimum and solar maximum of ozone mixing ratio. *Data source: ERA Interim reanalysis (https://apps.ecmwf.int/datasets/data/interim-full-daily/levtype=sfc/). From Kilifarska, N., 2017. Mechanism of Relation Between Cosmic Rays, Geomagnetic Field and Earth's Climate, (DSc thesis), Sofia, Bulgaria.*

A worldwide perspective of the decadal modulation of lower stratospheric O_3 density is given in Fig. 6.5, which compares the ozone spatial distribution in years of deep solar minimum (2009) and solar maximum (2001). The difference between the maps is shown in the bottom panel of Fig. 6.5 and reveals that in some regions, ozone density in solar minimum is higher than in solar maximum. This result is unexplainable within the concept for UV production of atmospheric ozone, and suggests the existence of additional ozone source in the lower stratosphere. The effect is stronger at northern middle and high latitudes, and undoubtedly could be attributed to the activation of electron-impact production of O_3 by the higher GCR flux during solar minimum. In the Southern Hemisphere, however, the cosmic ray effect is much smaller, focused mainly between the 50° and 70°S latitudes (see the bottom panel of Fig. 6.5). Interestingly, the equatorial O_3 over the maritime continent is also influenced by cosmic rays, despite their reduced intensity at these latitudes. Satellite measurements of energetic particles, detected on the low orbiting Russian station 'MIR' (379–410 km), shed a light on this particularity, measuring an enhancement of particles flux over the same region after the solar proton eruption on 29 September 1989 (Dachev et al., 1992). Consequently, ozone irregularity over the maritime continent could be related to the sporadic rise of energetic particles flux after solar proton events, or to some magnetospheric instabilities forcing particle precipitation from radiation belts.

An estimation of the spatial distribution of autocatalytically produced O_3 density over the northern middle and high latitudes is shown in Fig. 6.6. The latitude-longitude grid of the electron density in the Regener–Pfotzer maximum, needed as an input parameter (see Eqs 6.1 and 6.2), has been compiled from the results of Usoskin et al. (2004), Velinov et al. (2005), and Bazilevskaya et al. (2008). The calculated autocatalytically produced ozone density (contours) is compared with the 111-year climatology of winter O_3 mixing ratio at 70 hPa, taken from the ERA 20th century reanalysis (shading). Despite some uncertainties in conversion of concentration to a mixing ratio, related to the unknown depth of the Regener–Pfotzer maximum (i.e. the calculations have taken into account only the latitudinal dependence of ionization maximum, following Stozhkov et al., 1996, but not the longitudinal ones), the good correspondence between modelled and reanalysis O_3 distribution in the Northern Hemisphere is clear to see.

In resume: The results presented in this section indicate that ion-molecular reactions, initiated by the free electrons in the Regener–Pfotzer maximum, could serve as a mechanism for a local O_3 production in the extra-tropical lower stratosphere. It complements the findings of

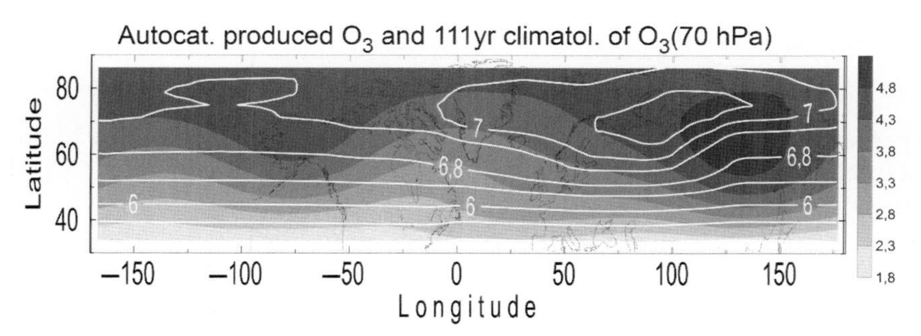

FIG. 6.6 Spatial distribution of autocatalytically produced ozone (ppmv) near the Regener–Pfotzer maximum (contours), in the latitudinal band 30–90°N, compared with the 111-year climatology of winter O_3 mixing ratio at 70 hPa, derived from ERA 20th century *(shading)*.

Grewe (2006), who shows that up to 70% of the winter lower stratospheric ozone at middle and high latitudes is produced locally and only about 30% are transported from the tropics. Finally, it is worth noting that activation of the O_3 producing autocatalytic cycle critically depends on the humidity of the ambient air. Due to the higher affinity of the O_2^+ ion to the H_2O molecule, the increased humidity prevents formation of the tetraoxygen (and consequently O_3), because of the formation of water clusters (de Petris, 2003). In the Northern Hemisphere, the Regener–Pfotzer maximum is placed in the driest levels, just above the tropopause, which provides two of the most important conditions triggering the autocatalytic cycle of O_3 production: high concentration of free low energy electrons and reduced water density.

6.2 Tropospheric O_3 destruction by GCRs

As pointed out in the previous section, the effect of GCRs on the lower stratospheric ozone depends critically on the depth of particles' penetration in the atmosphere. It has been shown that when ionization maximum is placed in a region with reduced humidity, the electron-impacted ion-neutral chemistry produces O_3. This section is devoted to the case when the level of maximum ionization is situated below the tropopause, in the wetter upper troposphere. In the denser upper troposphere, the lower energy electrons could be easily attached to the neutral molecules forming negative ions. Laboratory experiments with a mixture of O_3 (1%) and O_2 (99%), irradiated by an electron beam with energy 0–4 eV, reveal the formation of a negative ozone ions (Matejcik et al., 1996). Despite the initial large excess of oxygen molecules, surprisingly, the dominant attachment product from this experiment appears to be undissociated $(O_3^-)_n$ cluster ions, including the monomer O_3^-. The abundance of the O_3^- ions over the O_2^- ones has been more than one order of magnitude (Matejcik et al., 1996). The authors have explained this result not only with the considerably higher electron affinity of O_3 (2.103 eV against 0.44 eV of O_2), but also with the much higher condensation temperature of ozone (161 K as opposed to 55 K for O_2), making O_3 clustering more likely.

Another product from the above experiment is the tetraoxygen anion O_4^-, obviously due to its particular stability. For example, the binding energy of the extra electron in O_4^- is about twice as large as in O_2^-. However, the amount of the O_4^- ions decreases with an enhancement of the electrons' energy (see Figs 1 and 2 in Matejcik et al., 1996). In contrast, a continuous increase of O_3^- density and its clusters $(O_3^-)_n$ is detected at energies above ≈ 2 eV. Once formed in the atmosphere, the O_3^- anion reacts rapidly with CO_2 (Fehsenfeld and Ferguson, 1974) to form CO_3^- ions, i.e.:

$$O_3^- + CO_2 \rightarrow CO_3^- + O_2 \tag{XIX}$$

In the presence of H_2O, however, reaction (XIX) is strongly inhibited due to the formation of $O_3^-(H_2O)_n$ clusters, with O_2 as a stabilizing third body, through reaction (XX) (Fehsenfeld and Ferguson, 1974):

$$O_3^- + H_2O + O_2 \rightarrow O_3^-(H_2O) + O_2 \tag{XX}$$

The water abundance leads to a further increase of the attached H_2O molecules to the O_3^- ion:

$$O_3^-(H_2O)_{n-1}H_2O + O_2 \leftrightarrow O_3^-(H_2O)_n + O_2 \tag{XXI}$$

Another very fast reaction is the interaction with SO_2 (Fehsenfeld and Ferguson, 1974):

$$O_3^- + SO_2 \rightarrow SO_3^- + O_2$$

The reaction of charge exchange between the hydrated ozone anion and CO_2 molecule could be listed in addition, together with further reactions, all of them leading to ozone depletion in the upper troposphere under the impact of secondary electrons produced by GCRs.

Dissociative electron attachment to ozone, reported by Allan et al. (1996), Curran (1961), Cicman et al. (2007), Rangwala et al. (1999), etc., could be even more effective in ozone removal from the atmosphere. Laboratory experiments show that electrons of very low energies (less than 10 or even 5 eV) could be attached to the ozone molecule. The negative ozone ion O_3^- thus formed, dissociates subsequently in several channels:

$$O_3 + e \rightarrow O_3^- \rightarrow \begin{cases} O^- + O_2 \\ O_2^- + O \\ O^- + 2O \end{cases} \qquad \text{(XXII)}$$

The produced atomic oxygen could furthermore initiate some of the ozone destructive cycles (see Section 6.3), followed by a substantial depletion of the tropospheric ozone.

In resume: Beneath the tropopause, the secondary electrons (produced by cosmic rays in the lower atmosphere) are vastly thermalized, which favours their attraction to the heavier ozone molecule. Thus, the negative ozone ions, formed in such way, are either dissociated to products, urging further O_3 destructive cycles, or attracted by the water molecules, creating heavier water clusters. Consequently, the lower atmospheric ionization beneath the tropopause activates various ion-molecular reactions of ozone destruction.

6.3 Solar protons influence on the total O_3 density

6.3.1 Destruction of mesospheric and upper stratospheric O_3

The energetic particles of solar origin are less powerful and usually are unable to reach the lower stratosphere. The free electrons created by the energetic particles along their paths throughout the atmosphere (by a direct or dissociative ionization of the main atmospheric constituents) activate the HO_x and NO_x ozone destructive cycles. Electron impact dissociation of H_2O vapour demands relatively small energy of 9.19 eV (Müller et al., 1993), and thus formed atomic hydrogen and OH radicals effectively reduce O_3 density (e.g. through reactions XXIII–XXVI).

$$H_2O + e^- \rightarrow H + OH^* \qquad \text{(XXIII)}$$

$$H + O_2 + M \rightarrow HO_2 + M \qquad \text{(XXIV)}$$

$$HO_2 + O_3 \rightarrow 2O_2 + OH \qquad \text{(XXV)}$$

$$OH + O_3 \rightarrow O_2 + HO_2 \qquad \text{(XXVI)}$$

Dissociation of N_2 and O_2, however, is very demanding in terms of energy. The dissociation energy of N_2 is 50–60 eV (Itikawa, 2005), while that of O_2 is 33.5 eV (Cosby, 1993). Consequently, the N_2 and O_2 molecules are hardly dissociable by the bulk of secondary electrons, created by the solar energetic particles.

However, dissociatively excited oxygen fragments are well measured at energies near 14, 16, and 18 eV (Makarov et al., 2003). Moreover, the ionization potential of N_2 is 15.58 eV (Itikawa, 2005) and obtained N_2^+ ion dissociates easily on two nitrogen atoms, with energy between 1.06 and 5.82 eV, depending on the state of products (Banks and Kockarts, 1973). Consequently, the increased density of the free electrons at thermospheric and mesospheric levels (created by the energetic solar particles) is able to produce atomic nitrogen and oxygen. They could recombine by a three body reaction (XXVII), producing the ozone-depleting nitric oxide (NO). The simplest example of an ozone destructive cycle is illustrated by reactions (XXVIII)–(XXXI).

$$N + O + M \rightarrow NO + M + 6.505\,\text{eV} \quad \text{(XXVII)}$$

$$NO + O_3 \rightarrow NO_2 + O_2 + 0.653\,\text{eV} \quad \text{(XXVIII)}$$

$$NO_2 + O_3 \rightarrow NO_3 + O_2 + 1.128\,\text{eV} \quad \text{(XXIX)}$$

$$NO_3 + h\nu(\lambda \leq 620\text{nm}) \rightarrow NO_2 + O \quad \text{(XXX)}$$

$$NO_3 + h\nu(\lambda \leq 620\text{nm}) \rightarrow NO + O_2 \quad \text{(XXXI)}$$

The highly reactive nitrogen atom N interacts also with O_2, O_3, NO, NO_2, etc. molecules, creating more and more nitrogen oxides, thus destroying the odd oxygen (O and O_3) in the upper atmosphere. The ozone depletion during solar proton events is documented by many authors (e.g. Jackman et al., 1980, 2011; Rusch et al., 1981; Rohen et al., 2005; Krivolutsky et al., 2006; Hauchecorne et al., 2007; Seppälä et al., 2008) and is not the focus of this chapter.

6.3.2 Ozone self-restoration at lower atmospheric levels

Despite the destruction of the upper atmospheric ozone, the solar protons could influence the integral columnar ozone density indirectly, due to the unique ozone feature of restoring itself at lower levels, after being destroyed in mesosphere–upper stratosphere. The O_3 self-restoration is not easily understandable, because solar UV radiation reaching the middle stratosphere could not dissociate molecular oxygen O_2 (Banks and Kockarts, 1973). However, Slanger and co-workers (1988) noticed that the large UV continuum, known as the Hartley band (200–350 nm), is able to dissociate ozone, creating vibrationally excited molecular oxygen O_2^* (reaction XXXII). The latter is easily dissociated by the longer wave UV radiation, freely penetrating at middle stratosphere, creating two more oxygen atoms (reaction XXXIII). The latter rapidly interact with O_2 molecules, thus forming ozone, i.e.:

$$O_3 + hv(248\text{nm}) \rightarrow O_2^* + O \quad \text{(XXXII)}$$

$$O_2^* + hv(>300\text{nm}) \rightarrow 2O \quad \text{(XXXIII)}$$

$$\frac{O_2 + O + M \rightarrow O_3 + M}{\text{Net}: 1O_3 \rightarrow 3O_3} \quad \text{(XXXIV)}$$

In total, dissociation of one ozone molecule by solar UV radiation leads to the formation of three new ozone molecules (reaction XXXIV; Slanger et al., 1988). This mechanism for ozone production is known as Slanger's ozone formatting mechanism.

To estimate quantitatively the ozone *self-restoration* effect, ignoring dynamical influence, we have analysed the stationary budget of stratospheric ozone, examining its main sources and sinks (Kilifarska et al., 2013). The system of analysed chemical reactions is as follows:

$$O_2 + h\nu \rightarrow O + O \quad J_{21}$$

$$\begin{aligned} O_3 + h\nu &\rightarrow O_2 + O(^1P) \quad J_3 \\ &\rightarrow O_2 + O^*(^1D) \quad J_{3A} \\ &\rightarrow O_2^* + O(^1P) \quad J_{3B} \end{aligned}$$

$$O_2^* + h\nu \rightarrow O + O \quad J_{22}$$

$$O_2 + O + M \rightarrow O_3 + M \quad k_2$$

$$O_3 + O \rightarrow O_2 + O_2 \quad k_3$$

$$O^* + M \rightarrow O + M \quad k_6$$

$$O^* + H_2O \rightarrow OH + OH \quad k_7$$

$$O + OH \rightarrow H + O_2 \quad k_8$$

$$O + HO_2 \rightarrow OH + O_2 \quad k_{10}$$

$$O_3 + OH \rightarrow HO_2 + O_2 \quad k_{11}$$

$$OH + HO_2 \rightarrow OH + O_2 \quad k_{13}$$

$$NO_2 + O \rightarrow NO + O_2 \quad k_{16}$$

$$NO + O_3 \rightarrow NO_2 + O \quad k_{17}$$

$$O_3 + HO_2 \rightarrow OH + 2O \quad k_{36}$$

The balance between the main processes of ozone production and loss is described by the O_3 equilibrium Eq. (6.3). For simplicity, and bearing in mind that ozone self-restoration could be more effective at mid-latitudes, where the stratospheric temperatures do not favour the activation of ozone destructive halogen radicals, the latter are omitted in this analysis.

$$(J_{21} + J_{22}) \cdot [O_2] = A \cdot J_3 \cdot [O_3]^2 + (B \cdot J_3 + C) \cdot [O_3]^{1.5} + D \cdot J_3 \cdot [O_3] \quad (6.3)$$

where:

$$A = \frac{k_3}{k_2} \cdot \frac{1}{[O_2]^2}; \quad B = \beta \frac{k_8}{k_2} \cdot \left(\frac{k_7}{k_6 \cdot k_{13}} \right)^{0.5}; \quad C = \frac{\beta}{J_2} k_{11} \left(\frac{k_7}{k_6 \cdot k_{13}} \right)^{0.5};$$

and:

$$D = \frac{k_{16}}{k_2} \cdot \frac{[NO_2]}{[O_2]^2}; \quad \beta = \frac{1}{[O_2]} \cdot \sqrt{J_3 \cdot J_{3A} \frac{[OH]}{[O_2]}}$$

Here the square brackets denote concentrations, $\boldsymbol{J_i}$ are the photo-dissociation coefficients and $\boldsymbol{k_i}$ are the rate coefficients of the above-defined photochemical reactions. Differentiation of Eq. (6.3) with respect to O_3 and the substitution $[O_3]^{0.5} = \boldsymbol{X}$ transforms it into an easily solved square Eq. (6.4):

$$(2AJ_3) \cdot X^2 + 1.5(BJ_3 + C) \cdot X + DJ_3 = 0 \tag{6.4}$$

Given that quantum efficiency of $O(^1D)$ production at $\lambda = 248$ nm is much higher than that of $O(^1P)$ (Sander et al., 2000), the solution of Eq. (6.4) is written in the form:

$$X_{1,2} = \alpha \cdot \left\{ -1 \pm \sqrt{1 - \frac{D}{A} \cdot \frac{1}{\alpha^2}} \right\}, \quad \alpha = \frac{1.5(BJ_3 + C)}{4AJ_3} \tag{6.5}$$

The dissociation coefficients of O_2 and O_3 may be written as:

$$J_2 = C_2 \cdot \exp(-\tau_2(z)); \quad J_3 = C_3 \cdot \mathbf{exp}(-\tau_3(z))$$

where $\boldsymbol{C_2}$ and $\boldsymbol{C_3}$ are products of quantum efficiency, absorption cross-section, and solar radiation flux (see Eq. A.1 in the Appendix). After substitution of $\boldsymbol{J_3}$ and all other parameters in Eq. (6.5), $\boldsymbol{\alpha}$ is expressed in the form:

$$\alpha(z) = 1.125 \frac{k_2(T) \cdot k_{11}(T)}{k_3(T)} \cdot \left(\frac{k_7(T)}{k_6(T) \cdot k_{13}(T)} \right)^{0.5} \cdot$$

$$\left[\frac{k_8(T)}{k_2(T) \cdot k_{11}(T)} \cdot C_3(z) \cdot \exp(-\tau_3(z)) + \frac{1}{C_2(z) \cdot \exp(-\tau_2(z))} \right] \cdot [O_2(z)] \cdot \sqrt{0.1 \frac{[OH(z)]}{[O_3(z)]}} \tag{6.6}$$

Variables in Eq. (6.6) are the ozone optical depth ($\boldsymbol{\tau_3}$), the temperature-dependent photochemical reaction coefficients plus O_3, O_2, and OH densities. For a given wavelength, $\boldsymbol{\tau_3}$ depends on the O_3 absorption cross-section, the intensity, and the path length of the incident radiation (travelling from the upper atmospheric boundary to the level $\boldsymbol{z}$), as well as on the columnar ozone density above this level (see Eq. A.3 in the Appendix). Estimation of the term $\frac{D}{A} \cdot \frac{1}{\alpha^2}$ shows that it is very small and can be neglected. After variable transformation, we receive the equilibrium response of the O_3 profile, due to the decrease of its optical depth, in the form:

$$\frac{\Delta O_3(z)}{O_3(z)} = \left(\frac{\alpha_R(z)}{\alpha(z)} \right)^2 - 1 \tag{6.7}$$

The subscript $\boldsymbol{R}$ denotes reduced O_3 optical depth after intensive particle precipitations. Expression (6.7) allows us to estimate the altitude dependence of the self-restoration effect.

The vertical profile of the ozone self-restoration, shown in Fig. 6.7, is calculated by the use of Eq. (6.7), with the value of the O_3 absorption cross-section equal to $\mathbf{1.08e^{-17}}$ (cm^2) (Sander et al., 2000), and photon flux at the top of the atmosphere fixed to $\mathbf{1.0e^{17}}$ (photons cm^{-2} s^{-1}). The effect of the reduced optical depth has been estimated by assuming a 30% uniform reduction of the ozone profile above a given altitude. Allowing the negative ozone anomaly to reach different levels in the stratosphere (i.e. 35 km, 25 km, 20 km, and 15 km), the impact of the downward propagation of ozone depletion over the vertical O_3 profile is shown in Fig. 6.7 (Kilifarska et al., 2013). The standard ozone profile, used for comparison, is taken from the American Air Force Geophysics Laboratory AFGL (Anderson et al., 1986).

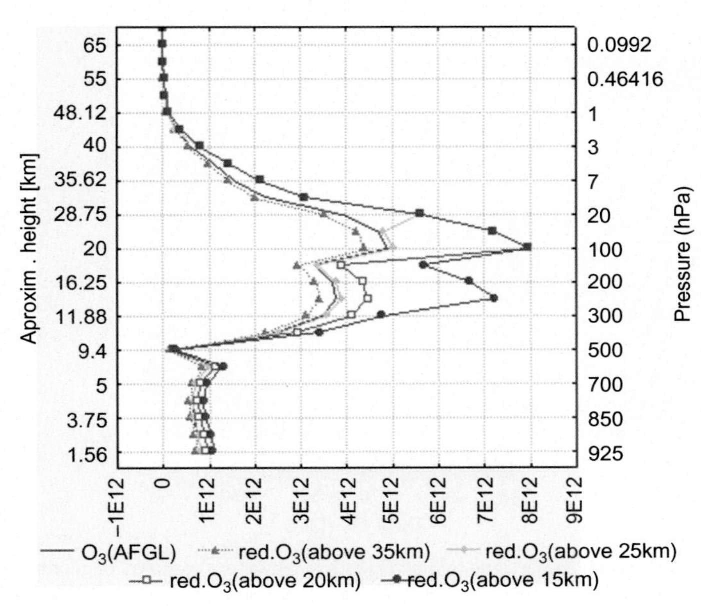

FIG. 6.7 Altitude dependence of the self-restoration effect on the reduction of ozone optical depth above 35, 25, 20, and 15 km, respectively. Note that the deeper penetration of a negative O_3 anomaly, formed at higher levels, increases the amplitude of the ozone self-restoration effect. *From Kilifarska, N.A., Bakhmutov, V.G., Melnyk, G.V., 2013. Energetic particles influence on the Southern Hemisphere ozone variability. C. R. Acad. Bulg. Sci. 66 (11), 1613–1622.*

It is clear that reduction of ozone density above 35 km leads to a depletion of the whole O_3 profile. However, when the negative ozone anomaly reaches lower levels, the self-restoration effect becomes to play a role, and its amplitude increases with downward propagation of the ozone disturbance (e.g. compare profiles with 25 km and 15 km depth of the negative O_3 anomaly). Consequently, the deeper the ozone depletion penetrates, the stronger the self-restoration effect is.

In resume: The less energetic solar protons are rarely able to reach the deepest atmospheric levels, so the majority of their effect on the atmospheric ozone is concentrated at the thermosphere-mesosphere levels. Due to the activation of the HO_x and NO_x ozone destructive cycle, these particles reduce the O_3 density in the upper atmosphere for several days after strong proton events on the Sun. In addition to this effect, which is well-elucidated in the scientific literature, an enhancement of the lower stratospheric ozone density could appear, after a strong depletion of O_3 density at upper levels. The manifestation of this reaction depends on the depth of downward penetration of the negative O_3 anomaly (from the upper atmosphere to the stratosphere). In other words, the deeper the penetration, the stronger the self-restoration of the lower level ozone is.

6.4 Hemispherical and longitudinal asymmetry of ozone response to energetic particles forcing

Analysis of the 111-year climatology of total O_3 density reveals its high hemispherical asymmetry (see Fig. 6.8). The bifocal distribution of the winter ozone (shading) matches fairly well the corresponding distribution of the geomagnetic field intensity (contours) in the Northern Hemisphere (compare Fig. 6.8 with the Hovmoller diagram, presented in Fig. 4.4). In the

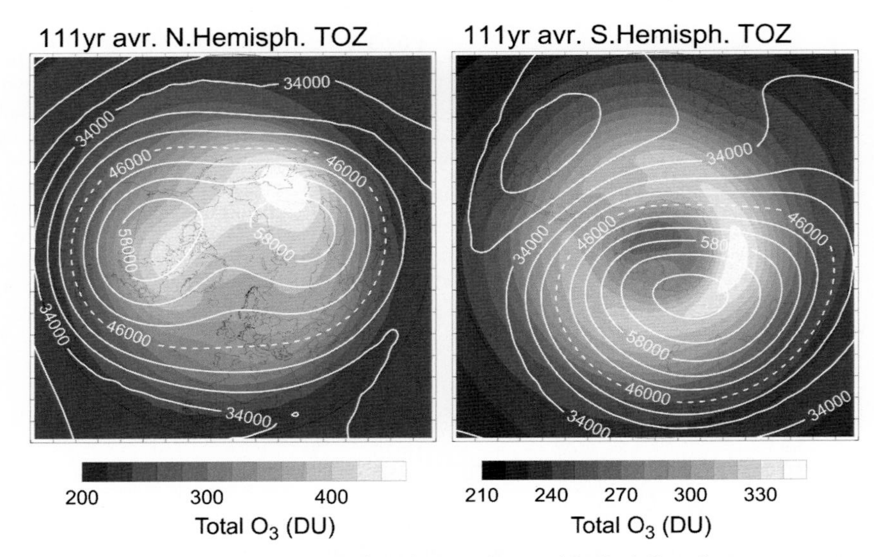

FIG. 6.8 Centennial climatology (1900–2010) of total ozone density *(shading)*. Overdrawn are contours of geomagnetic field intensity, averaged over the same period. The median value of geomagnetic field intensity (i.e. 46,000 nT) is marked by a *dashed curve*. Note the fairly good correspondence between spatial distributions of the ozone and geomagnetic field intensity maxima. *Data sources: ERA 20th century reanalysis for ozone data (https://apps.ecmwf.int/datasets/data/cera20c/levtype=sfc/type=an/) and IGRF (International Reference Geomagnetic field), http://wdc.kugi.kyoto-u.ac.jp/igrf/index.html, for geomagnetic field data. From Kilifarska, N., 2017. Mechanism of Relation Between Cosmic Rays, Geomagnetic Field and Earth's Climate, (DSc thesis), Sofia, Bulgaria.*

Southern Hemisphere, however, there is a single maximum in ozone spatial distribution, corresponding fairly well to the single maximum of geomagnetic field intensity. Moreover, the ozone in the Northern Hemisphere is much thicker than that of the Southern one, a fact that is currently attributed to the stronger winter polar vortex in the Southern Hemisphere, and conditions favouring severe O_3 destruction during Antarctic winter (e.g. IPCC, 2007).

The asymmetrical spatial distribution of total ozone, including its bidecadal variability, has also been noticed by Hood and Zaff (1995), Peters et al. (2008), and Steblova (2001). The first and second papers attribute these O_3 features to the upper troposphere–lower stratosphere planetary wave structure. However, these authors were not able to explain the reason for the suggested interdecadal variations of the large-scale wave structures. Moreover, the maximal amplitude of the stationary planetary waves is found at ~300 hPa (Hood and Zaff, 1995), while the highest amplitude of O_3 longitudinal variations in ERA Interim reanalysis is placed near 150–70 hPa (refer to Fig. 6.9). These and some other problems (e.g. Pan et al., 1997, 2000) suggest that other factor(s) (e.g. energetic particles) may have an important influence on the spatial and interannual variability of the extra-tropical near tropopause O_3.

6.4.1 Response of total O_3 density to GCR impact

Analysing the north–south asymmetry of total ozone density, Steblova (2001) goes to the conclusion that it is somehow related to the intensity of galactic cosmic rays, modulated by

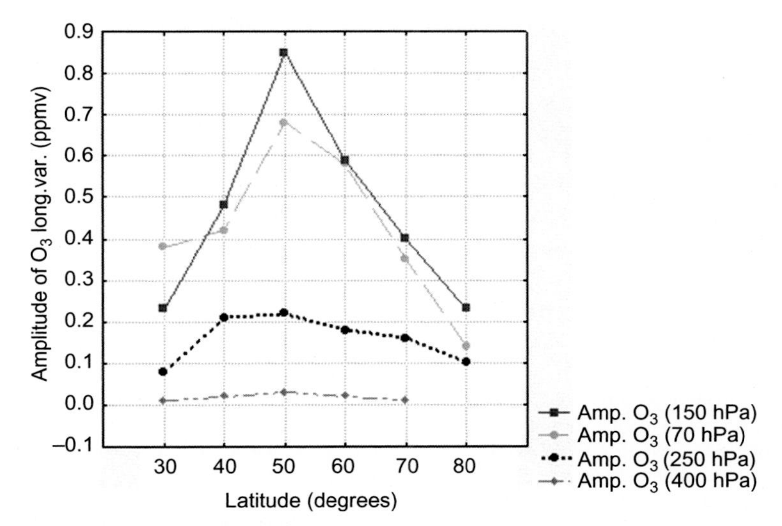

FIG. 6.9 Latitudinal–altitudinal dependence of the ozone longitudinal variations in the Northern Hemisphere, calculated from ERA Interim reanalysis. *From Kilifarska, N., 2017. Mechanism of Relation Between Cosmic Rays, Geomagnetic Field and Earth's Climate, (DSc thesis), Sofia, Bulgaria.*

the solar variability. Similarly, Krivolutsky and co-workers (2002) found an in phase covariation between the total ozone density and the decadal variations of GCRs, measured on balloon flights at several regions in Russia. The question about the mechanism of such interdependence has remained open until recently. Two mechanisms capable of explaining such a relation were described in Sections 6.1 and 6.2. This section therefore focuses on the statistical estimation of a degree of connectivity between GCRs and total ozone density and its spatial heterogeneity, determined for the period 1900–2010.

Fig. 6.10 shows a map of correlation coefficients between GCRs and total ozone density (TOZ), created from the highest, statistically significant at 2σ level lagged correlation coefficients, calculated in each point of the analysed grid, with latitudinal and longitudinal increments of 10 degrees. To reduce the uncertainty of comparison between correlations with different time lags, the correlation coefficients have been multiplied by the autocorrelation function of the forcing factor (Kenny, 1979). This procedure, which substantially reduces correlation coefficients with longer time lags, is based on the reasonable assumption that the effect of the applied forcing in a given moment of time decreases with moving away from this moment.

Analysis of Fig. 6.10 reveals that at polar latitudes, GCRs and total ozone covariate synchronously with high correlation coefficients. At tropical and extra-tropical latitudes, however, the correlation becomes negative, which implies that centennial decrease of GCR intensity (see Fig. 6.12) is accompanied by a rise in the total ozone density at these latitudes. The exceptional is the O_3 variability over maritime continent, having a well-pronounced negative centennial trend (Kilifarska et al., 2017) and correspondingly positive correlation with GCRs. The ozone response to the galactic particles forcing is without delay at the Northern Hemisphere high latitudes, and lagged by 4–7 years over the tropics (see right-hand panel in Fig. 6.10). The time delay rises to 10 years or more over the White Continent. An overlay of the GCR–TOZ correlation map on the centennial climatology of total ozone density, shown in

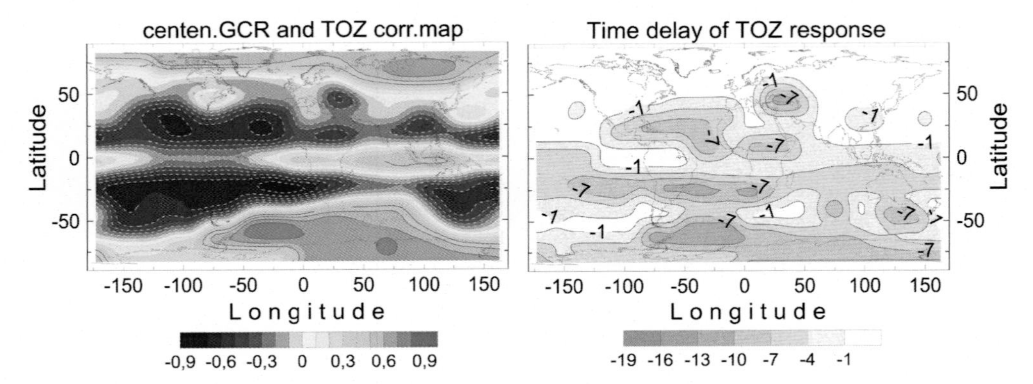

FIG. 6.10 *(left)* Correlation maps of GCRs and winter total ozone density (TOZ) corrected for the time lags; *(right)* time delay of TOZ response to GCR forcing in years. *Data sources: GCR are from Usoskin, I.G., et al., 2008. Cosmic Ray Intensity Reconstruction. IGBP PAGES/World Data Center for Paleoclimatology, Data Contribution Series # 2008-013. NOAA/NCDC Paleoclimatology Program, Boulder, CO; total* O_3 *density is taken from ERA 20th century reanalysis (https://apps.ecmwf.int/datasets/data/cera20c/levtype=pl/type=an/). From Kilifarska, N.A., Bakhmutov, V.G., Melnyk, G.V., 2017. Galactic cosmic rays and tropical ozone asymmetries. C. R. Acad. Bulg. Sci. 70 (7), 1003–1010.*

Fig. 6.11, clarifies that the positive correlation coefficients correspond fairly well to the regions with a thicker TOZ density. In contrast, the areas with a thinner column of ozone coincide with the negative correlation coefficients. This means that, unlike the high-latitude ozone layer, which has been continuously decreasing during the 20th century, the thinner tropical one increases slowly.

Different centennial trends of polar and tropical ozone are clearly visible in the TOZ time series, shown in Fig. 6.12 in two longitudinal sectors and at two latitudes.

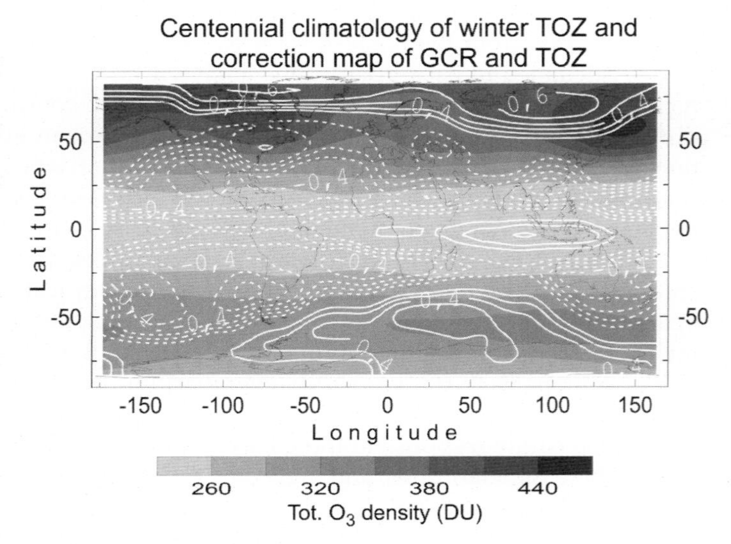

FIG. 6.11 Comparison of the centennial climatology of winter total ozone density TOZ *(shading)* with GCR–TOZ correlation map, identical to that in Fig. 6.10 (contours: *continuous curves* depict positive correlation, *dashed ones* negative). *Data sources: Total* O_3 *density is taken from ERA 20th-century reanalysis, while GCR are provided by Usoskin, I.G., et al., 2008. Cosmic Ray Intensity Reconstruction. IGBP PAGES/World Data Center for Paleoclimatology, Data Contribution Series # 2008-013. NOAA/NCDC Paleoclimatology Program, Boulder, CO. Based on Kilifarska, N., 2017. Mechanism of Relation Between Cosmic Rays, Geomagnetic Field and Earth's Climate, (DSc thesis), Sofia, Bulgaria.*

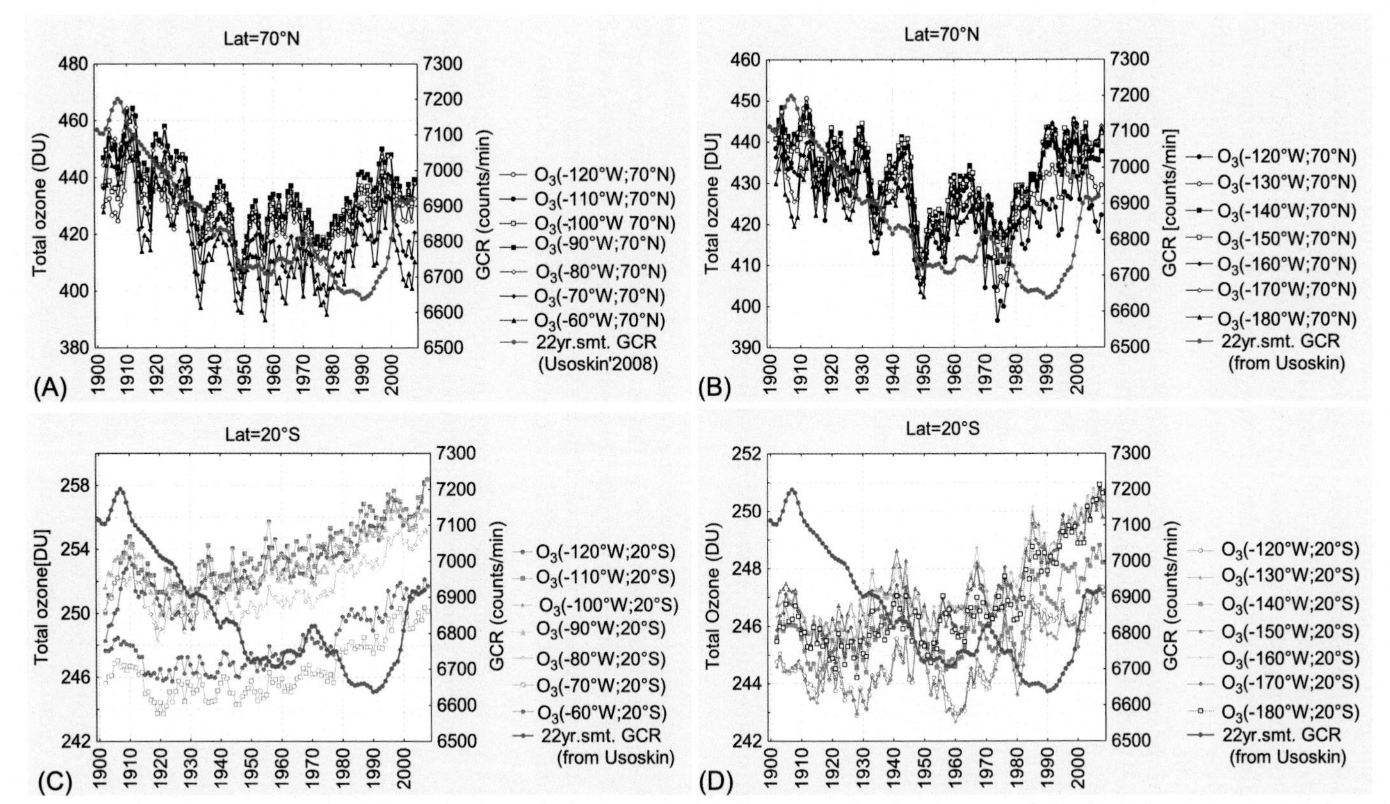

FIG. 6.12 Time series of total ozone density at 70°N (A and B panels) and at 20°S latitudes (C and D panels). Note the different tendencies in TOZ centennial evolution at polar and tropical latitudes. *Data source: ERA 20th century reanalysis. From Kilifarska, N., 2017. Mechanism of Relation Between Cosmic Rays, Geomagnetic Field and Earth's Climate, (DSc thesis), Sofia, Bulgaria.*

The A and B panels in Fig. 6.12 illustrate the centennial evolution of winter TOZ at 70°N latitude, in the longitudinal sectors, where the maxima of TOZ spatial distribution over the globe are located. It is easily seen that high latitude ozone follows quite well the temporal evolution of GCRs, during the examined time interval, 1900–2010, steadily decreasing up to 1980, and gradually increasing after that. However, the tropical total ozone at 20°S latitude shows a completely different tendency, increasing continuously during the entire century (Fig. 6.12 C and D).

The consistent positive trend of the tropical-subtropical total ozone density is still unexplained. The World Meteorological Organization (WMO, 2014) suggests that O_3 recovery (detected since the beginning of the 21st century) follows the depletion of ozone destructive substances, after the entry into force of the Montreal protocol in 1989. This explanation is certainly not applicable for the continuous enhancement of the tropical ozone density throughout the entire 20th century. Other possible explanation could be based on the centennial rise of solar activity during the 20th century, and on the more intensive photochemical ozone production in the upper tropical stratosphere. However, solar activity has started to decrease in the 21st century, while the weak but stable rise of the tropical total ozone density, visible in Fig. 6.12, continues. Another explanation of this puzzling ozone behaviour will be given in Section 6.4.3.

6.4.2 GCR influence on the lower stratospheric O_3

Long ago it was noticed that the greatest impact on the surface temperature belongs not to the changes of stratospheric O_3 maximum, but rather to the variations of its minimal density in the lower stratosphere (Wang et al., 1993; Hansen et al., 1997; de F. Forster and Shine, 1997; Aghedo et al., 2011). However, until recently it was unclear what the factors are that drive the variability of the lower stratospheric ozone. For this reason, the current and following sections focus on the detection and attribution of factors controlling temporal and spatial variability of the lower stratospheric O_3.

At the altitudes of lower stratosphere, the most promising factors able to affect ozone density and its spatial variability are energetic particles (see Section 6.1). The estimated connectivity between ozone mixing ratio at 70 hPa and galactic cosmic rays (smoothed by a 22-point moving window) is shown in Fig. 6.13. Their in phase covariance at tropical and polar latitudes is clearly visible in the left-hand panel of Fig. 6.13. This means that systematic decrease of GCR intensity during the 20th century (see Fig. 6.12) is accompanied by a persistent decrease of the lowermost stratospheric O_3 in these regions. In contrast, negative correlation found in the northern extra-tropic (near 50°N latitude) and over the southernmost edge of Latin America suggests a centennial ozone enhancement in these regions.

This longitudinal and hemispherical asymmetry of connectivity between GCRs and O_3 could be attributed to the dramatic longitudinal variability of the strength of geomagnetic mirror in the Southern Hemisphere (see Chapter 5, Section 5.4). Thus, the strongest field in the region of the geomagnetic North Pole (situated near south Australia) confines many particles in Earth's radiation belts, which the weaker field in the Northern Hemisphere could not hold within the geomagnetic trap, i.e. in radiation belts. As noted in Section 5.4, these particles leave the geomagnetic trap in the Northern Hemisphere, especially in regions with a

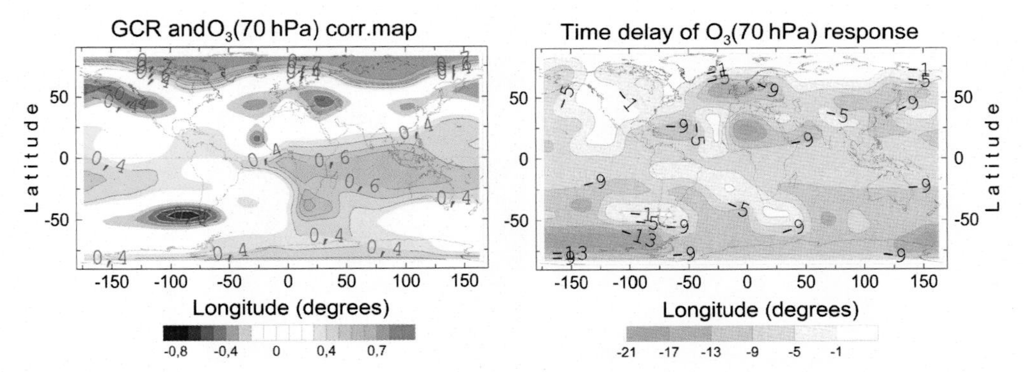

FIG. 6.13 *(left)* Correlation coefficients of galactic cosmic rays with winter ozone at 70 hPa, statistically significant at 95% and corrected for the time lag; *(right)* time delay of O_3 response to GCR forcing in years. *Data sources: GCRs are from Usoskin, I.G., et al., 2008. Cosmic Ray Intensity Reconstruction. IGBP PAGES/World Data Center for Paleoclimatology, Data Contribution Series # 2008-013. NOAA/NCDC Paleoclimatology Program, Boulder, CO; O_3 volume mixing ratio at 70 hPa is taken from ERA 20th century reanalysis. From Kilifarska, N., 2017. Mechanism of Relation Between Cosmic Rays, Geomagnetic Field and Earth's Climate, (DSc thesis), Sofia, Bulgaria.*

stronger longitudinal gradient of the geomagnetic field, producing the extended atmospheric showers of secondary ionization in the lower atmosphere. This ionization, as explained in Section 6.1, is able to activate the autocatalytic production of ozone in the lower stratosphere. Thus, the inexplicable positive trend found in the lower stratospheric ozone near the 50°N latitude (see also Figs 4.10B and 7.6C) could be related to the geomagnetic lensing of charged particles in the upper troposphere–lower stratosphere (see Chapter 5, Section 5.5.2).

The situation is similar in the region of the South Atlantic Magnetic anomaly, the weaker geomagnetic field of which could not hold the particles confined in the Northern Hemisphere (see Fig. 6.3). Over the South Pacific Ocean, near the southernmost edge of Latin America, the secondary ionizations from the extended atmospheric showers obviously meets the conditions necessary for activation of the autocatalytic ozone production. This reasoning offers a physically based explanation of the positive centennial ozone trend (see Fig. 4.10d), and of the strong antiphase correlation between ozone and GCRs found in the region (see Fig. 6.13).

An estimation of the hypothesized geomagnetic–cosmic rays coupled imprint on the lower stratospheric O_3 is shown in Fig. 6.14, presenting the lagged correlation maps of ozone at 70 hPa with geomagnetic field intensity (shading) and GCR (contours, corresponding to the correlation map in Fig. 6.13). Note that the stronger imprint of geomagnetic field on the ozone centennial variability in the Northern Hemisphere (shading) corresponds fairly well to the belt of antiphase covariance between GCRs and ozone (dashed contours). The opposite sign of geomagnetic–ozone correlations over the North Atlantic–Europe, and North America–North Pacific is due to the different centennial trends of the geomagnetic field (positive in Euro-Asia, and negative over the American continent and the Pacific Ocean; see Fig. 5.11).

Analysis of the time lags of ozone response to geomagnetic and GCR forcing reveals that the cosmic rays–ozone coupling is without time delay in the Northern Hemisphere polar cap

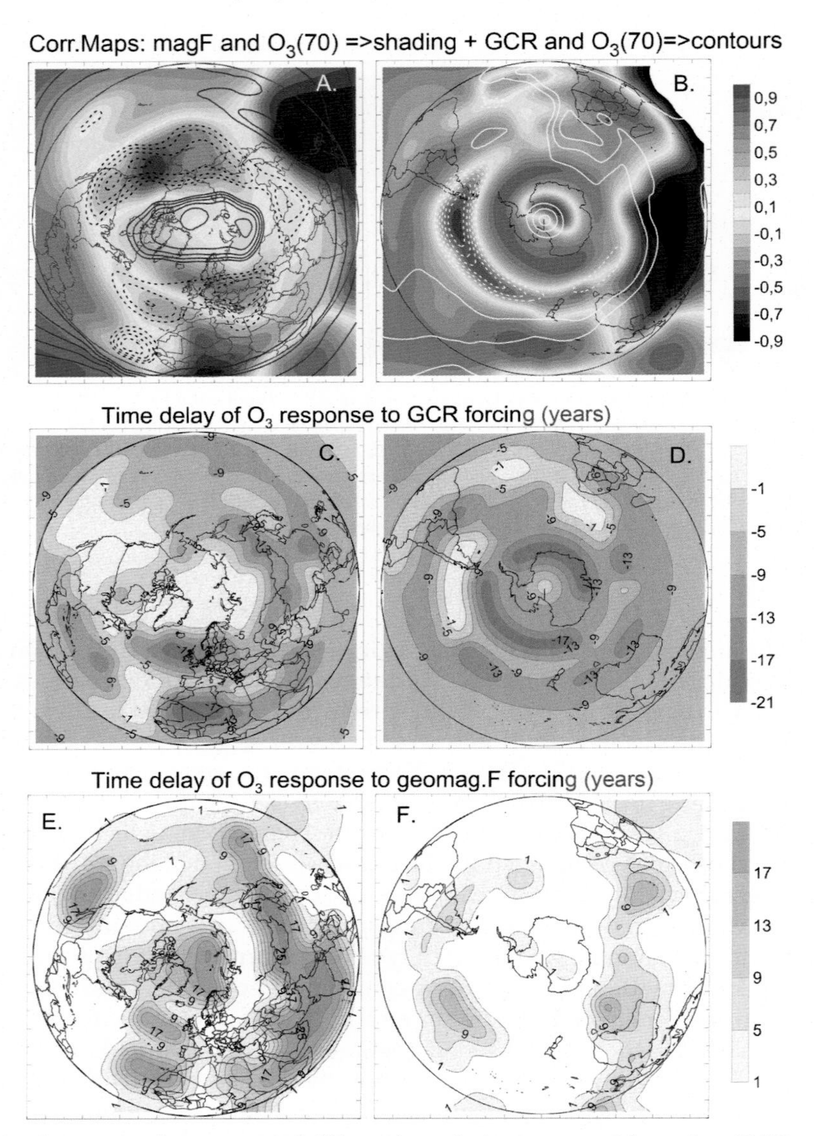

FIG. 6.14 Correlation map of geomagnetic field intensity and winter ozone mixing ratio at 70 hPa calculated over the period 1900–2010 *(shading)*, for the northern (A) and southern (B) hemispheres. Overdrawn are contours of correlation between ozone and GCRs, calculated for the same period (*continuous contours* denote positive correlation, and *dashes* negative one). Middle and bottom panels illustrate the time lag of ozone responce to GCR (C, D) and geomagnetic field (E, F) forcings. *Data sources: IGRF 2010 model for the geomagnetic field and ERA 20th century reanalysis for the ozone.*

(refer to Figs 6.13 and 6.14C). In extra-tropics, and more specifically within the belt 45–60°N latitudes, the ozone responds to particles forcing with a delay of 1–9 years. At the same time, the temporal synchronization between geomagnetic and ozone variations, at these latitudes, is without or with a minimal delay of ozone response (Fig. 6.14E). This indicates that the centennial rise of the lower stratospheric ozone at these latitudes should be related to the particles trapped in the Van Allen radiation belts (e.g. Lazutin, 2010; Dachev et al., 2015a,b), whose resident time could be 10 years or more (Selesnick et al., 2007). In the part of Earth closest to their trajectories, these particles are redistributed irregularly over the globe by the heterogeneous geomagnetic field.

In the Southern Hemisphere, a synchronous antiphase covariation of the lower stratospheric ozone with both GCRs and geomagnetic field intensity is found only over the South Pacific Ocean and the southernmost edge of Latin America (Fig. 6.14B). Analysis of the time lag of ozone response to GCR forcing shows that it is severely delayed in the entire Southern Hemisphere, with the exception of the above-mentioned region (Fig. 6.14D). At the same time, the regional ozone response to slowly varying geomagnetic forcing is without or with a small time delay (Fig. 6.14F). This is additional evidence that coupling between ozone, GCRs and the geomagnetic field is largely determined by the hemispherical asymmetry of the geomagnetic mirror. As noted earlier (see Chapter 5, Section 5.4), particles confined by the geomagnetic field that are successfully reflected by the Northern Hemispheric mirror point cannot be held in the magnetic trap by the weaker geomagnetic field in the South Atlantic–South American region, and precipitate in the atmosphere, producing (where possible) ozone. Consequently, the centennial ozone enhancement in this particular area could be attributed to the surplus of ionization in the lower atmosphere.

It is important to stress that the correlation between ozone and GCR intensity has been derived by the ground-based measurements of GCRs, which also contain the impact of trapped and quasitrapped particles, and the most energetic solar protons able to reach the Earth's surface. In this context, the regions of negatively correlated ozone and GCRs reveal the impact of the trapped radiation on the lower stratospheric ozone density. As described in Chapter 5 (Sections 5.2 and 5.4) the trapped radiation (in the lower part of their cyclotron orbits around magnetic field lines) is more sensitive to the geomagnetic deviations from a dipolar magnetic moment. The longitudinal variability of the geomagnetic field favours particles' precipitation in regions with an increased cross-longitudinal gradient, where two main processes are initiated in the lower stratosphere: (i) ozone production by the lower energetic electrons in the Regener–Pfotzer maximum and (ii) nuclear reactions between products of GCR interaction with atmospheric molecules (e.g. π-mesons) and the heavier O_3 molecule. The effects of these processes are with opposite sign, and the domination of one of them could increase or decrease O_3 density in the lower stratosphere.

The influence of the trapped and quasitrapped radiation in the Van Allen radiation belts on the total ozone density (TOZ) is clearly visible in Fig. 6.10. The figure reveals that the impact of these particles on TOZ is much stronger than those found in ozone at 70 hPa (Fig. 6.13), the former being maximal in the region of the South Atlantic magnetic anomaly. This result could be attributed to the altitude decrease of the Regener–Pfotzer maximum at this region of severely weak geomagnetic field. Consequently, in the region of question the ozone at or beneath the tropopause is impacted by the lower atmospheric ionization (note the slight increase of winter ozone profile near 200 hPa, shown in Fig. 6.2F). In the latitudinal interval 45–60°N, the height of the Regener–Pfotzer maximum is near to 70 hPa (i.e. close to the ozone layer maximum at these latitudes), which could explain the stronger influence of trapped radiation on the ozone at this level.

6.4.3 Solar protons and stratospheric O_3

In recent decades, the problem of solar protons influence on the atmospheric O_3 has been a subject of intensive investigations. The short-term strato-mesospheric–thermospheric response to different solar proton events has been studied by many authors (e.g. Jackman et al., 1980, 2011; Rusch et al., 1981; Krivolutsky et al., 2006; Hauchecorne et al., 2007; Seppälä et al., 2008). The particles' effect on the upper atmospheric O_3 is well elucidated in scientific literature, so this subsection focuses instead on the long-term (centennial) effect of solar protons on the total ozone density (TOZ), and ozone in the lower stratosphere. The potential impact of solar protons during the period 1900–2010 is estimated by the use of a lagged correlation analysis with TOZ data taken from the ERA 20th-century reanalysis. The time record of solar proton flux (SPF), with energies 10 MeV or higher, is compiled from several sources: (i) historical reconstruction of the huge solar proton events between 1561 and 1950 (McCracken et al., 2001); (ii) published data for solar proton fluencies during the period 1955–86 (Shea and Smart, 1990); and (iii) satellite data for solar proton events affecting Earth's environment since 1976 (courtesy of NOAA data records). Correlation maps, prepared in a similar way as in GCR–ozone analysis, are shown in Fig. 6.15.

Fig. 6.15 shows an absence of persistent relation between solar protons and total ozone density during the examined century. A small exception is visible near the Antarctic Peninsula, indicating a systematic decrease of TOZ, when solar proton fluxes are intensified. This region of increased TOZ sensitivity corresponds fairly well to the increased flux of energetic particles, measured on board various satellites orbiting around the Earth (e.g. Fiandrini et al., 2004; Dachev et al., 1992, 2015a,b). This means that higher energetic solar protons, penetrating systematically deeper into the region of the South Atlantic Anomaly, due to the weaker stopping power of the geomagnetic field there, reduce the O_3 density through activation of ozone destructive cycles at mesospheric–thermospheric levels.

On the other hand, examination of connectivity between solar proton flux and the ozone mixing ratio at 70 hPa reveals a widely spread negative correlation in the tropics (see Fig. 6.16). The tropical O_3 response could hardly be explained with intense particle

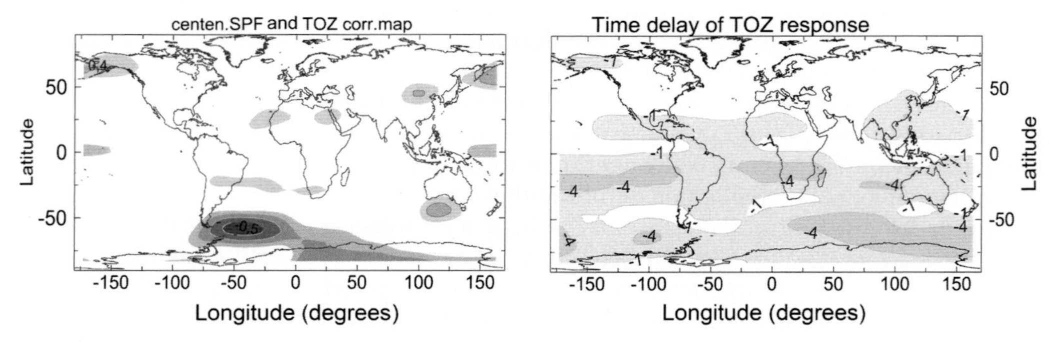

FIG. 6.15 *(left)* Centennial correlation maps of winter total ozone density (TOZ) and solar protons with E $\geq$10 MeV (contours); *(right)* time delay of TOZ response to the solar protons forcing, in years. *From Kilifarska, N., 2017. Mechanism of Relation Between Cosmic Rays, Geomagnetic Field and Earth's Climate, (DSc thesis), Sofia, Bulgaria.*

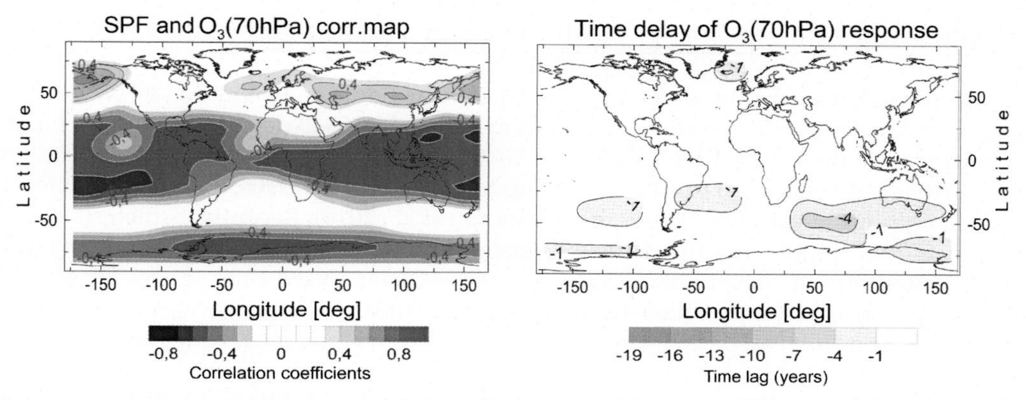

FIG. 6.16 *(left)* Correlation coefficients of solar protons (SPF) with winter's ozone volume mixing ratio at 70 hPa, statistically significant at 95% and corrected for time lag; *(right)* time delay of O_3 response to SPF forcing in years. *From Kilifarska, N., 2017. Mechanism of Relation Between Cosmic Rays, Geomagnetic Field and Earth's Climate, (DSc thesis), Sofia, Bulgaria.*

precipitation, because the probability for such a deep penetration in the entire equatorial region is very small. Very likely it could be attributed to the Forbush decrease of GCRs, due to their deflection by the stronger (in periods of solar proton events) solar wind. This conclusion is supported also by Fig. 6.13, which shows that GCRs and tropical ozone at 70 hPa correlate positively.

In resume: A lot of statistical evidence shows that asymmetrical distribution of O_3 over the globe could be related to geomagnetic modulation of energetic particles of different origin entering Earth's atmosphere. The effect of solar protons on the mesospheric and upper stratospheric ozone is well-known, but their influence on the centennial evolution of total ozone density appears to be very small. Over short-term periods, however, solar protons could contribute to the lower stratospheric O_3 density indirectly, through activation of the mechanism for its self-restoration, after being destroyed at mesospheric–upper stratospheric levels (see Section 6.3). Although less effective, compared to autocatalytic ozone production, this mechanism could have its impact in some periods, and in some regions of Earth's atmosphere.

The impact of galactic cosmic rays (arriving along the open geomagnetic field lines) on the lower stratospheric ozone are fixed mainly in the polar regions, although their effect is also traceable at equatorial latitudes (see Fig. 6.13). The temporal covariation between ozone and these particles is in-phase. In some regions over the world, however, the lower stratospheric ozone and total ozone density covariate in antiphase with the temporal variations of cosmic ray intensity (measured by ground-based detectors). Detailed analysis reveals that this is an ozone response to trapped particles in Earth's radiation belts, which are subject to geomagnetic lensing by the nondipolar part of the geomagnetic field. Thus, the irregularly distributed ionization near the tropopause, with its ability to activate ozone production or destruction in certain regions, could be a reasonable explanation of the observed longitudinal variations, and hemispherical asymmetry, of ozone density distribution globally.

6.5 Statistical evidence for GCR influence on the near tropopause water vapour

Balloon measurements of water vapour, taken near Boulder, Colorado (40°N, 105.25°W), along with zonally averaged satellite measurements in the 35–45 degrees latitudinal band (by Aura-MLS instrument, UARS-HALOE and SAGE II), show a steady increase of the lower stratospheric H_2O vapour at 82 hPa, lasting up to the end of the 1990s, followed by a decrease (or small amplitude variations) after that (Solomon et al., 2010). Recently, Schieferdecker and co-workers (2015) have identified the existence of quasidecadal signal in the lower stratospheric water vapour, despite the limited length of the analysed time series. These studies raise the question whether longer-term variations of near tropopause water vapour could be found in the centennial data records of ERA 20th century, and if so, what the driver of such variations is.

The examination of the time series of a specific humidity, at different latitudes and longitudes, confirms that there is a gradual moistening of the atmosphere at 150 hPa over the entire 20th century. The comparison with a smoothed time series of GCRs shows pronounced antiphase coherence at almost all examined latitudes (Fig. 6.17A, C, and D). The temporal evolution of water vapour over Antarctica is exceptional, where a phase-lagged covariance with that of GCR centennial variations is clearly visible (see Fig. 6.17B).

Taking into account that H_2O molecules are relatively easily dissociated by secondary electrons, produced by GCRs (see Section 6.3.1), it is reasonable to look for a potential relation between GCRs and humidity. The correlation map (based on the statistically significant at 2σ level coefficients) of GCRs and specific humidity at 150 hPa reveals that the overall covariance between them is in antiphase, with the exception of the Antarctic continent, where a relatively weak in-phase relation is found (Fig. 6.18, left panel). Bearing in mind that GCR intensity persistently decreased during the 20th century, this result suggests a global (with exception of Antarctica) increase of the water vapour at said level. This estimation is in good agreement not only with the time series shown in Fig. 6.17, but also with the recent instrumental measurements of the water vapour (Roscoe and Rosenlof, 2011; Solomon et al., 2010).

On the other hand, many authors have noticed the antiphase covariance between ozone and water vapour in the upper troposphere–lower stratosphere (UTLS) region (e.g. Pan et al., 1997, 2000; Hegglin et al., 2009; Tilmes et al., 2010). This reasonably raises the question whether this synchronicity is due to the simultaneous influence of GCRs on the ozone and H_2O vapour. Taking into account the heterogeneous distribution of secondary ionization in the lower atmosphere (due to the geomagnetic lensing of trapped energetic particles), the relatively homogeneous map of connectivity between GCRs and H_2O vapour (see Fig. 6.18) seems unrelated to the direct impact of particles on the water vapour density.

In resume: The continuous rise of water vapour at 150 hPa throughout the entire 20th century (Fig. 6.17), which opposes the long-term decrease of GCR intensity, suggests that secondary ionization created by GCRs in the lower stratosphere could directly drive the long-term variability of the water vapour. However, the heterogeneous distribution of the lower atmospheric ionization (projected on the ground-based neutron monitors' measurements; see Fig. 5.5) implies that the overall negative correlation between GCRs and water vapour is not a result of direct GCR influence.

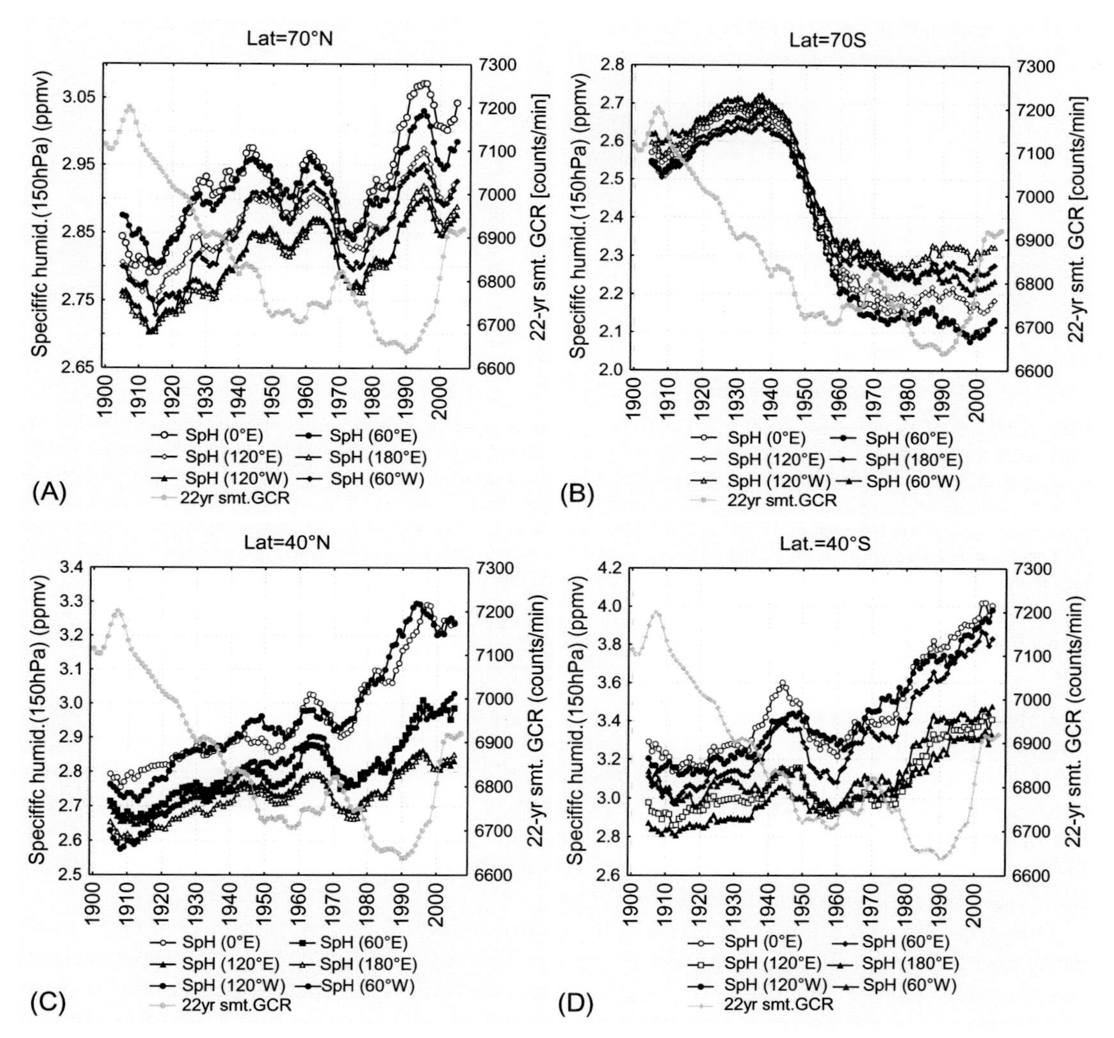

FIG. 6.17 Time series of specific humidity at 150 hPa, shown at four latitudes and six longitudes *(black symbols)*, smoothed by an 11-year moving window; (A) at 70°N latitude; (B) at 70°S latitude; (C) at 40°N latitude; (D) at 40°S latitude. Long-term variations of GCRs, i.e. annual mean values smoothed by a 22-year moving window *(grey curve)*, are shown for comparison with the water vapour temporal evolution. *Data source: ERA 20th century for specific humidity, and Usoskin, I.G., et al., 2008. Cosmic Ray Intensity Reconstruction. IGBP PAGES/World Data Center for Paleoclimatology, Data Contribution Series # 2008-013. NOAA/NCDC Paleoclimatology Program, Boulder, CO for GCR. Reproduced with permission from Kilifarska, N.A., 2017. Hemispherical asymmetry of the lower stratospheric* O_3 *response to galactic cosmic rays forcing. ACS Earth Space Chem. 1, 80–88. DOI: https://doi.org/10.1021/acsearthspacechem.6b00009.*

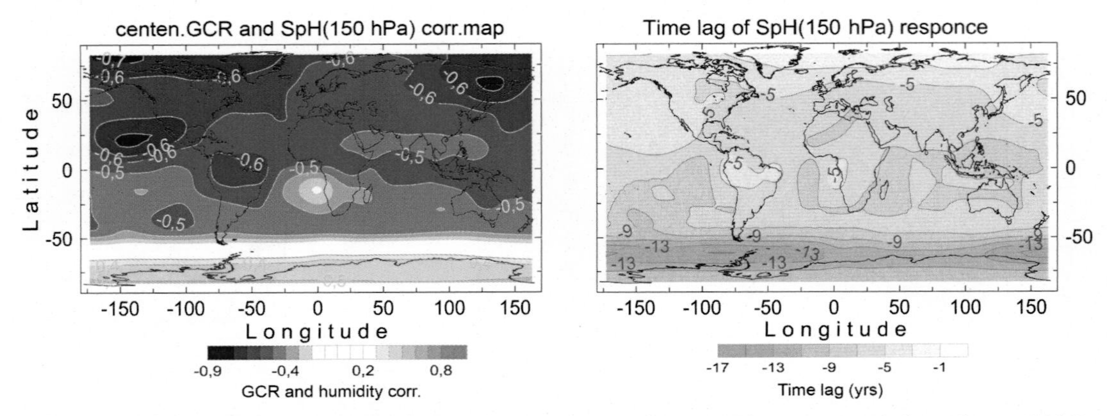

FIG. 6.18 Statistically significant (at 2σ level) correlation maps of GCRs and winter-specific humidity at 150 hPa, calculated for the period 1900–2010 and corrected for a time lag *(left)*. Time delay of H_2O vapour response to GCR forcing is shown in the *right-hand panel. Data source: ERA 20th century for specific humidity, and historical reconstruction of GCR intensity by Usoskin, I.G., et al., 2008. Cosmic Ray Intensity Reconstruction. IGBP PAGES/World Data Center for Paleoclimatology, Data Contribution Series # 2008-013. NOAA/NCDC Paleoclimatology Program, Boulder, CO. Reproduced with permission from Kilifarska, N.A., 2017. Hemispherical asymmetry of the lower stratospheric O_3 response to galactic cosmic rays forcing. ACS Earth Space Chem. 1, 80–88. DOI: https://doi.org/10.1021/acsearthspacechem.6b00009.*

6.6 Interplay between near tropopause ozone and water vapour

It has long been known that the products of H_2O molecule dissociation by UV radiation or energetic particles destroy ozone, a phenomenon known in the scientific literature as the HO_x destructive cycle of ozone (Banks and Kockarts, 1973; Brasseur and Solomon, 2005). The lower stratospheric O_3 could also affect the H_2O density in the UTLS layer, through an influence on the upper tropospheric static stability (see Section 7.2.2). Therefore, the interaction between O_3 and water vapour in time and space needs to be investigated carefully, in the context of their possible influence on climate variability.

One approach for such an analysis is the examination of the strength of O_3–H_2O connectivity and its spatial distribution over the globe. The connectivity ozone–vapour covariation has been estimated by the use of the lagged cross-correlation technics, applied over the ozone and water vapour time series, preliminary smoothed by 11-year moving window, in each point of a grid with 10-degree increments in latitude and longitude. The spatial distribution of the covariance is analysed by the use of correlation maps, derived from the statistically significant at 2σ level cross-correlation coefficients. It is useful to note that in a lot of grid points, simultaneous correlation has been found (i.e. without a time lag between applied forcing and examined response). At others, synchronization between ozone and humidity variations occurs with some delay in time. For this reason, two types of maps have been drawn, showing simultaneous covariances (Fig. 6.19) and lagged correlation maps (Figs 6.20 and 6.21).

Analysis of Fig. 6.19 shows that simultaneous coupling between ozone and water vapour is persistently found in the Northern Hemisphere extra-tropics (around 50°N latitude), being stronger during the winter season. Comparison with centennial climatology of dynamical surface temperature anomalies (coloured shading in Fig. 6.19) reveals that the circle of

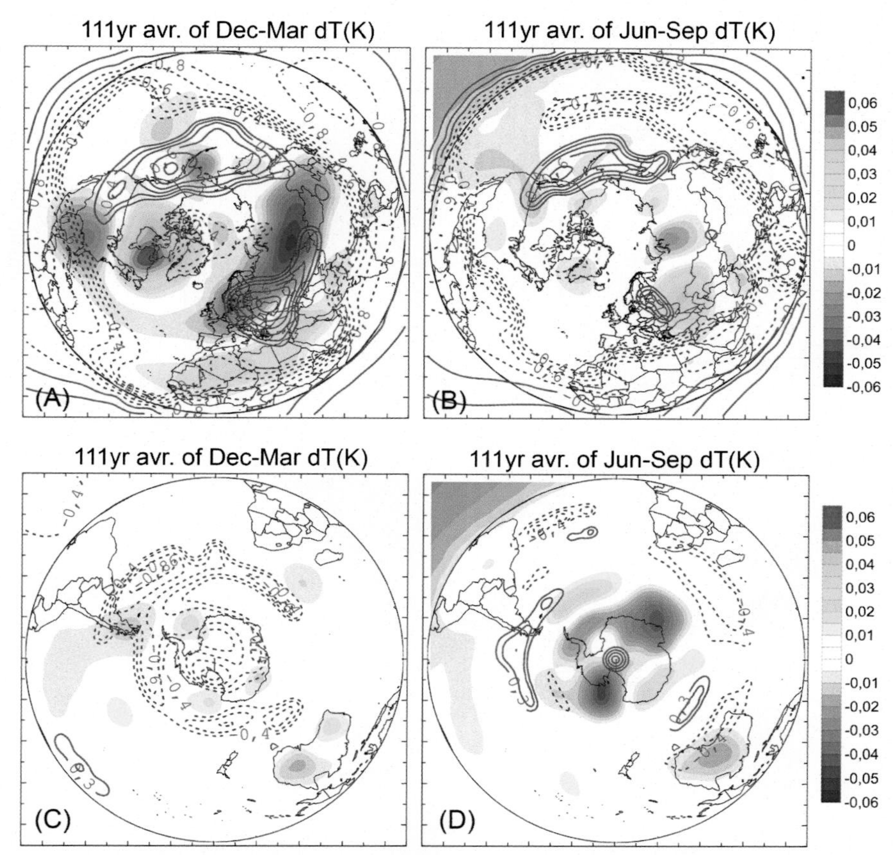

FIG. 6.19 Simultaneous correlation maps of O_3 mixing ratio at 70 hPa and water vapour at 150 hPa (*continuous contours* denote positive correlation, and *dashed contours* negative) in Dec–Mar *(left column)* and Jun–Sep *(right column)*. The *coloured background* presents the centennial climatology of the near-surface air temperature dynamical anomalies (i.e. temperature deviation from its dynamically varying decadal means) for Dec–Mar and Jun–Sep, respectively. A and B panels correspond to the Northern Hemisphere, C and D to the Southern Hemisphere. *Data source: ERA 20th-century reanalysis.*

negative winter temperature anomalies at mid-latitudes matches fairly well the regions with in-phase covariances of ozone and humidity (continuous contours). In the Southern Hemisphere, the ozone–water vapour connectivity is weaker, and relations with surface temperature anomalies are not found. The strongest antiphase synchronization of ozone and humidity variations (without time delay) is found in tropical regions in both examined periods: Dec–Mar and Jun–Sep.

The simultaneous correlation does not allow, however, distinction of dependent from forcing variable. Such a possibility is provided by lagged correlation maps. Two possibilities have been explored: (i) ozone is a driving factor for humidity variations; and (ii) changing humidity initiates ozone variations. Correspondingly, two types of map have been created, with leading (independent factor) O_3 and with H_2O vapour as independent variable. Due to the high

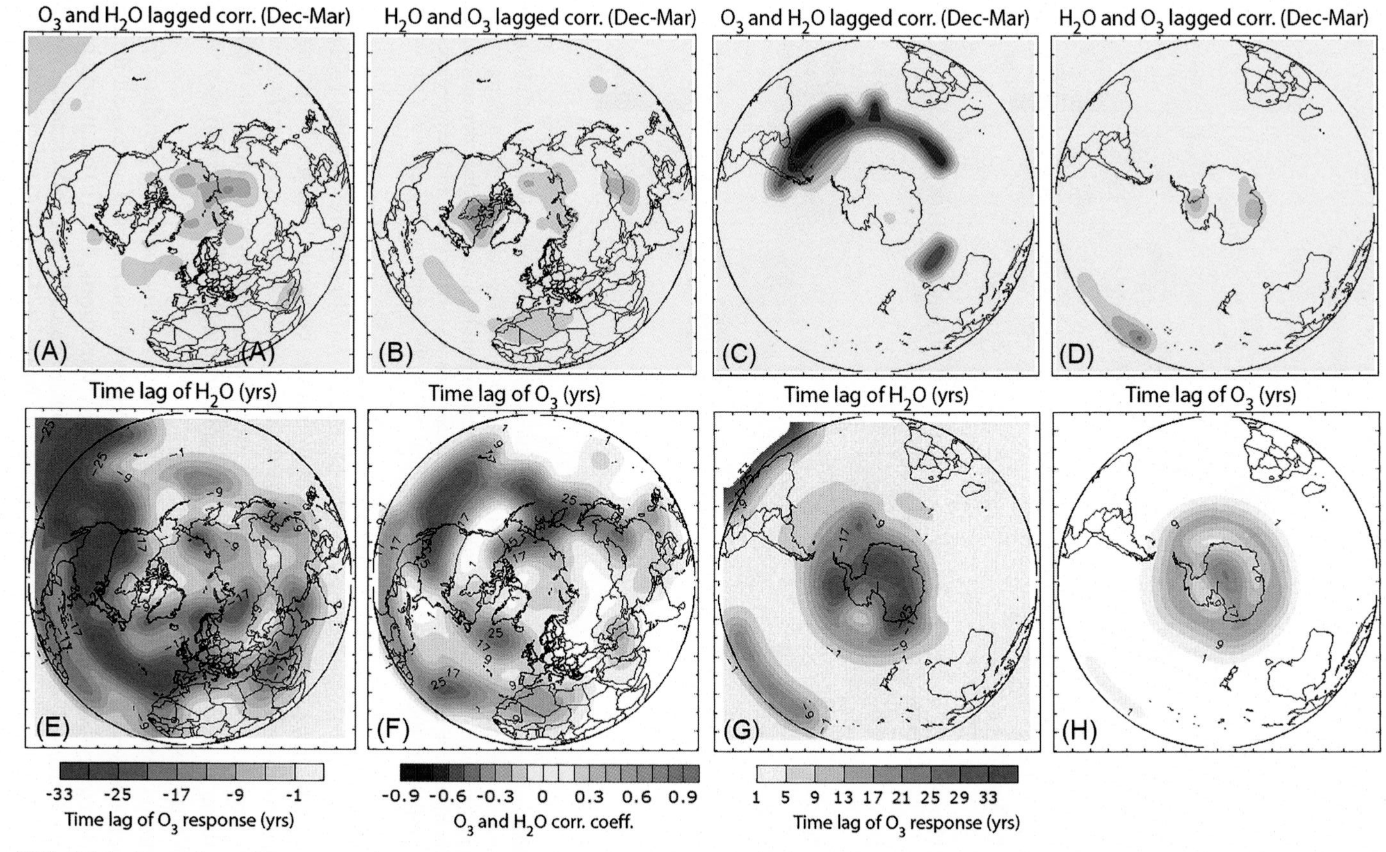

FIG. 6.20 Lagged correlation map of O_3 at 70 hPa and H_2O vapour at 150 hPa *(coloured shading)* calculated for Dec–Mar, during the period 1900–2010. All maps are created from the preliminary weighted correlation coefficients by the autocorrelation function of forcing factor, taken for a given time lag (specific for each grid point). Panels A and C illustrate the ozone influence on the water vapour, while B and D show the humidity influence on the ozone. The bottom panels illustrate the time delay of the H_2O response to ozone forcing (panels E and G) or O_3 response to humidity forcing (panels F and H). *Data source: ERA 20th-century reanalysis.*

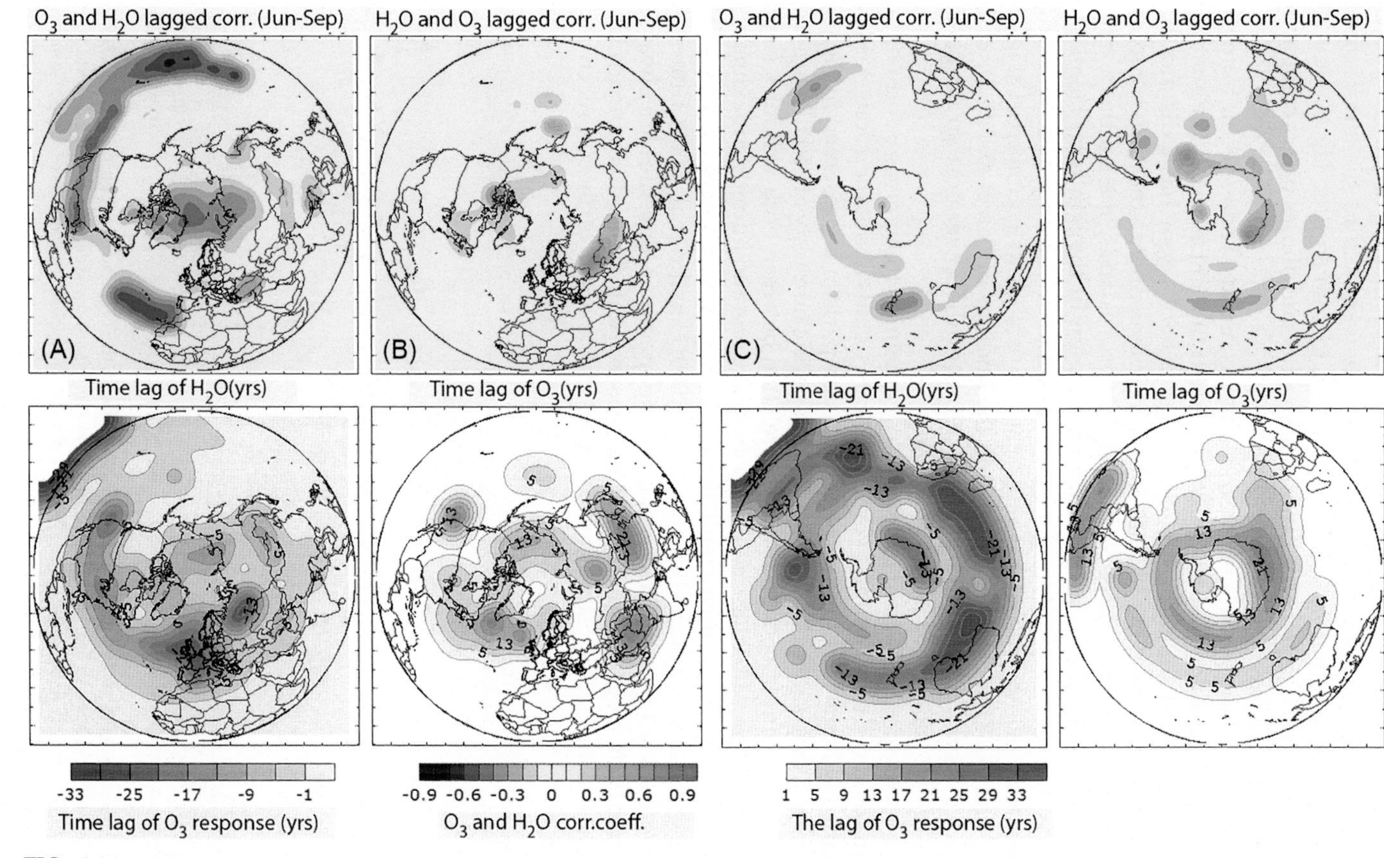

FIG. 6.21 As for Fig. 6.20, but calculated for Jun–Sep. *Data source: ERA 20th-century reanalysis.*

variability of time lags between grid points, and between maps with different forcing factor, the lagged correlation coefficients have been initially weighted by the autocorrelation function of independent (forcing) factor, corresponding to a given time lag (Kenny, 1979). This procedure reduces the delayed correlation coefficients and allows direct comparison of correlation with different time lags.

Figs 6.20 and 6.21 show the impact of ozone (panels A and C) and specific humidity (panels B and D) in their lagged coupled variability. It is clear that in Dec–Mar (the austral summer), and Jun–Sep (the boreal summer), the lower stratospheric ozone has a distinct influence on the water vapour at 150 hPa. In the Northern Hemisphere, the polar cap is strongly influenced, as are several tropical regions: the Azores Islands, the southern part of North America, and the Central Pacific. The impacted region in the Southern Hemisphere is the Southern Ocean, with a clear resemblance to the Antarctic convergent zone. During winter, none of the ozone or humidity has a significant impact on its counterpart (Figs 6.20A and B and 6.21C and D). This result is in line with the findings of Pan et al. (1997), who reported a hemispherical asymmetry in seasonal variations of the lowermost stratospheric O_3 and H_2O, based on the analysis of the SAGE II data.

The cross-correlation coefficient of time series containing trend gives information about the similarity in trends of the examined variables. More details about their long-term variability give comparison of analysed time series. Such a comparison is presented in Fig. 6.22, where zonally averaged time series of O_3 at 70 hPa and specific humidity at 150 hPa are compared in the regions with a simultaneous positive correlation between both variables (see Fig. 6.19). A pronounced tendency for phase shifting between ozone and humidity maxima and minima is clearly seen (see Fig. 6.22). Such a tendency has been previously noticed in the extra-tropical transition layer by other authors (Pan et al., 1997, 2000; Hegglin et al., 2009; Tilmes et al., 2010). They discovered that high extra-tropical near tropopause O_3 densities are always accompanied by the reduced specific humidity. On the seasonal data basis, this anticorrelation is usually attributed to the strongest winter downward transport of the O_3-reach air from the

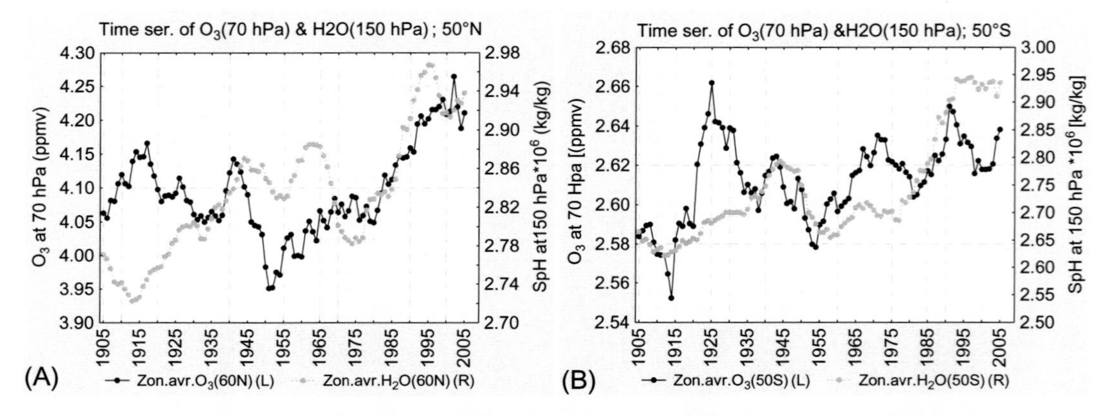

FIG. 6.22 Time series of zonally averaged winter (Dec–Mar in the NH and Jun–Sep in the SH) O_3 mixing ratio at 70 hPa *(black symbols)* and specific humidity at 150 hPa *(grey symbols)*, shown for 50°N latitude (A) and 50°S latitude (B). *Data source: ERA 20th century. From Kilifarska, N., 2017. Mechanism of Relation Between Cosmic Rays, Geomagnetic Field and Earth's Climate, (DSc thesis), Sofia, Bulgaria.*

stratosphere, and to a direct transport of the O_3-poor air from the upper troposphere across the warmer extra-tropical summer tropopause (Pan et al., 1997, 2000). However, the maintenance of this tendency in the long-term variations of the near tropopause O_3 and water vapour requires an explanation. Therefore, Chapter 7 will focus on the clarification of this problem and the coupled O_3–H_2O vapour influence on climate.

In resume, the synchronized variations of the upper troposphere-lower stratospheric ozone and water vapour raise a question about the reason for such synchronization. Statistical analysis indicates that this covariance is not a synchronous response to an external forcing (e.g. cosmic rays ionization created in the Regener–Pfotzer maximum), because the fingerprints of GCRs on ozone and humidity are quite different. The synchronization is more likely to be due to the continuous interaction between ozone and humidity. Thus in the summer extra-tropics, the triggering factor seems to be O_3, while in the tropics, the instantaneous interdependence between ozone and humidity does not allow the separation of the triggering factor. The differentiation of the leading factor in the SH polar region also seems impossible (at least with the applied method).

Appendix

List of photochemical reaction rates used in the ozone equilibrium equation (6.3); units are $cm^6 molecule^{-2} s^{-1}$ for $\boldsymbol{k_2}$, otherwise $cm^3 molecule^{-1} s^{-1}$ (JPL, 2000).

$$k_2 = 6.10^{-34}\cdot(T/300)^{-2.4} \qquad k_{10} = 3.10^{-11}\exp(250/T)$$
$$k_3 = 8.10^{-12}\exp(-2060/T) \qquad k_{11} = 1.5\cdot10^{-12}\exp(-880/T)$$
$$k_{6a} = 1.8\cdot10^{-11}\exp(110/T) \qquad k_{13} = 4.8\cdot10^{-11}\exp(250/T)$$
$$k_{6b} = 3.2\cdot10^{-12}\exp(70/T) \qquad k_{16} = {}^{16}5.6\cdot10^{-12}\exp(180/T)$$
$$k_7 = 2.2\cdot10^{-10} \qquad k_{17} = 3.10^{-12}\exp(-1500/T)$$
$$k_8 = 2.2\cdot10^{-11}\exp(120/T) \qquad k_{36} = 1.1\cdot10^{-14}\exp(-500/T)$$

The photo-dissociation rate of oxygen molecules and ozone is defined as follows:

$$\begin{aligned} J_{O_2,O_3}(\lambda,z) &= \varepsilon_{O_2,O_3}(\lambda)\cdot\sigma_{O_2,O_3}(\lambda)\cdot I(z,\lambda)\cdot\exp\langle-\tau_{O_2,O_3}(z,\lambda)\rangle \\ &= C_{2,3}\cdot\exp\langle-\tau_{O_2,O_3}(z,\lambda)\rangle \end{aligned} \tag{A.1}$$

$$\sigma(\lambda,z) = \frac{\ln(I_o) - \ln[I(z)]}{n(z)} \tag{A.2}$$

$$\tau(\lambda,z) = \int_z^{\infty}\sigma(\lambda,z)\cdot n(s)\cdot l(s)\cdot ds \tag{A.3}$$

where $J_{O_2,\,O3}(\lambda,z)$ is the rate of dissociation, resulting from a flux $\boldsymbol{I(\lambda)}$ in the wavelength interval $\boldsymbol{\lambda+d\lambda}$ entering the atmosphere, $\varepsilon_{O_2,\,O3}(\lambda)$ is the photo-dissociation quantum efficiency; $\boldsymbol{\sigma(\lambda)}$ in ($cm^2 molecule^{-1}$) is the absorption cross section, $\boldsymbol{n(z)}$ is the total concentration of O_2 or O_3 in (molecules cm^{-3}) above the level z, and $\boldsymbol{l(z)}$ is the path length in (cm) (JPL, 2000); $\boldsymbol{\tau(\lambda,z)}$ is the optical depth of the atmosphere for penetrating radiation with a given wavelength.

References

Aghedo, A.M., Bowman, K.W., Worden, H.M., Kulawik, S.S., Shindell, D.T., Lamarque, J.F., Faluvegi, G., Parrington, M., Jones, D.B.A., Rast, S., 2011. The vertical distribution of ozone instantaneous radiative forcing from satellite and chemistry climate models. J. Geophys. Res. Atmos. 116https://doi.org/10.1029/2010JD014243.

Allan, M., Asmis, K.R., Popovic, D.B., Stepanovic, M., Mason, N.J., Davies, J.A., 1996. Resonances in collisions of low-energy electrons with ozone: experimental elastic and vibrationally inelastic differential cross sections and dissociative attachment spectra. J. Phys. B Atomic Mol. Phys. 29, 4727. https://doi.org/10.1088/0953-4075/29/20/024.

Anderson, G.P., Clough, S.A., Kneizys, F.X., Chetwynd, J.H., Shettle, E.P., 1986. AFGL Atmospheric Constituent Profiles (0-120 km). AFGL-TR-86-0110, U.S. Air Force Geophysics Laboratory, Optical Physics Division.

Banks, P.M., Kockarts, G., 1973. Aeronomy—Part A. Academic Press, New York and London.

Bazilevskaya, G.A., Usoskin, I.G., Flückiger, E.O., Harrison, R.G., Desorgher, L., Bütikofer, R., Krainev, M.B., Makhmutov, V.S., Stozhkov, Y.I., Svirzhevskaya, A.K., Svirzhevsky, N.S., Kovaltsov, G.A., 2008. Cosmic ray induced ion production in the atmosphere. Space Sci. Rev. 137, 149–173. https://doi.org/10.1007/s11214-008-9339-y.

Brasseur, G.P., Solomon, S., 2005. Aeronomy of the Middle Atmosphere: Chemistry and Physics of the Stratosphere and Mesosphere, third ed. Atmospheric and Oceanographic Sciences Library. Springer, Netherlands.

Cacace, F., de Petris, G., Troiani, A., 2001. Experimental detection of tetraoxygen. Angew. Chem. Int. Ed. 40, 4062–4065. https://doi.org/10.1002/1521-3773(20011105)40:21<4062::AID-ANIE4062>3.0.CO;2-X.

Cacace, F., de Petris, G., Rosi, M., Troiani, A., 2002. Formation of O3+ upon ionization of O2: the role of isomeric O4+ complexes. Chem. A Eur. J. 8, 3653–3659. https://doi.org/10.1002/1521-3765(20020816)8:16<3653::AID-CHEM3653>3.0.CO;2-1.

Chapman, S.A., 1930. A theory of upper-atmospheric ozone. Mem. R. Meteorol. Soc. 3 (26), 103–125.

Cicman, P., Skalny, J.D., Fedor, J., Mason, N.J., Scheier, P., Illenberger, E., Mark, T.D., 2007. Dissociative electron attachment to ozone at very low energies revisited. Int. J. Mass Spectrom. 260, 85–87.

Cosby, P.C., 1993. Electron-impact dissociation of oxygen. J. Chem. Phys. 98, 9560–9569. https://doi.org/10.1063/1.464387.

Curran, R.K., 1961. Negative ion formation in ozone. J. Chem. Phys. 35, 1849–1851. https://doi.org/10.1063/1.1732155.

Dachev, T.P., Matviichuk, Y.N., Bankov, N.G., Semkova, J.V., Koleva, R.T., Ivanov, Y.J., Tomov, B.T., Petrov, V.M., Shurshakov, V.A., Bengin, V.V., Machmutov, V.S., Panova, N.A., Kostereva, T.A., Temny, V.V., Ponomarev, Y.N., Tykva, R., 1992. 'Mir' radiation dosimetry results during the solar proton events in September–October 1989. Adv. Space Res. 12, 321–324. https://doi.org/10.1016/0273-1177(92)90122-E.

Dachev, T.P., Tomov, B.T., Matviichuk, Y.N., Dimitrov, P.G., Bankov, N.G., Shurshakov, V.V., Ivanova, O.A., Häder, D.-P., Schuster, M.T., Reitz, G., Horneck, G., 2015a. 'BION-M' No. 1 spacecraft radiation environment as observed by the RD3-B3 radiometer-dosimeter in April–May 2013. J. Atmos. Sol. Terr. Phys. 123, 82–91. https://doi.org/10.1016/j.jastp.2014.12.011.

Dachev, T.P., Tomov, B., Matviichuk, Y., Dimitrov, P., Bankov, N., Häder, D.P., Horneck, G., Reitz, G., 2015b. ISS radiation environment as observed by liulin type-r3dr2 instrument in October-November 2014. Aerosp. Res. Bulg. 27, 17–41.

de F. Forster, P.M., Shine, K.P., 1997. Radiative forcing and temperature trends from stratospheric ozone changes. J. Geophys. Res.-Atmos. 102, 10841–10855. https://doi.org/10.1029/96JD03510.

de Petris, G., 2003. Atmospherically relevant ion chemistry of ozone and its cation. Mass Spectrom. Rev. 22, 251–271. https://doi.org/10.1002/mas.10053.

Dorman, L.I., Kovalenko, V.A., Milovidova, N.P., 1967. The latitude distribution, integral multiplicities and coupling coefficients for the neutron, total ionizing and hard components of cosmic rays. Nuovo Cimento B (1965-1970) 50, 27–39. https://doi.org/10.1007/BF02710680.

Dütsch, H.U., 1971. Photochemistry of atmospheric ozone. Adv. Geophys. 15, 219–315.

Fehsenfeld, F.C., Ferguson, E.E., 1974. Laboratory studies of negative ion reactions with atmospheric trace constituents. J. Chem. Phys. 61, 3181–3193. https://doi.org/10.1063/1.1682474.

Fiandrini, E., Esposito, G., Bertucci, B., Alpat, B., Ambrosi, G., Battiston, R., Burger, W.J., Caraffini, D., Masso, L.D., Dinu, N., Ionica, M., Ionica, R., Pauluzzi, M., Menichelli, M., Zuccon, P., 2004. High-energy protons, electrons, and positrons trapped in Earth's radiation belts. Space Weather. 2. https://doi.org/10.1029/2004SW000068.

Funke, B., López-Puertas, M., Gil-López, S., von Clarmann, T., Stiller, G.P., Fischer, H., Kellmann, S., 2005. Downward transport of upper atmospheric NOx into the polar stratosphere and lower mesosphere during the Antarctic 2003 and Arctic 2002/2003 winters. J. Geophys. Res.-Atmos. 110. https://doi.org/10.1029/2005JD006463.

Garcia, R.R., 1992. Transport of thermospheric NOX to the stratosphere and mesosphere. Adv. Space Res. 12, 57–66. https://doi.org/10.1016/0273-1177(92)90444-3.

Grewe, V., 2006. The origin of ozone. Atmos. Chem. Phys. 6, 1495–1511. https://doi.org/10.5194/acp-6-1495-2006.

Hansen, J., Sato, M., Ruedy, R., 1997. Radiative forcing and climate response. J. Geophys. Res.-Atmos. 102, 6831–6864. https://doi.org/10.1029/96JD03436.

Hauchecorne, A., Bertaux, J.-L., Lallement, R., 2007. Impact of solar activity on stratospheric ozone and No2 observed by GOMOS/ENVISAT. In: Calisesi, Y., Bonnet, R.-M., Gray, L., Langen, J., Lockwood, M. (Eds.), Solar Variability and Planetary Climates. In: Space Sciences Series of ISSI, Springer New York, New York, NY, pp. 393–402. https://doi.org/10.1007/978-0-387-48341-2_31.

Hegglin, M.I., Boone, C.D., Manney, G.L., Walker, K.A., 2009. A global view of the extratropical tropopause transition layer from atmospheric chemistry experiment Fourier transform spectrometer O3, H2O, and CO. J. Geophys. Res.-Atmos. 114. https://doi.org/10.1029/2008JD009984.

Hood, L.L., Zaff, D.A., 1995. Lower stratospheric stationary waves and the longitude dependence of ozone trends in winter. J. Geophys. Res.-Atmos. 100, 25791–25800. https://doi.org/10.1029/95JD01943.

Intergovernmental Panel on Climate Change (IPCC), 2007. Climate change 2007. In: Solomon, S. et al., (Eds.), The Physical Science Basis. Cambridge University Press, New York, p. 996.

Itikawa, Y., 2005. Cross sections for electron collisions with nitrogen molecules. J. Phys. Chem. Ref. Data Monogr. 35, 31–53. https://doi.org/10.1063/1.1937426.

Jackman, C.H., Frederick, J.E., Stolarski, R.S., 1980. Production of odd nitrogen in the stratosphere and mesosphere: an intercomparison of source strengths. J. Geophys. Res. Oceans 85, 7495–7505. https://doi.org/10.1029/JC085iC12p07495.

Jackman, C.H., Marsh, D.R., Vitt, F.M., Roble, R.G., Randall, C.E., Bernath, P.F., Funke, B., López-Puertas, M., Versick, S., Stiller, G.P., Tylka, A.J., Fleming, E.L., 2011. Northern Hemisphere atmospheric influence of the solar proton events and ground level enhancement in January 2005. Atmos. Chem. Phys. 11, 6153–6166. https://doi.org/10.5194/acp-11-6153-2011.

JPL, 2000. Chemical Kinetics and Photochemical Data for Use in Stratospheric Modelling. In: Evaluation No. 13. NASA, Jet Propulsion Lab., California Institute of Technology, Pasadena, California, USA.

Kenny, D.A., 1979. Correlation and Causality. John Wiley & Sons Inc., New York.

Kilifarska, N.A., 2013. An autocatalytic cycle for ozone production in the lower stratosphere initiated by Galactic Cosmic rays. C. R. Acad. Bulg. Sci. 66 (2), 243–252.

Kilifarska, N.A., 2017. Hemispherical asymmetry of the lower stratospheric O_3 response to galactic cosmic rays forcing. ACS Earth Space Chem. 1, 80–88. https://doi.org/10.1021/acsearthspacechem.6b00009.

Kilifarska, N.A., Bakhmutov, V.G., Melnyk, G.V., 2013. Energetic particles influence on the Southern Hemisphere ozone variability. C. R. Acad. Bulg. Sci. 66 (11), 1613–1622.

Kilifarska, N.A., Bakhmutov, V.G., Melnyk, G.V., 2017. Geomagnetic fingerprint on the Lower stratospheric O_3 found in ERA 20 century reanalysis. C. R. Acad. Bulg. Sci. 70 (7), 1003–1010.

Krivolutsky, A., Bazilevskaya, G., Vyushkova, T., Knyazeva, G., 2002. Influence of cosmic rays on chemical composition of the atmosphere: data analysis and photochemical modelling. Phys. Chem. Earth A/B/C 27, 471–476. https://doi.org/10.1016/S1474-7065(02)00028-1.

Krivolutsky, A.A., Klyuchnikova, A.V., Zakharov, G.R., Vyushkova, T.Y., Kuminov, A.A., 2006. Dynamical response of the middle atmosphere to solar proton event of July 2000: three-dimensional model simulations. Adv. Space Res. 37, 1602–1613. https://doi.org/10.1016/j.asr.2005.05.115. Particle acceleration; Space plasma physics; solar radiation and the earth's atmosphere and climate.

Lazutin, L.L., 2010. Belt of protons 1–20 MeV at L = 2. Cosm. Res. 48, 108–112. https://doi.org/10.1134/S0010952510010090.

Makarov, O.P., Kanik, I., Ajello, J.M., 2003. Electron impact dissociative excitation of O2: 1. Kinetic energy distributions of fast oxygen atoms. J. Geophys. Res. Planets. 108. https://doi.org/10.1029/2000JE001422.

Matejcik, S., Cicman, P., Kiendler, A., Skalny, J.D., Illenberger, E., Stamatovic, A., Märk, T.D., 1996. Low-energy electron attachment to mixed ozone/oxygen clusters. Chem. Phys. Lett. 261, 437–442. https://doi.org/10.1016/0009-2614(96)01005-6.

McCracken, K.G., Dreschhoff, G.A.M., Zeller, E.J., Smart, D.F., Shea, M.A., 2001. Solar cosmic ray events for the period 1561–1994: 1. Identification in polar ice, 1561–1950. J. Geophys. Res. Space Physics 106, 21585–21598. https://doi.org/10.1029/2000JA000237.

Meraner, K., Schmidt, H., 2016. Transport of nitrogen oxides through the winter mesopause in HAMMONIA. J. Geophys. Res. Atmos. 121, 2556–2570. https://doi.org/10.1002/2015JD024136.

Müller, U., Bubel, T., Schulz, G., 1993. Electron impact dissociation of H2O: emission cross sections for OH*, OH+*, H*, and H2O+* fragments. Z. Phys. D: At. Mol. Clusters 25, 167–174. https://doi.org/10.1007/BF01450171.

NIST Chemical Kinetics Database, 2002. Standard Reference Database 17, Version 7.0 (Web Version), Release 1.6.5, Data Version 2012.02.

Pan, L., Solomon, S., Randel, W., Lamarque, J.-F., Hess, P., Gille, J., Chiou, E.-W., McCormick, M.P., 1997. Hemispheric asymmetries and seasonal variations of the lowermost stratospheric water vapor and ozone derived from SAGE II data. J. Geophys. Res. Atmos. 102, 28177–28184. https://doi.org/10.1029/97JD02778.

Pan, L.L., Hintsa, E.J., Stone, E.M., Weinstock, E.M., Randel, W.J., 2000. The seasonal cycle of water vapor and saturation vapor mixing ratio in the extratropical lowermost stratosphere. J. Geophys. Res. Atmos. 105, 26519–26530. https://doi.org/10.1029/2000JD900401.

Peters, D.H.W., Gabriel, D.H.W., Entzian, G., 2008. Longitude-dependent decadal ozone changes and ozone trends in boreal winter months during 1960–2000. Ann. Geophys. 26, 1275–1286. https://doi.org/10.5194/angeo-26-1275-2008.

Porter, H.S., Jackman, C.H., Green, A.E.S., 1976. Efficiencies for production of atomic nitrogen and oxygen by relativistic proton impact in air. J. Chem. Phys. 65, 154–167. https://doi.org/10.1063/1.432812.

Rangwala, S.A., Kumar, S.V.K., Krishnakumar, E., Mason, N.J., 1999. Cross sections for the dissociative electron attachment to ozone. J. Phys. B Atomic Mol. Phys. 32, 3795. https://doi.org/10.1088/0953-4075/32/15/311.

Rohen, G., von Savigny, C., Sinnhuber, M., Llewellyn, E.J., Kaiser, J.W., Jackman, C.H., Kallenrode, M.-B., Schröter, J., Eichmann, K.-U., Bovensmann, H., Burrows, J.P., 2005. Ozone depletion during the solar proton events of October/November 2003 as seen by SCIAMACHY. J. Geophys. Res. Space Phys. 110. https://doi.org/10.1029/2004JA010984.

Roscoe, H.K., Rosenlof, K.H., 2011. Revisiting the lower stratospheric water vapour trend from the 1950s to 1970s. Atmos. Sci. Lett. 12, 321–324. https://doi.org/10.1002/asl.339.

Rosen, J.M., Hofmann, D.J., 1981. Balloon-borne measurements of electrical conductivity, mobility, and the recombination coefficient. J. Geophys. Res. Oceans 86, 7406–7410. https://doi.org/10.1029/JC086iC08p07406.

Rusch, D.W., Gérard, J.-C., Solomon, S., Crutzen, P.J., Reid, G.C., 1981. The effect of particle precipitation events on the neutral and ion chemistry of the middle atmosphere—I. Odd nitrogen. Planet. Space Sci. 29, 767–774. https://doi.org/10.1016/0032-0633(81)90048-9.

Sander, S.P., Friedl, R.R., DeMore, W.B., Ravishankara, A.R., 2000. NASA Panel for Data Evaluation: 13. Jet Propulsion Lab., California Institute of Technology, Pasadena, CA.

Schieferdecker, T., Lossow, S., Stiller, G.P., von Clarmann, T., 2015. Is there a solar signal in lower stratospheric water vapour? Atmos. Chem. Phys. 15, 9851–9863. https://doi.org/10.5194/acp-15-9851-2015.

Selesnick, R.S., Looper, M.D., Mewaldt, R.A., 2007. A theoretical model of the inner proton radiation belt. Space Weather. 5. https://doi.org/10.1029/2006SW000275.

Seppälä, A., Clilverd, M.A., Rodger, C.J., Verronen, P.T., Turunen, E., 2008. The effects of hard-spectra solar proton events on the middle atmosphere. J. Geophys. Res. Space Physics. 113. https://doi.org/10.1029/2008JA013517.

Shea, M.A., Smart, D.F., 1990. A summary of major solar proton events. Sol. Phys. 127, 297–320. https://doi.org/10.1007/BF00152170.

Slanger, T.G., Jusinski, L.E., Black, G., Gadd, G.E., 1988. A new laboratory source of ozone and its potential atmospheric implications. Science 241, 945–950. https://doi.org/10.1126/science.241.4868.945.

Solomon, S., Rosenlof, K.H., Portmann, R.W., Daniel, J.S., Davis, S.M., Sanford, T.J., Plattner, G.-K., 2010. Contributions of stratospheric water vapor to decadal changes in the rate of global warming. Science 327, 1219–1223. https://doi.org/10.1126/science.1182488.

Steblova, R.C., 2001. North-south asymmetry of the ozone layer. Rep. Russ. Acad. Sci. 379 (5), 675–679.

Stozhkov, Y.I., Bazilevskaya, G.A., Pokrevsky, P.E., Svirzhevsky, N.S., Martin, I.M., Turtelli, A., 1996. Cosmic rays in the atmosphere: north–south asymmetry. J. Geophys. Res. Space Physics 101, 2523–2528. https://doi.org/10.1029/95JA03317.

Tilmes, S., Pan, L.L., Hoor, P., Atlas, E., Avery, M.A., Campos, T., Christensen, L.E., Diskin, G.S., Gao, R.-S., Herman, R.L., Hintsa, E.J., Loewenstein, M., Lopez, J., Paige, M.E., Pittman, J.V., Podolske, J.R., Proffitt, M.R., Sachse, G.W., Schiller, C., Schlager, H., Smith, J., Spelten, N., Webster, C., Weinheimer, A., Zondlo, M.A., 2010. An aircraft-based upper troposphere lower stratosphere O3, CO, and H2O climatology for the Northern Hemisphere. J. Geophys. Res. Atmos. 115. https://doi.org/10.1029/2009JD012731.

US Standard Atmosphere, 1976. NOAA, NASA. US Air Force, Washington DC.

Usoskin, I.G., Gladysheva, O.G., Kovaltsov, G.A., 2004. Cosmic ray-induced ionization in the atmosphere: spatial and temporal changes. J. Atmos. Sol. Terr. Phys. 66, 1791–1796. https://doi.org/10.1016/j.jastp.2004.07.037.

Velinov, P.I.Y., Mateev, L., Kilifarska, N., 2005. 3-D model for cosmic ray planetary ionisation in the middle atmosphere. Ann. Geophys. 23, 3043–3046. https://doi.org/10.5194/angeo-23-3043-2005.

Wang, W.-C., Zhuang, Y.-C., Bojkov, R.D., 1993. Climate implications of observed changes in ozone vertical distributions at middle and high latitudes of the northern hemisphere. Geophys. Res. Lett. 20, 1567–1570. https://doi.org/10.1029/93GL01318.

World Meteorological Organization (WMO), 2007. Scientific assessment of ozone depletion: 2006. Global Ozone Research and Monitoring Project, Report No. 50, Geneva.

World Meteorological Organization (WMO), 2014. Scientific assessment of ozone depletion: 2014. World Meteorological Organization, Geneva Global Ozone Research and Monitoring Project—Report No. 55, 416 pp.

Zhaunerchyk, V., Geppert, W.D., Österdahl, F., Larsson, M., Thomas, R.D., Bahati, E., Bannister, M.E., Fogle, M.R., Vane, C.R., 2008. Dissociative recombination dynamics of the ozone cation. Phys. Rev. A 77, 022704.

CHAPTER

7

Mechanisms of geomagnetic influence on climate

OUTLINE

7.1 Brief review of existing mechanisms

Despite the existing controversy in scientific literature regarding the relation between geomagnetic field and climate, several hypotheses have attempted to explain such a relation (Dickinson, 1975; Pudovkin and Babushkina, 1991; Pudovkin and Veretenenko, 1992; Pudovkin and Raspopov, 1992; Tinslev, 1996; Svensmark and Friis-Christensen, 1997; Svensmark, 1998). Almost all of them, however, are inevitably related to the charged energetic

The Hidden Link Between Earth's Magnetic Field and Climate
https://doi.org/10.1016/B978-0-12-819346-4.00007-3

© 2020 Elsevier Inc. All rights reserved.

particles (mainly protons and electrons), continuously or sporadically precipitating in Earth's atmosphere. The differentiation between suggested mechanisms consists in the different mediating factors that are supposed to translate the particles' impact down to the surface.

The authors of the 'optical' mechanism suggest that atmospheric transparency to the incoming solar irradiance (Pudovkin and Raspopov, 1992) is altered by energetic particles' precipitation, which excite series of ion-molecular reactions in the middle atmosphere. More specifically, they suggest that increased amounts of NO_2 could substantially reduce the intensity of solar radiation reaching the troposphere and Earth's surface. The effect should be stronger at high latitudes, where the intensity of precipitating particles is higher. The surface temperature and pressure are expected to respond to the differential solar heating with stronger meridional gradients (Pudovkin and Veretenenko, 1992), which should be followed by changes of atmospheric circulation (Pudovkin and Babushkina, 1991). The main deficiency of this hypothesis is that it expects zonally symmetric climate changes, which should depend mainly on latitude, due to the strongest latitudinal dependence of precipitating particles. The long-term variations of the surface temperature show, however, that at multidecadal and centennial timescales the climate variations have a predominantly regional character (e.g. Shindell et al., 2003; Schneider et al., 2004; Kobashi et al., 2011).

According to the 'clouds' hypothesis, the geomagnetic influence on weather and climate is attributed to the cosmic rays influence on the amount of clouds. Two mechanisms are proposed for the impact of cosmic rays (CRs) over the clouds microphysics. The *first* one, proposed by Dickinson (1975), the secondary ionization (induced by CRs near the tropopause) favours the formation of sulphate aerosols, supposedly serving as cloud condensation nuclei. The *second* one, proposed by Tinslev (1996), suggests that the vertical electric current J_z stimulates the rate of ice nucleation. The J_z is associated with the electric potential difference between the ionosphere and the Earth's surface, continuously maintained by the global thunderstorm activity.

However, the hypothesized CR influence on the cloud cover received its greatest popularity due to the empirical study of Svensmark and Friis-Christensen (1997) and Svensmark (1998). The presented evidence for existing tight correlation between cloud cover and CR intensity, despite their not offering any deterministic mechanism by which CRs might influence the clouds' formation. However, most satellite and ground-based measurements of clouds' microphysical parameters and aerosols do not support the speculative CRs' influence on the cloud formation (Kristjánsson et al., 2008; Sloan and Wolfendale, 2008; Calogovic et al., 2010; Kulmala et al., 2010, etc.). The hypothesis has been tested in a specially designed experiment in the CERN laboratory (CLOUD), aimed at studying the ion-induced nucleation or ion-ion recombination as sources of aerosol particles. The first published results were somewhat controversial (see Duplissy et al., 2010; Kirkby et al., 2011). However, the latest publication of Dunne et al. (2016) shows undoubtedly that "variations in cosmic ray intensity does not significantly affect climate via nucleation in the present day atmosphere."

The main deficiency of the 'cloud' hypothesis (as a potential mediator of CR influence on the weather and climate) is the fact that clouds' contribution to greenhouse effect is only ~25% (Schmidt et al., 2010). At the same time, the water vapour impact on greenhouse warming varies between ~50% (in model experiments with clouds) and ~67% (in clear-sky experiments). Bearing in mind that 90% of the H_2O vapour greenhouse effect belongs to the upper tropospheric humidity (Inamdar et al., 2004), it becomes clear that in addition to the other uncertainties, the clouds' impact on Earth's radiation budget is substantially smaller than that of the water vapour.

Recently a new 'exotic' mechanism has been suggested, according to which the solubility of the CO_2 in the seawater decreases when the magnetic field weakens. Pazur and Winklhofer surmised that geomagnetic field variations modulate the carbon exchange between atmosphere and ocean, thus altering the greenhouse impact of CO_2 on the climate (Pazur and Winklhofer, 2008). Dissenting studies were published almost immediately by other authors, which stressed that such a conclusion contradicts the results from oceanic field programs, marine chemistry, palaeomagnetic reconstructions, and carbon cycle modelling (Köhler et al., 2009).

Conclusively, none of the proposed up to now mechanisms is able to explain the detected geomagnetic 'signal' in climate variability. On the other hand, there is a great deal of evidence suggesting that the long-term and spatial variations of climate could be related to the corresponding changes of the geomagnetic field (see Chapter 4 and references therein).

This fact is a good motivation to propose and explore another mechanism transferring particles' impact on the lowermost stratosphere (i.e. changes in ozone density), down to the Earth's surface.

7.2 Ozone as a mediator of geomagnetic influence on climate

Chapter 5 reveals that besides well-accepted influence of particles penetrating along the open magnetic field lines in polar regions, the geomagnetically trapped particles in Earth's radiation belts also have their imprint on the chemical composition and thermo-dynamic balance near the tropopause. The trapped particles are subject to geomagnetic lensing (in the lowest part of their trajectories, closest to the planetary surface), being focused in some regions over the globe and defocused at others. In addition, Chapter 6 shows that in certain conditions, i.e. high density of low energy electrons and dry atmosphere, an autocatalytic production of ozone could be activated in the lower stratosphere. The ionization in the Regener–Pfotzer maximum is, however, quite irregularly distributed over the globe and this irregularity is projected on the near tropopause ozone density. This chapter will describe the mechanism through which ozone irregularities created in this way are projected furthermore on the upper tropospheric humidity and surface temperature.

7.2.1 Tropopause temperature adjustment to changes in the lower stratospheric ozone density

Ozone influence on the equilibrium atmospheric temperature profile and climate has been observed for many decades (Manabe and Strickler, 1964; Manabe and Wetherald, 1967; Ramanathan et al., 1976; de Forster and Shine, 1997; Stuber et al., 2001, etc.). The efficiency of this influence depends on the thickness of the ozone layer as well as on the altitude of its maximum (Manabe and Wetherald, 1967), and is well-illustrated in Fig. 7.1. The figure compares decadal mean temperature profiles for the 1960s and 1990s, taken from ERA 40 reanalysis, at four different latitudes on the Greenwich Meridian. Note that the thinner tropical ozone layer and the higher level of its peak density have much less influence on

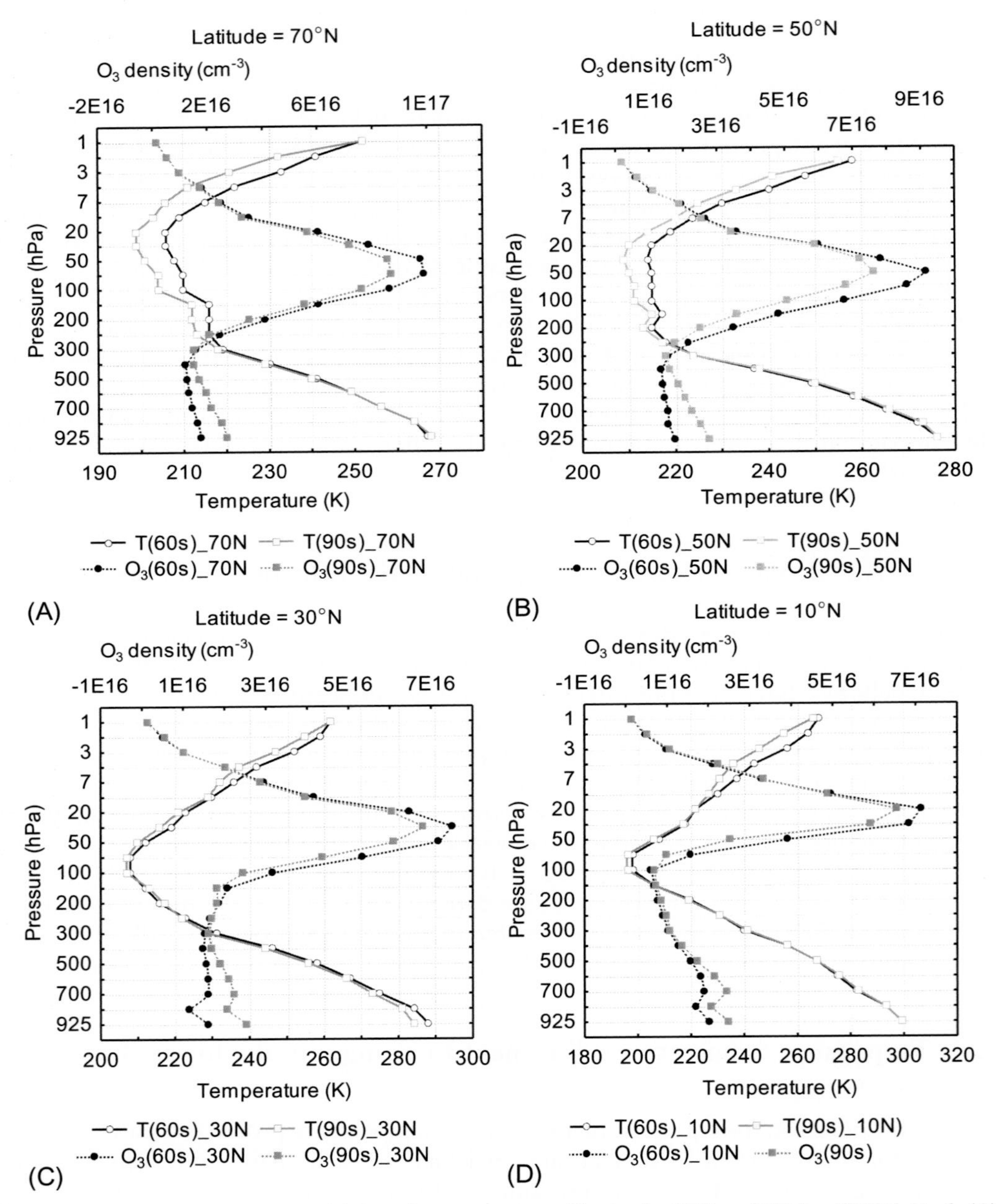

FIG. 7.1 Comparison between decadal O_3 and temperature profiles for the 1960s and 1990s; at 70°N latitude (A); at 50°N latitude (B); at 30°N latitude (C); and at 10°N latitude (D). Note that the effect of the severe ozone depletion in the 1990s on stratospheric temperature is much stronger in the extra-tropics than at tropical latitudes.

the decadal mean temperature profiles (in the upper troposphere and stratosphere) than the thicker layer with a peak placed at lower altitudes, i.e. the O_3 profiles at latitudes poleward of 30°N. Thus, Fig. 7.1 shows that the global ozone depletion in the 1990s led to a substantial decrease of stratospheric temperature at middle and high latitudes, having little impact in the tropics.

Changes of ozone density near the O_3 layer maximum mostly influences the *shortwave* radiative forcing due to the highest values of the ozone absorption cross-section in the most intensive Hartley and Huggins bands of solar UV radiation (195–350 nm). The much lower efficiency of the stratospheric ozone absorption in the infrared *atmospheric window* (where the atmosphere is practically transparent to the Earth's long-wave radiation) is due to the *tropospheric* O_3 absorption in the 9600 nm band (Ramanathan et al., 1976; Clough and Iacono, 1995). Interestingly, the greatest effect on the surface temperature is not the changes in the stratospheric O_3 maximum, but the variation of its minimum density near the tropopause O_3 (Hansen et al., 1997; de Forster and Shine, 1997; Aghedo et al., 2011).

Many authors have suggested that climate sensitivity to the variations of near tropopause ozone density is due to the ozone impact on the planetary radiation balance. However, the fact that the greatest part of greenhouse warming belongs to the upper tropospheric water vapour (Spencer and Braswell, 1997; Inamdar et al., 2004; Schmidt et al., 2010) implies that the radiative forcing of ozone is a trigger, but not the main factor influencing surface temperature. In fact, the high efficiency of ozone to absorb electromagnetic radiation directly affects the tropopause temperature (Wirth, 1993; de Forster and Tourpali, 2001; Seidel and Randel, 2006; Gauss et al., 2006; Randel et al., 2007), and correspondingly the upper tropospheric lapse rate. The latter, in turn, influences the amount of water vapour near the tropopause, but details of these relations are given in the following section.

7.2.2 Response of the upper tropospheric static stability and water vapour density to changes in tropopause temperature

The frequent mixing of the tropospheric and stratospheric air masses is a characteristic feature of the near tropopause region (known also as an extra-tropical transition layer). Different processes contribute to the air mixing in the region, e.g. turbulence, diffusion, small-scale vertical or horizontal instabilities, convection, stirring and stretching of large-scale flow. The mass exchange across the tropopause is bi-directional, i.e. from stratosphere to troposphere, and from troposphere to stratosphere, with different intensities and variabilities (Sprenger and Wernli, 2003). This section focuses on the troposphere–stratosphere mixing, and the ability of the lowermost stratospheric O_3 to control the cross-tropopause transport of the upper tropospheric H_2O vapour, e.g. through influence on the upper tropospheric static stability.

The variations of tropopause temperature affect the local vertical temperature gradient, altering the static stability of the upper tropospheric air (Young, 2003). Thus unstable air masses could propagate upward, moistening the upper troposphere and lower stratosphere (UTLS). The stably stratified air, on the other hand, prevents upward propagation of the water vapour.

According to North and Eruhimova (2009), the unsaturated air becomes unstable if the environmental lapse rate becomes greater than the dry adiabatic gradient. The situation with saturated air is more complicated, however, because its stability depends on both dry and moist (known also as wet or saturated) adiabatic temperature gradients. The latter depends on the environmental temperature, and on the water vapour mixing ratio (Eq. 7.1):

$$\Gamma_w = g \cdot \frac{1 + \dfrac{H_v \cdot r_v}{R_{sd} \cdot T}}{c_{pd} + \dfrac{H_v^2 \cdot \varepsilon \cdot r_v}{R_{sd} \cdot T^2}} \tag{7.1}$$

where $\boldsymbol{\Gamma_w}$ is the wet adiabatic lapse rate (K m^{-1}), $\boldsymbol{g}$ is the Earth's gravitational acceleration (i.e. 9.8076) (m s^{-2}), $\boldsymbol{H_v}$ is the heat of vaporization of water (2.501*10^6) (J kg^{-1}), $\boldsymbol{r_v}$ is the water vapour mixing ratio (i.e. ratio of the mass of H_2O vapour to the mass of dry air) (kg kg^{-1}), $\boldsymbol{R_{sd}}$ is the specific gas constant of dry air (287 J kg^{-1} K^{-1}), $\boldsymbol{\varepsilon} = 0.622$ is the molecular weight of water to that of dry air, $\boldsymbol{T}$ is the temperature of the saturated air (K), and $\boldsymbol{c_{pd}}$ is the specific heat capacity of dry air at constant pressure (J kg^{-1} K^{-1}).

Numerical experiments with a temperature profile of Terra Nova Bay uniformly increased and decreased by 10 K (taken from Tomasi et al., 2004), in the altitude interval 400–430 hPa (where temperature variations, initiated by the ozone changes, are expected), show that cooling of the near tropopause region reduces the wet adiabatic lapse rate, while its warming increases the $\boldsymbol{\Gamma_w}$ value (see Fig. 7.2). The H_2O vapour mixing ratio profile remained unchanged in these calculations (Kilifarska, 2012).

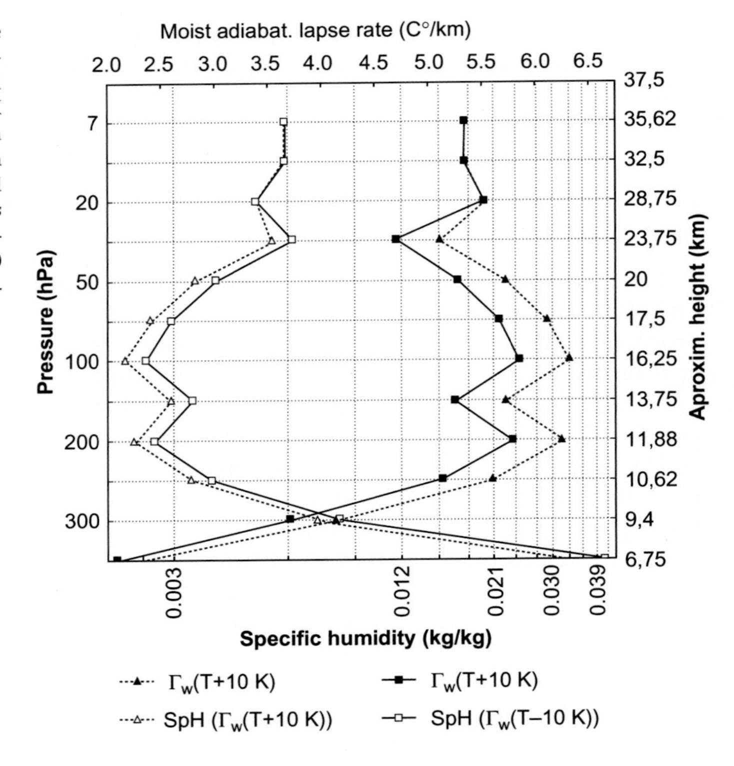

FIG. 7.2 Moist adiabatic lapse rate ($\boldsymbol{\Gamma_w}$) profiles calculated for reduced *(continuous line with full squares)* or increased *(dotted line with full triangles)* by 10 K temperature profile from Terra Nova Bay (Tomasi et al., 2004). Also given are specific humidity (***SpH***) calculated with decreased temperature *(continuous line with open squares)*, or increased temperature *(dotted line with open triangles)* and corresponding for these temperatures $\boldsymbol{\Gamma_w}$.

In addition, the sensitivity of air masses specific humidity (***SpH***) to changes in $\boldsymbol{\Gamma}_w$ (due to temperature changes) has also been examined. A couple of calculations have been made, with $\boldsymbol{\Gamma}_w(\boldsymbol{T} + 10\ \mathrm{K})$ and $\boldsymbol{\Gamma}_w(\boldsymbol{T} - 10\ \mathrm{K})$ corresponding to the temperature profile being increased and reduced by 10 K. Results are also shown in Fig. 7.2 (the open symbols). The inverse dependence between ***SpH*** and $\boldsymbol{\Gamma}_w$ is clear, meaning that cooling of the lower stratosphere moistens the UTLS layer, and vice versa.

Finally, the combined effect of the reduced temperature (by 15 K) and increased humidity (by 20%) on the moist adiabatic lapse rate is shown in Fig. 7.3. The resulted $\boldsymbol{\Gamma}_w(\boldsymbol{T} - 15\ \mathrm{K};$ ***SpH*** +20%) profile is compared with undisturbed $\boldsymbol{\Gamma}_w$, as well as with lapse rate profiles corresponding to only cooler $\boldsymbol{\Gamma}_w(\boldsymbol{T} - 15\ \mathrm{K})$ and only wetter conditions $\boldsymbol{\Gamma}_w$(***SpH*** + 20%) in the troposphere-stratosphere system. It can be clearly seen that a drop in temperature or growth of humidity causes reduction of $\boldsymbol{\Gamma}_w$, while the combined effect of both is a stronger reduction of the saturated lapse rate. The decreased wet adiabatic lapse rate *increases* the possibility for local, or transported from outside, air masses to fit the requirement for conditional instability (i.e. $\boldsymbol{\Gamma}_w < \boldsymbol{\Gamma} < \boldsymbol{\Gamma}_d$), which allows them to propagate upwards.

Reaching the lowest stratosphere, the water vapour starts to reinforce the initial effect of the ozone decrease, through additional cooling of the region (Randel et al., 2007). The further cooling could be interrupted by an enhancement of the lower stratospheric ozone, which will increase the static stability of the upper troposphere, thus reducing the water supply from the lower levels. Such a cyclic antiphase relation between near tropopause O_3 and H_2O vapour is illustrated in Fig. 7.4, based on the data from the merged ERA 40 and ERA Interim reanalyses. This comprehensive coupling between near tropopause O_3 and water vapour could be

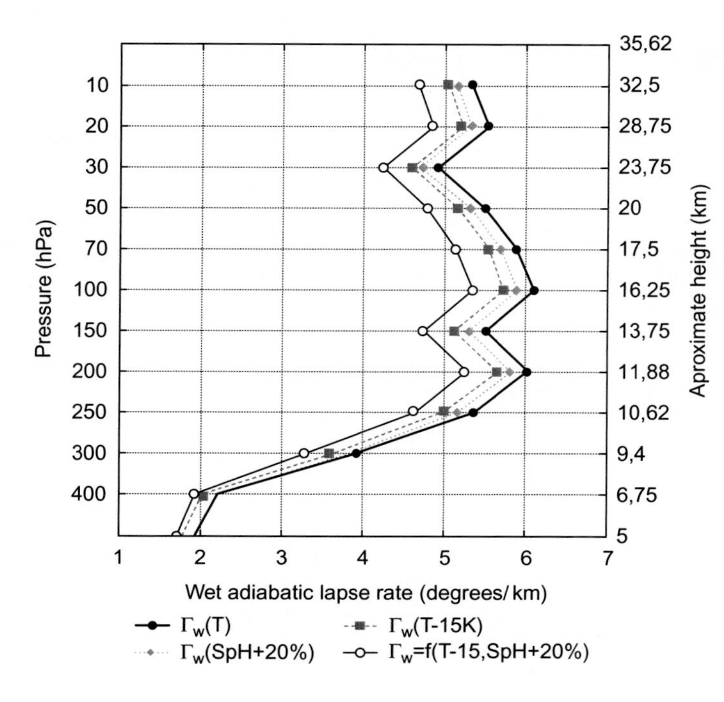

FIG. 7.3 Temperature and humidity dependence of saturated lapse rate ($\boldsymbol{\Gamma}_w$): (i) $\boldsymbol{\Gamma}_w$ corresponding to *observed* values of temperature and water vapour mixing ratio from Terra Nova Bay *(thick black line with dots)*; (ii) $\boldsymbol{\Gamma}_w$ corresponding to a reduced ***T*** profile by 15 K *(dark grey dashed line with full squares)*; (iii) $\boldsymbol{\Gamma}_w$ corresponding to an increased humidity by 20% *(pale dotted line with full diamonds)*; and (iv) $\boldsymbol{\Gamma}_w$ corresponding to cooler by 15 K and wetter by 20% near tropopause region *(thin black line with circles)*.

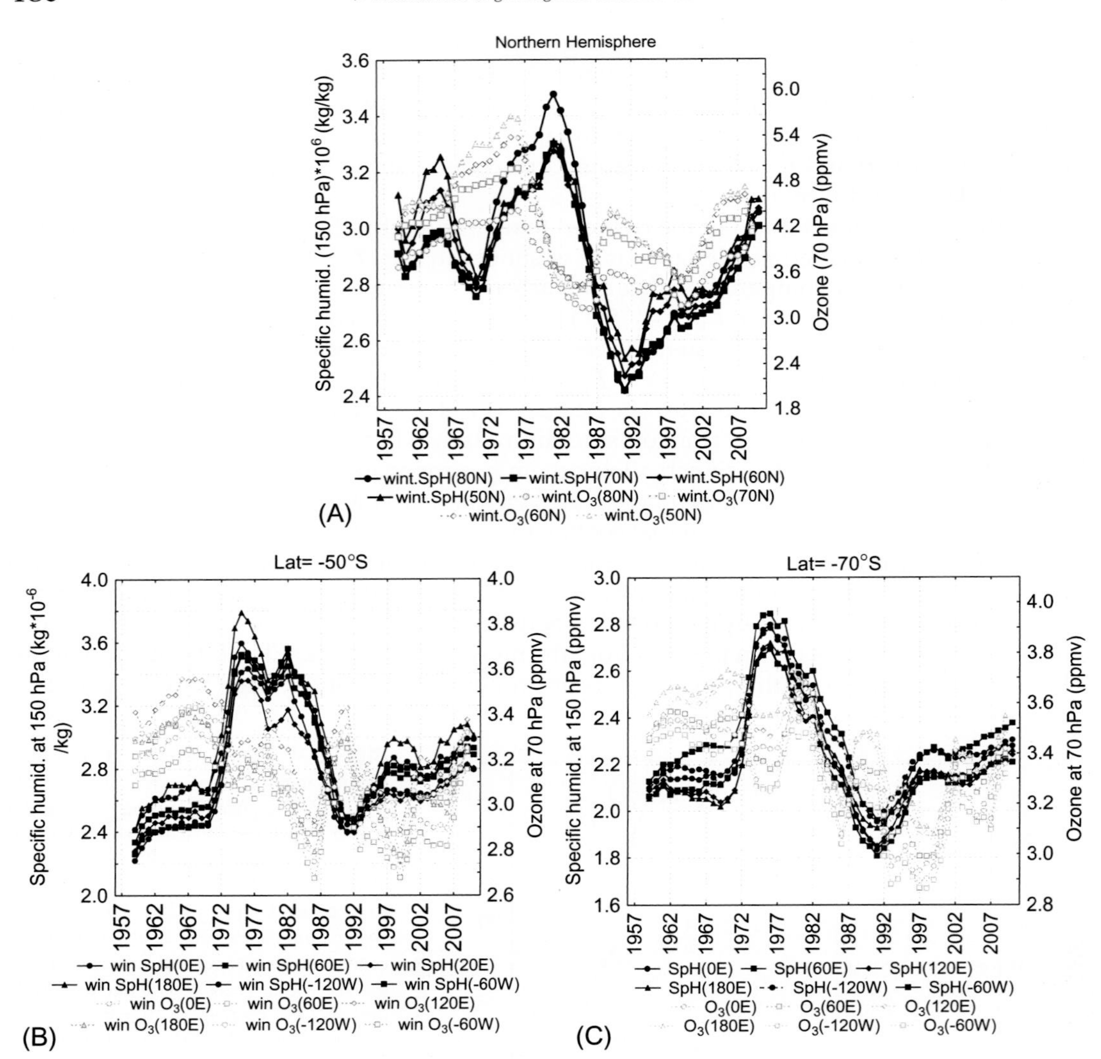

FIG. 7.4 (A) Time series of a zonally averaged winter O_3 mixing ratio at 70 hPa and specific humidity at 150 hPa, taken at 80-, 70-, 60-, and 50-degree latitudes (data are smoothed by 5-point running average); (B and C) Southern Hemisphere O_3 (70 hPa) and H_2O vapour (150 hPa) time series at 50°S and 70°S latitude and different longitudes. *Data source: ERA-40 and ERA Interim reanalyses, merged at January 2000 (long time averages of two reanalyses have been equalized before merging).*

attributed to the imprint of Schwabe (~11-year) and Hale (~22-year) solar cycles on the lower stratospheric O_3 density (Kilifarska, 2011a,b; Hood and Soukharev, 2012). This possibility for local moistening of the upper troposphere and the lowermost stratosphere could be a natural explanation for the patchy character of the humidity spatial distribution at these levels (Schaeler et al., 2009).

7.2.3 Surface temperature response to the variations in the greenhouse effect of the near tropopause ozone and water vapour

The role of water vapour in Earth's radiation balance has long been recognized (Manabe and Wetherald, 1967). The importance of the water absorption in far infrared and its significant influence on the greenhouse effect was pointed out by Clough et al. (1992) and Sinha and Allen (1994). Moreover, it has been found that the efficiency of the water vapour absorption depends strongly on altitude. Thus Sinha and Allen (1994) showed that the lower tropospheric water vapour absorbs mainly in the so-called 'window' region (8–12 μm wavelength). In contrast, the perturbation in the upper tropospheric humidity generally absorb in the water vapour pure rotational bands (i.e. wavelengths > 12 μm), followed by vibration-rotational bands (wavelengths < 8 μm). In addition, Sinha and Harries (1995) pointed out that the far infrared water vapour absorption produces the majority of its greenhouse forcing in the subarctic winter atmosphere (53%), which drops to 19% in the standard tropical atmosphere. Later, the numerical experiments of Spencer and Braswell (1997) proved the key role of the upper tropospheric water vapour in the Earth's radiation balance. These findings were confirmed later by the satellite observations of the greenhouse effect (Inamdar et al., 2004), who showed that ~90% of the greenhouse effect of the whole atmospheric water vapour is ensured by the rotational–vibrational and pure rotational absorption of the upper tropospheric water vapour in the *nonwindow* long-wave radiation band. They also reported that only 10% of the water vapour greenhouse effect belongs to the lower tropospheric humidity, due to its absorption in the *window* band (i.e. wavelengths 5–12 μm).

Sensitivity experiments with radiative transfer model performed by Randel et al. (2007) clarified additionally that besides ozone and H_2O vapour, none of the well-mixed greenhouse gases, high stratospheric clouds, or aerosols has an important radiative influence near the tropopause. Consequently, the radiation emitted from the lower troposphere (at higher *effective temperature*) is absorbed by the upper tropospheric water vapour and lower stratospheric ozone, which have much smaller *effective temperatures* near the tropopause (i.e. a lower emissivity). Due to the strong vertical gradient, O_3 at these levels has a mostly local radiative effect, warming the lowermost stratosphere down to the tropopause (Randel et al., 2007). However, the lower emissivity of the extremely cold upper tropospheric H_2O vapour, which traps the long-wave radiation inside the troposphere, has a major impact on Earth's radiation balance (e.g. Inamdar et al., 2004).

Cooling of the tropopause increases the static instability of the upper troposphere, favouring the upward propagation of the moister air across the extra-tropical tropopause (see Section 7.2.2). When reaching the lowermost stratospheric level, the water vapour starts to cool it, due to the small optical thickness of H_2O vapour at these levels (see e.g. de Forster and Shine, 2002), thus amplifying the effect of the initial tropopause cooling forced by the ozone depletion. Consequently, the variations of the lower stratospheric ozone affect simultaneously temperature and humidity at levels where the sensitivity of Earth's radiation budget to the water vapour density is highest and nonlinear (Spencer and Braswell, 1997). The latter means that even a small increase of humidity could enhance strongly the amount of the trapped long-wave radiation, i.e. the greenhouse effect.

The spatial distribution of the statistical relation between water vapour at 150 hPa and ozone at 70 hPa, with the air temperature at 2 m above the surface, is shown in Figs 7.5

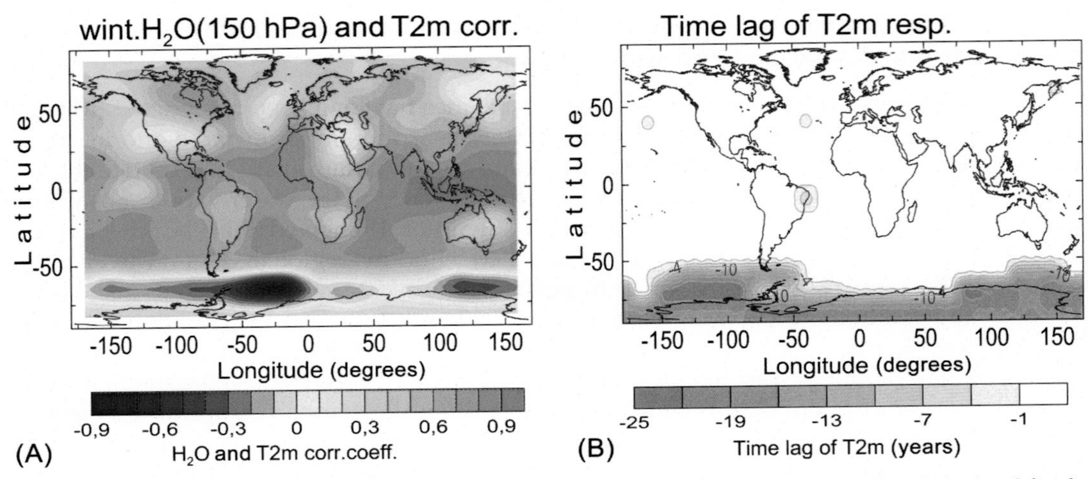

FIG. 7.5 (A) Correlation map between water vapour at 150 hPa and surface air temperature (calculated for the period 1900–2010); (B) time lag of temperature response to water vapour changes. *Data source: ERA 20th-century reanalysis.*

and 7.7. The lagged correlation maps has been created from the statistically significant at 2σ level correlation coefficients, which have been preliminarily weighted by the autocorrelation function of the forcing variable, corresponding to a given time lag. Fig. 7.5 reveals that the strongest in phase relation between H_2O vapour and temperature is found over the maritime continent (Indonesia) and tropical regions (reaching higher latitudes in the Southern Hemisphere). The single place with antiphase relation between H_2O vapour and temperature is found around the 60°S latitude. This change of the phase relation is due to the centennial drying of the whole Antarctic continent and the strongest surface temperature rise at these latitudes (see Fig. 7.6A and B). However, Antarctic temperature response to water vapour changes is significantly delayed (Fig. 7.5B).

Fig. 7.6 illustrates the difference between decadal means of the 21st and 20th centuries' first decades of the three examined variables: surface air temperature, water vapour at 150 hPa, and O_3 at 70 hPa. Note that unlike the tropical upper troposphere, the polar regions in the Southern Hemisphere are notably dried, while the temperature rise roughly marks the Antarctic convergent zone (Fig. 7.6A and B). Similarly, the strongest warming during the 20th century, visible in the Northern Hemisphere high latitudes, corresponds to negligible moistening, despite the good correspondence between the 'hot' spot in north-eastern Canada and the high vapour–temperature correlation (compare Figs 7.5 and 7.6A and B). On the other hand, the strongest moistening of the upper troposphere over Indonesia, and the high correlation coefficients between water vapour and temperature, correspond to a very weak surface warming (compare Figs 7.5 and 7.6A). These results indicate that the long-term variations of the upper tropospheric humidity are not indicative of the changes of near-surface temperature.

The relation between lower stratospheric ozone and surface air temperature is presented in Fig. 7.7. The correlation map is compared with dynamical temperature and ozone anomalies, averaged over the period 1900–2010 (Fig. 7.7A and B), and their centennial changes, i.e. the difference between the first decades of the 21st and 20th centuries (Fig. 7.7C and D). Note that the strongest positive correlation between ozone and temperature corresponds fairly well to

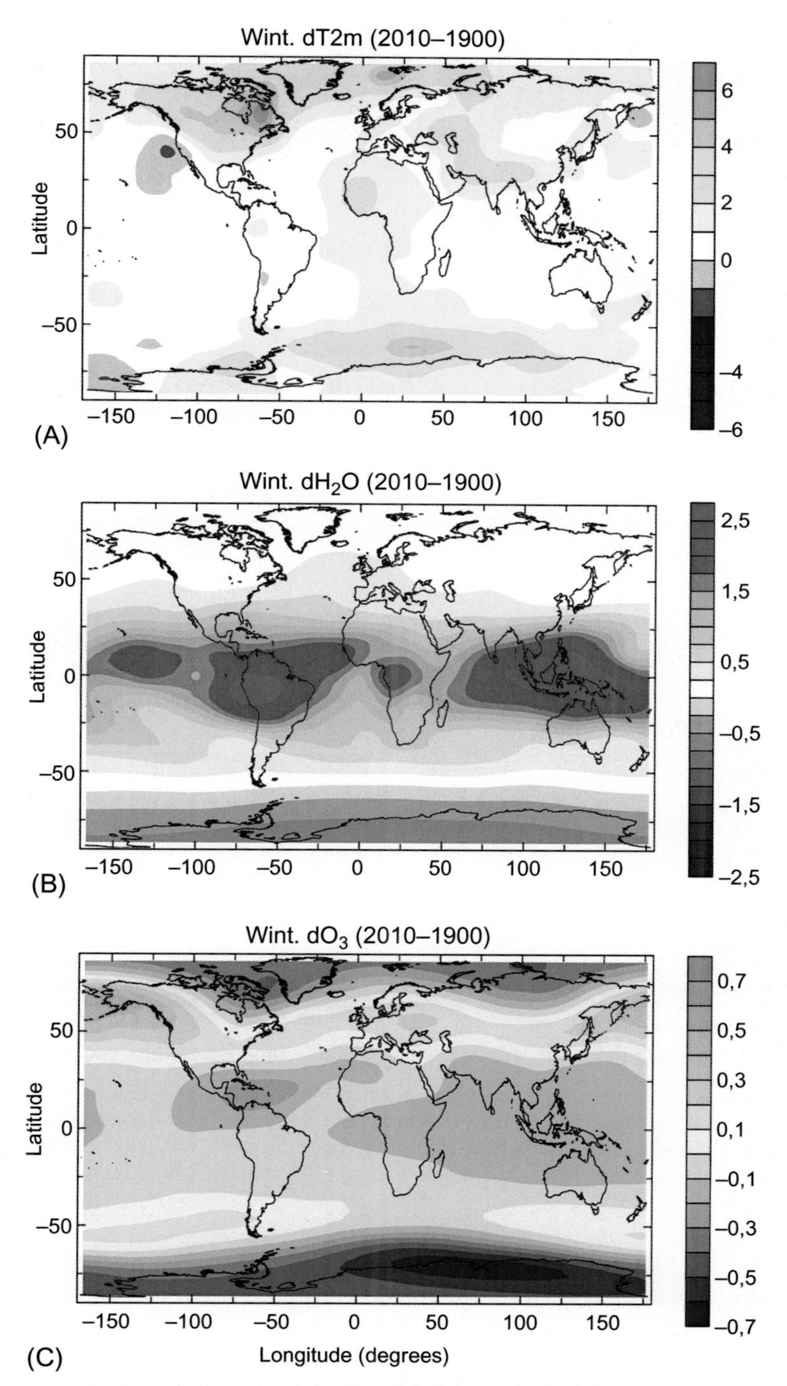

FIG. 7.6 Difference between first decades of the 21st and 20th centuries of air temperature at 2 m above the surface (A), H_2O at 150 hPa (B), and O_3 at 70 hPa (C). *Data source: ERA 20th century reanalysis.*

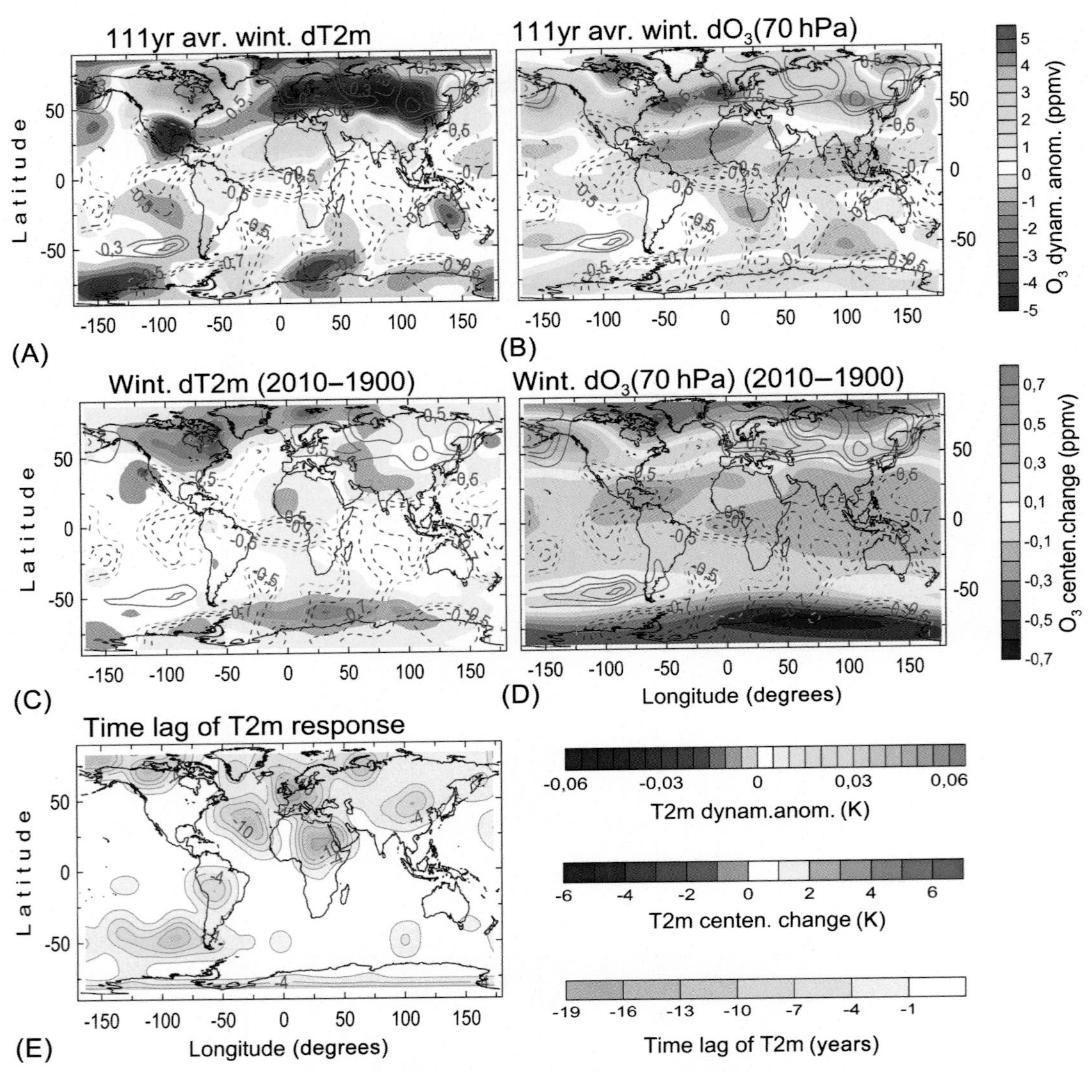

FIG. 7.7 Comparison of correlation map between ozone at 70 hPa and air temperature at 2 m above the surface (contours) with: (A and B) centennial climatology of temperature and ozone dynamical anomalies; (C and D) difference between the first decades of the 21st and 20th century, for temperature and ozone, respectively. (E) Time lag of near surface temperature response to centennial changes of ozone at 70 hPa. *Data source: ERA 20th century reanalysis.*

the regions with persistent negative temperature and positive ozone anomalies in the Northern Hemisphere extra-tropics. In addition, Fig. 7.7D shows that the coherent in-phase variations of ozone and temperature correspond to the latitudinal belts of the centennial ozone enhancement between the 20th and 21st centuries: around 50-degree latitudes in both hemispheres. This result contradicts the proposed mechanism for O_3 influence on climate (described in Section 7.2.2), according to which surface temperature cooling should be expected in periods of near tropopause O_3 enhancement. This discrepancy should be attributed to the fact that the regional ozone increase is unable to change the global positive trend of surface temperature. On shorter timescales, however, the positive ozone deviations from its centennial nonlinear trend are projected on the surface temperature by the mechanism described in Section 7.2.2. This relation is clear in Fig. 7.7, where the positive ozone anomalies (Fig. 7.7B) match fairly well the negative surface temperature anomalies (Fig. 7.7A) in both hemispheres.

Fig. 7.7D shows in addition that in the tropics, where the lower stratospheric ozone density is weakly depleted, its correlation with the surface temperature is negative. Comparison with Fig. 7.7A and C reveals that surface temperature in these regions is either unchanged or slightly raised during the 20th century, despite the strongest rise of the upper tropospheric humidity (e.g. over Amazonia and Indonesia; see Fig. 7.6). This result could be explained by the work of Sinha and Harries (1995), who have shown that the absorption of the upper tropospheric water vapour in vibrational–rotational and pure rotational bands drops from 53% in the subarctic to 19% in the tropical atmosphere. Fig. 7.7 confirms their result, illustrating that in subarctic and subantarctic regions, where the lower stratospheric ozone is markedly decreased (see Fig. 7.7D), the negative O_3–temperature correlation is indicative of significant surface warming (Fig. 7.7C). This is particularly visible in north-eastern Canada and the Southern Ocean where the ‘hot’ spots in the centennial surface temperature change Fig. 7.7C) coincide with the strongest centennial ozone depletion (Fig. 7.7D).

Finally, the imprint of the *coupled* near tropopause ozone and water vapour on the surface temperature is shown in Fig. 7.8, where the ozone–vapour correlation is overdrawn on the centennial climatology of surface temperature dynamical anomalies for winter and summer seasons. All statistically significant (at 2σ level) correlation coefficients, used for map drawing, have been preliminary weighted by the ozone autocorrelation function, corresponding to the humidity time lag in each grid point. Such a weighting allows comparison of the strength of correlations with different time lags (Kenny, 1979). It can be easily seen that antiphase covariance between ozone and water vapour in tropical regions exists year-round. However, their impact on the surface temperature is quite small (see also Fig. 7.6C). The less efficiency of the tropical ozone–humidity influence of the surface temperature should be attributed to the weaker water vapour absorption in the vibrational and rotational bands of its far infrared absorption spectrum, i.e. with wavelengths $>12\ \mu m$ or $<8\ \mu m$ (Sinha and Harries, 1995). On the other hand, the in phase coherence between ozone and water vapour, found in extratropics, matches very well the negative surface temperature dynamical anomalies (i.e. temperature deviations from its nonlinear centennial trend).

Analysis of the water vapour time lag to the ozone forcing, shown in Fig. 7.8C and D, shows that their year-round synchronous variability at tropical latitudes is without time delay. There are two other regions where the near tropopause vapour response to the ozone forcing is without, or with minimal time delay. These are Antarctica and Northern Hemisphere extratropics—particularly the North Pacific Ocean and Eastern Europe up to the Ural Mountains. The water response over the eastern North America, North Atlantic, and Western Europe, as well as Eastern Asia, is significantly delayed. A reference to Fig. 7.6B reveals that these latter regions are characterized by systematically positive ozone anomalies during the 20th century. This is an indication that the ozone–water imprint on the surface temperature could be explained by the mechanism described in Section 7.2.2. The mechanism is more effective in winter, due to the higher ozone density in this season.

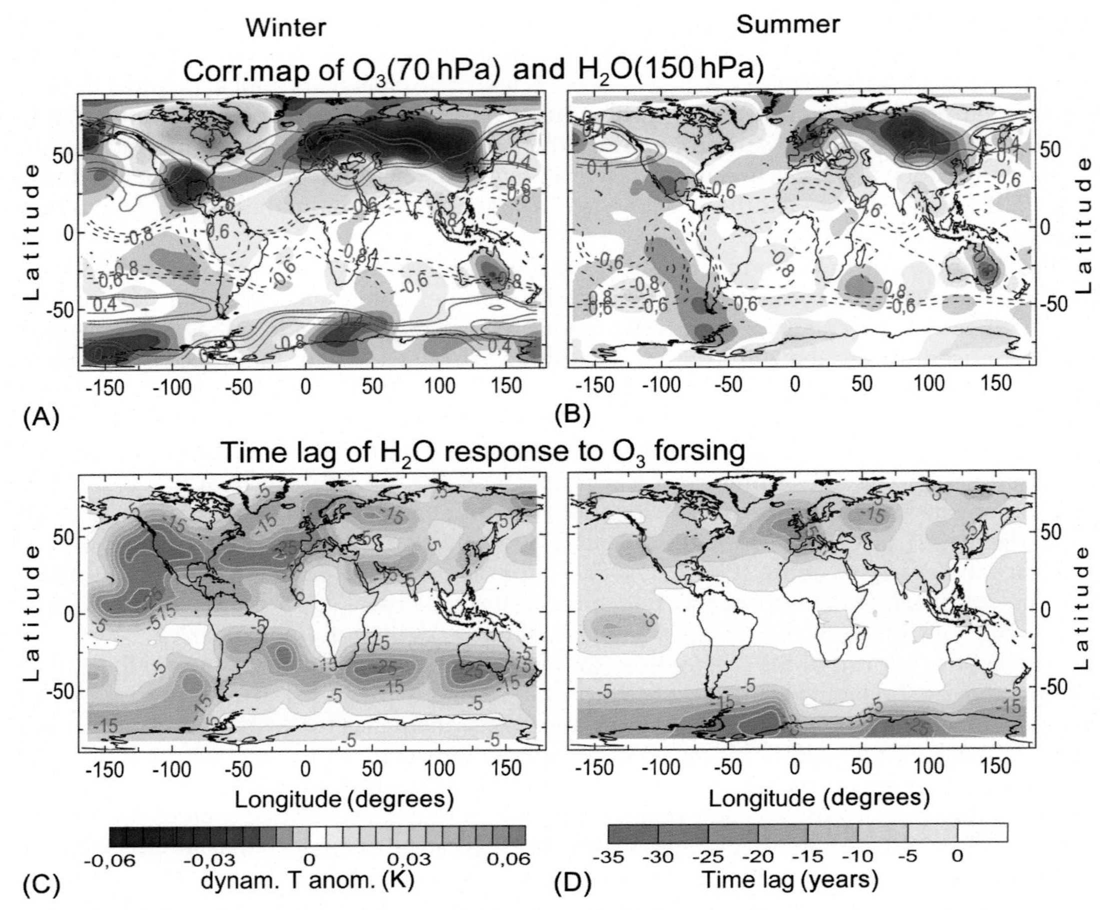

FIG. 7.8 Lag corrected correlation maps between ozone at 70 hPa and water vapour at 150 hPa, calculated over the period 1900–2010, for winter (A) and summer (B). Time lag of humidity response to ozone changes is shown in (C) and (D) panels. *Data source: ERA 20th century reanalysis.*

7.2.4 GCR projection on the ozone–water vapour coupling in the upper troposphere–lower stratosphere

Chapter 6 elucidated the influence of highly energetic galactic cosmic rays on the lower stratospheric ozone balance. It showed that depending on the level of particles absorption in the atmosphere, and on the humidity of the ambient air, they are able to activate production or destruction of the near tropopause ozone. Consequently, the spatial and temporal variations of GCR intensity are projected on the near tropopause ozone and, as shown in Section 7.2.2, on the near tropopause humidity. An estimation of the persistence and spatial distribution of such relations is shown in Fig. 7.9, which compares the correlation maps of:

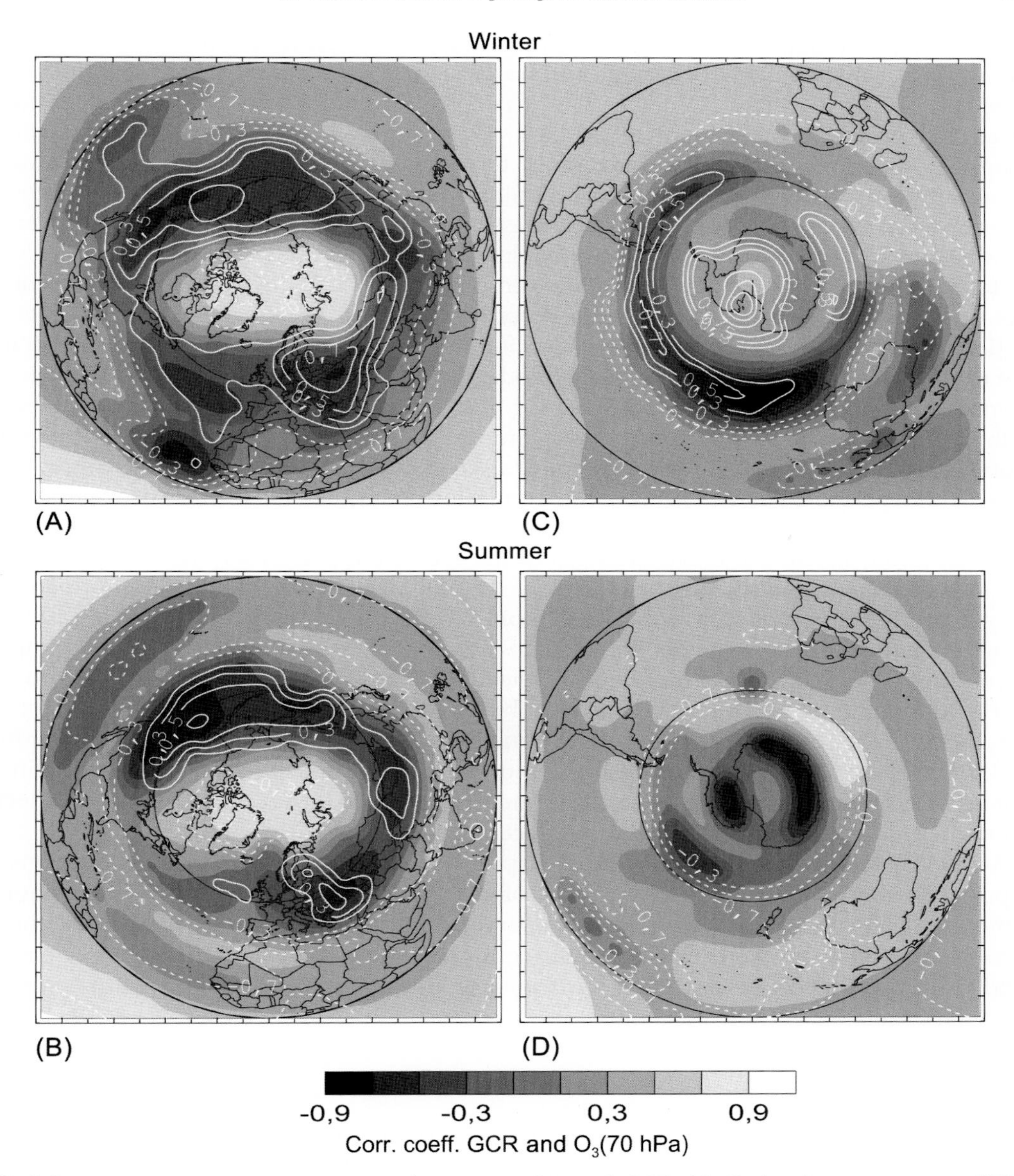

FIG. 7.9 Comparison of correlation maps of ozone at 70 hPa with GCRs *(shading)* and water vapour at 150 hPa (contours), for winter (A) and (C), and summer (B) and (D) panels. *Data source: ERA 20th century reanalysis, and historical reconstruction of GCR intensity by Usoskin, I.G., et al., 2008. Cosmic Ray Intensity Reconstruction. IGBP PAGES/ World Data Center for Paleoclimatology, Data Contribution Series # 2008-013. NOAA/NCDC Paleoclimatology Program, Boulder, CO, USA.*

(i) GCR and ozone mixing ratio at 70 hPa; and (ii) ozone and humidity near the tropopause. Before the maps were drawn, all correlation coefficients were weighted by autocorrelation functions of the forcing variables (i.e. GCRs and ozone), as described by Kenny (1979).

Note that latitudinal band of antiphase correlation between GCRs and ozone (dark shading), and *in-phase* correlation between ozone and water vapour (continuous contours), coincide impressively. In the Northern Hemisphere, it is persistent round the year, although slightly reduced in summer. In the winter Southern Hemisphere, it is narrower and practically disappears in summer (Fig. 7.9D). In summer, the galactic cosmic rays' influence on the ozone is concentrated mainly over the Antarctica, followed by disappearance of the in phase covariance between ozone and water vapour over the Antarctic convergent zone. As noted earlier, the extra-tropics around 50 degrees are the only places over the world that have a rise in the lower stratospheric ozone density during the 20th century (e.g. see Fig. 7.6C). This enhancement has been attributed to the impact of the charged particles trapped in Van Allen radiation belts (see Section 6.4.2) and is a reasonable explanation for the regional specificity of climate variability. The results presented in Fig. 7.9 are a good indication that ozone–humidity variations, which are projected down to Earth's surface (see Fig. 7.8), are actually related to GCR variability.

In *conclusion*, this section introduces a new mechanism through which the near tropopause ozone could influence the surface temperature. According to this mechanism, changes of the lower stratospheric O_3 density trigger corresponding changes in the tropopause temperature and respectively in the near tropopause humidity, by the virtue of alteration of the wet lapse rate. The latter, in turn, alter the strength of the greenhouse warming inserted by the near tropopause water vapour. Statistical analysis of the efficiency of O_3–humidity influence on the surface temperature shows that it is more effective at extratropical to polar latitudes (in agreement with the previous studies of Sinha and Harries, 1995), where their imprint on the surface temperature is more easily detectable (Fig. 7.8). The ozone–water vapour impact is stronger in winter, perhaps due to the higher ozone density in extra-tropics. In the tropics, the efficiency of the near tropopause ozone–humidity impact on the surface temperature is weakened, due to the reduced absorption of the upper tropospheric water vapour in the vibrational-rotational and pure rotational bands (Sinha and Harries, 1995), which is reflected in its smaller impact on the greenhouse effect (see Fig. 7.8).

According to the analyses provided in Chapter 6, the near tropopause ozone variations are imposed by the spatial-temporal variations of galactic cosmic rays (GCRs). Juxtaposition of correlation maps between GCRs and ozone, and ozone and water vapour, illustrates the existence of tight connections between them in certain regions of the world (see Fig. 7.9). This is a good indication that ozone–humidity variations, which are projected down to Earth's surface, are actually related to GCR variability.

7.3 Model simulation of the proposed mechanism for lower stratospheric ozone influence on climate

The slowdown of global warming of surface temperature, also known as hiatus, is characterized by a cooling trend of Eurasia (Cohen et al., 2012; Li et al., 2015; Kug et al., 2015; Sun et al., 2018). Fig. 7.10 compares decadal variations of surface temperature in two consecutive solar cycles: the decades 2002–12 and 1991–2001. The negative winter temperature anomalies are easily noticeable. For unknown reasons, the cooling is observed dominantly in winter, which has motivated some authors to relate it to Arctic warming (Kug et al., 2015) or a stratospheric polar vortex (Garfinkel et al., 2017).

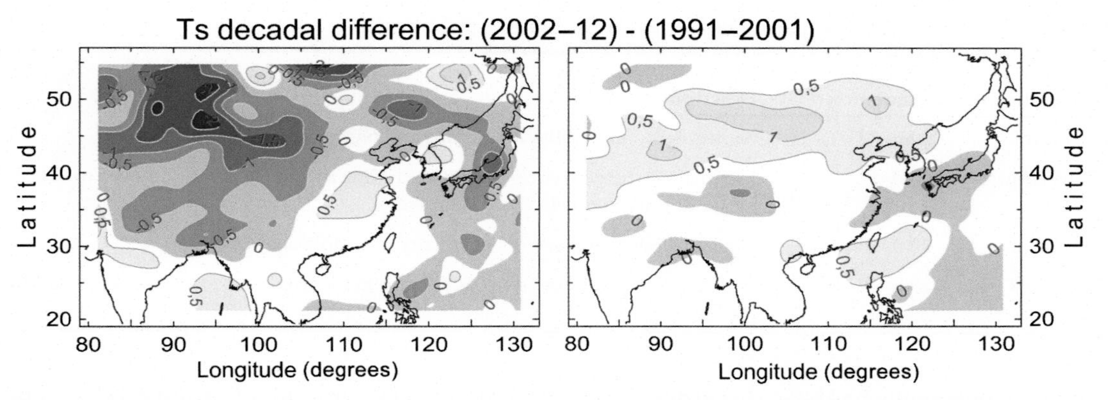

FIG. 7.10 Decadal variability of surface temperature over China domain for winter *(left)* and summer seasons *(right)*, derived from ERA Interim monthly mean data. *From Kilifarska, N., Wang, T., Ganev, K., Xie, M., Zhuang, B., Li, S., 2018. Decadal cooling of East Asia—the role of aerosols and near tropopause ozone forcing. C. R. Acad. Bulg. Sci. 71 (6), 937–944.*

Our initial guess for this surface cooling consists of two hypotheses: an enhanced aerosol concentration and an increased ozone density in the lowermost stratosphere. The cooling effect of aerosols (consisting of scattering, reflection, or absorption of solar radiation, which reduce the heat received by the planet) is well-known and will not be described here. The mechanism of near tropopause O_3 influence on the surface temperature is described in Section 7.2. To test our hypotheses, we have used the regional climate model RegCM, version 4.6.0. This is a limited area model, developed for long-term regional climate simulations (Giorgi et al., 2012). RegCM is a compressible, sigma-p vertical coordinate model, working in hydrostatic or nonhydrostatic mode. Radiation transfer of the model takes into account short-wave and infrared parts of the spectrum, including the absorption, scattering, and emission from the atmospheric gases (i.e. H_2O, CO_2, O_3, CH_4, N_2O, and CFCs). The calculation of scattering and absorption of solar radiation by aerosols is based on the aerosol optical properties. The optical characteristics of clouds are determined on the basis of the amount of their liquid water, prognostically calculated by the model, while the clouds' fractional cover is calculated diagnostically. In addition, the models' radiative scheme diagnostically calculates a fraction of cloud ice as a function of temperature. In the infrared spectrum, the cloud emissivity is calculated as a function of cloud liquid/ice water path and cloud infrared absorption cross-sections, depending on effective radii for the liquid and ice phases.

The model experiments have been conducted by the use of the MIT-Emanuel convection scheme (Emanuel, 1991; Emanuel and Zivkovic-Rothman, 1999). The land surface processes have been described via the Biosphere-Atmosphere Transfer Scheme (BATS) of Dickinson et al. (1993). The sea surface temperature has been prescribed every 6 h from the temporally interpolated weekly sea surface NOAA products.

We have conducted four numerical experiments, summarized in Table 7.1. The control run, the experiments with increased O_3 just above the tropopause, and those with reduced H_2O below the tropopause cover the period 2001–12. The experiment with active aerosol chemistry covers the period 2001–10, due to the limitations in the available emissions data. The first year

TABLE 7.1 Numerical experiments carried out with the RegCM model.

Experiments	Period of simulation	Description
Control	2001–12	No aerosols, no additional O_3; fully interactive H_2O vapour
abvTropO_3	2001–12	Increased O_3 between tropopause (excluding it) and 50 hPa, by amount produced by GCR; fully interactive H_2O vapour
Aerosols	2001–10	Active tropospheric aerosol chemistry; no additional O_3
redH_2O	2001–12	Reduced water vapour from 350 hPa up to the tropopause; standard model's O_3 profile

of all experiments has been discarded as a spin-up period. The amount of ozone produced by GCRs is calculated following parameterization of Kilifarska (2013, 2017), while the spatial distribution of ionization in the Regener–Pfotzer maximum has been compiled from the works of Usoskin et al. (2004), Velinov et al. (2005), and Bazilevskaya et al. (2008).

7.3.1 Impact of aerosol chemistry and near tropopause O_3 in surface cooling of China

The aerosols optical thickness over China is quite substantial (Luo et al., 2014). Therefore, our first suggestion for the explanation of the decadal cooling of the domain has been related to the enhanced aerosol loading in the atmosphere. For this reason, all aerosol sources (i.e. dust, sea salt, organic and black carbon, together with SO_2 and SO_4) have been activated in the RegCM-4.6.0 model, taking into account their emissions, transport, and removal from the atmosphere. The deviation of surface temperature, calculated by the run with active aerosol chemistry, from it values derived by the *control* run is shown in Fig. 7.11A and B, for winter and summer seasons. The substantial cooling of the surface, especially strong during the winter, can be easily seen in the figure. The annually average effect during the examined 2002–10 period is in the order of −3.0 to −3.5 K.

The other possible reason for the observed surface temperature cooling could be a higher O_3 density just above the tropopause, produced by the more intensive GCRs in the period 2002–12 (see Section 6.1). The unusually low 24th solar cycle, and correspondingly weaker heliomagnetic field, allows more particles (with galactic and extragalactic origin) to penetrate the heliosphere (e.g. Jokipii, 1989). As a consequence, more charged particles precipitate in Earth's atmosphere, producing more ozone just above the tropopause (see Chapter 6). The tropopause temperature readjusts vastly to the higher ozone density (e.g. de Forster and Tourpali, 2001; Seidel and Randel, 2006) and in accordance with our hypothesis, the warmer tropopause will stabilize the upper troposphere, making it drier. Consequently, the impact of the vibrational–rotational and pure rotational water vapour bands (i.e. <8 μm and >12 μm) in the greenhouse effect will be reduced, followed by a surface cooling.

In the *abvTropO*$_3$ experiment, we have added the estimated amount of O_3, produced by GCRs near the Regener–Pfotzer maximum, to the regular ozone profile calculated by the model at all levels between tropopause (excluding it) and 50 hPa (the top of the model). It

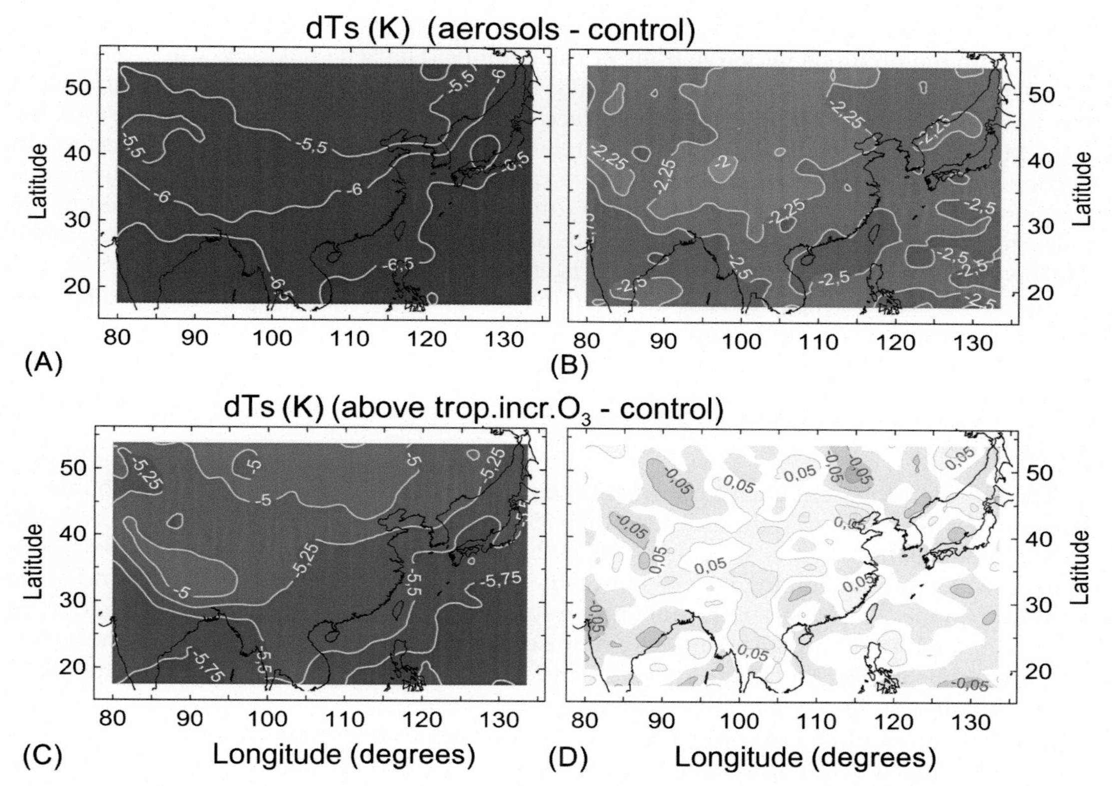

FIG. 7.11 Surface temperature difference between: (A and B) run with *active aerosol chemistry* and *Control* runs; (C and D) run with *increased* O_3 above the tropopause and *Control* runs. The *left-hand column* presents response of the *winter* surface temperature (Ts), while the *right-hand column* presents that of the *summer* Ts. *From Kilifarska, N., Wang, T., Ganev, K., Xie, M., Zhuang, B., Li, S., 2018. Decadal cooling of East Asia—the role of aerosols and near tropopause ozone forcing. C. R. Acad. Bulg. Sci. 71 (6), 937–944.*

is worth stressing that ozone produced by the low energy electrons near the Regener–Pfotzer maximum is nonuniformly distributed over the globe, due to the corresponding irregularities in the lower atmospheric ionization (see Section 6.4).

The surface temperature response to the ozone forcing is shown in Fig. 7.11C and D, which presents the temperature difference between *abvTropO*$_3$ and *Control* runs. The surface cooling (especially strong in winter) is clear. This is arguably due to a reduction of the solar radiation reaching the surface, due to its absorption by the increased ozone density in the lowermost stratosphere. However, it is important to recall that the greatest part of solar UV radiation (in the most intensive radiation band 0.195–0.35 μm) is absorbed in the ozone layer maximum. At latitudes of China, the maximum ozone layer is substantially higher than 50 hPa. Therefore, the ozone density between the tropopause and 50 hPa is too low to have a noticeable impact on the shortwave radiative forcing of the planetary surface. Its *long-wave* radiative forcing is also small, due to the fact that within the atmospheric *window* the infrared radiation is captured mainly by the tropospheric ozone, in a 9.6 μm absorption band (Clough and Iacono, 1995).

The RegCM response to the increased density of the lowermost stratospheric O_3 is inconsistent with the previous radiative transfer calculations (Hansen et al., 1997), as well as with the chemistry–climate models experiments (Gauss et al., 2006), predicting a *positive* impact of the lower stratospheric ozone on the surface temperature. The opposite response of the RegCM model suggests that it should be related to the lapse rate feedback (described in Section 7.2), which overwhelms the positive radiative forcing of ozone. According to this mechanism, the surface cooling obtained by the RegCM model should be attributed to the warmed tropopause, increased wet lapse rate, and reduced near tropopause humidity, when the lowermost stratospheric ozone density is enhanced.

Therefore, the next experiment is intended to simulate the climate response to variations in the upper tropospheric H_2O vapour.

7.3.2 Impact of the upper tropospheric water vapour on surface temperature variability

The fourth of the numerical experiments conducted investigates the effect of a double reduction of water vapour in the altitude range 350 hPa to tropopause. The difference between *redH₂O* and *Control* runs is shown in Fig. 7.12. The seasonal character of the upper tropospheric water vapour forcing is very noticeable, with dominant cooling in winter and some warming (in the central part of the domain) in summer. This result indicates that the radiative scheme of the RegCM model adequately represents the impact of the upper tropospheric water vapour absorption (in the far infrared band of Earth's radiation spectrum) into the planetary radiation budget, especially in winter.

Comparison of model experiments with reduced upper tropospheric water vapour (Fig. 7.12) and reduced lower stratospheric ozone (Fig. 7.11C and D) shows a striking similarity between them, although the amplitude of the response is different. Such a similarity is an indirect confirmation of our hypothesis that the lower stratospheric ozone variability acts as a modulator of the near tropopause humidity.

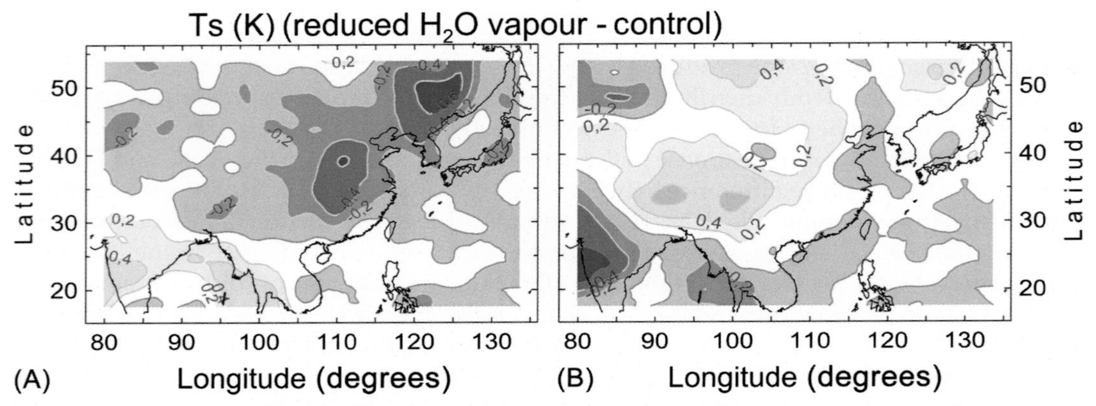

FIG. 7.12 Surface temperature response to a reduced water vapour between 350 hPa and tropopause level shown for winter *(left)* and summer *(right)*. *From Kilifarska, N., Wang, T., Ganev, K., Xie, M., Zhuang, B., Li, S., 2018. Decadal cooling of East Asia—the role of aerosols and near tropopause ozone forcing. C. R. Acad. Bulg. Sci. 71 (6), 937–944.*

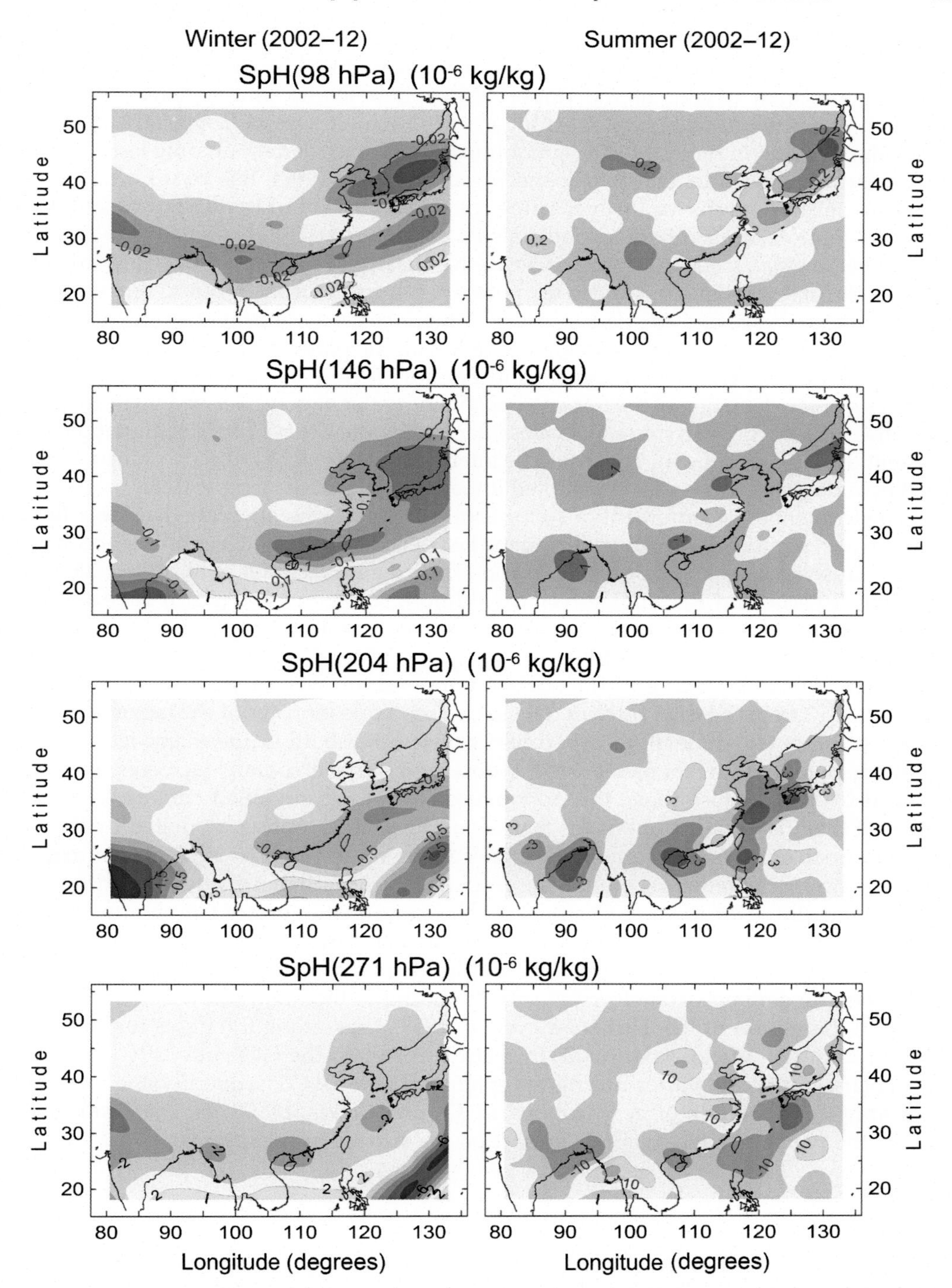

FIG. 7.13 Water vapour response to an increased O_3 density just above the tropopause, i.e. its deviations from the control experiment; the *left-hand column* is for winter and the *right-hand one* for summer. *From Kilifarska, N., Wang, T., Ganev, K., Xie, M., Zhuang, B., Li, S., 2018. Decadal cooling of East Asia—the role of aerosols and near tropopause ozone forcing. C. R. Acad. Bulg. Sci. 71 (6), 937–944.*

Additional analysis of the water vapour difference between *enhanced* near tropopause ozone (*abvTropO*$_3$) and *Control* runs at the four highest model levels are shown in Fig. 7.13. Note the persistent drying of the upper troposphere-lower stratosphere (UTLS) in the *abvTropO*$_3$ experiment, being much stronger in the winter season (especially at the upper two levels). Bearing in mind the strong nonlinear sensitivity of the outgoing long-wave radiation to the water vapour density (Spencer and Braswell, 1997), this result offers excellent support to our hypothesis for the mediating role of water vapour in the experiment with enhanced lowermost stratospheric ozone density.

Moreover, the *abvTropO*$_3$ experiment reveals that the negative feedback of an increased upper tropospheric lapse rate on the surface temperature is a result of the reduced amount of water vapour at these levels, and to a lesser extent of the increased atmospheric radiation from the warmer tropopause. This result corresponds well with the previous studies of Sinha and Allen (1994), Sinha and Harries (1995), Spencer and Braswell (1997), and Inamdar et al. (2004).

Analysis of the deviation of the upper tropospheric water vapour, derived by the experiment with active aerosol chemistry, from its values derived by the *Control* run, does not indicate such persistent drying (not shown). This could be interpreted as another indirect confirmation of the hypothesized ozone–water vapour coupled impact on the surface temperature.

The instrumental measurements also confirm our hypothesis, showing that the *reduction* of the lowermost stratospheric ozone during the last decades of the 20th century is accompanied by tropopause cooling and *moistening* of the UTLS region (Randel et al., 2004, 2009; Soden et al., 2005; Roscoe and Rosenlof, 2011). In addition, Solomon et al. (2010) have shown that partial ozone recovery at the beginning of the 21st century is accompanied by about 10% decrease of the water vapour at 82 hPa.

In resume: Decadal cooling of the surface temperature over China, obtained during the period 2002–12 from the ERA Interim data set, raises a question about the factors responsible for such cooling. Our modelling experiments reveal that both of the major factors initially suggested, potentially responsible for the observed surface cooling (aerosols loading and enhanced O_3 density just above the tropopause), affect the modelled climate. The effect of the two of them is stronger in winter than in summer. The cooling modelled by RegCM is, however, much exaggerated compared to the cooling reproduced by ERA interm reanalysis. Especially strong is the modelled climate response to the enhanced aerosol loading.

The fingerprint of the lowermost stratospheric ozone on the surface temperature resembles better the decadal variations found in ERA Interim data, particularly the summer warming in the centre of the domain. The nonrealistic amplitude of the surface response could be attributed to the exaggerated ozone forcing, corresponding to the highest level of GCR intensity, reached in the minim of the 23rd solar cycle in 2009. This strong forcing is imposed persistently through the whole experiment. In reality, however, the GCR intensity is modulated by the solar activity and varies during the examined period. Consequently, the introduction of a temporal modulation of near tropopause O_3 production, by GCRs, could substantially improve the paired O_3–H_2O vapour impact on the surface temperature.

References

Aghedo, A.M., Bowman, K.W., Worden, H.M., Kulawik, S.S., Shindell, D.T., Lamarque, J.F., Faluvegi, G., Parrington, M., Jones, D.B.A., Rast, S., 2011. The vertical distribution of ozone instantaneous radiative forcing from satellite and chemistry climate models. J. Geophys. Res. Atmos. 116. https://doi.org/10.1029/2010JD014243.

Bazilevskaya, G.A., et al., 2008. Cosmic ray induced ion production in the atmosphere. Space Sci. Rev. 137, 149–173. https://doi.org/10.1007/s11214-008-9339-y.

Calogovic, J., Albert, C., Arnold, F., Beer, J., Desorgher, L., Flueckiger, E.O., 2010. Sudden cosmic ray decreases: no change of global cloud cover. Geophys. Res. Lett. 37. https://doi.org/10.1029/2009GL041327.

Clough, S.A., Iacono, M.J., 1995. Line-by-line calculation of atmospheric fluxes and cooling rates: 2. Application to carbon dioxide, ozone, methane, nitrous oxide and the halocarbons. J. Geophys. Res. Atmos. 100, 16519–16535. https://doi.org/10.1029/95JD01386.

Clough, S.A., Iacono, M.J., Moncet, J.-L., 1992. Line-by-line calculations of atmospheric fluxes and cooling rates: application to water vapor. J. Geophys. Res. Atmos. 97, 15761–15785. https://doi.org/10.1029/92JD01419.

Cohen, J.L., Furtado, J.C., Barlow, M., Alexeev, V.A., Cherry, J.E., 2012. Asymmetric seasonal temperature trends. Geophys. Res. Lett. 39. https://doi.org/10.1029/2011GL050582.

de Forster, P.M.F., Shine, K.P., 1997. Radiative forcing and temperature trends from stratospheric ozone changes. J. Geophys. Res. Atmos. 102, 10841–10855. https://doi.org/10.1029/96JD03510.

de Forster, P.M., Shine, K.P., 2002. Assessing the climate impact of trends in stratospheric water vapour. Geophys. Res. Lett. 29, 1086. https://doi.org/10.1029/2001GL013909.

de Forster, P.M.F., Tourpali, K., 2001. Effect of tropopause height changes on the calculation of ozone trends and their radiative forcing. J. Geophys. Res. Atmos. 106, 12241–12251. https://doi.org/10.1029/2000JD900813.

Dickinson, R.E., 1975. Solar variability and the lower atmosphere. Bull. Am. Meteorol. Soc. 56, 1240–1248. https://doi.org/10.1175/1520-0477(1975)056<1240:SVATLA>2.0.CO;2.

Dickinson, R.E., Henderson-Sellers, A., Kennedy, P., 1993. Biosphere–atmosphere transfer scheme (BATS) version 1e as coupled to the NCAR community climate model. Tech. Rep., National Centre for Atmospheric Research Tech. Note, NCAR/TN-387+STR. NCAR, Boulder, CO.

Dunne, E.M., Gordon, H., Kürten, A., Almeida, J., Duplissy, J., Williamson, C., Ortega, I.K., Pringle, K.J., Adamov, A., Baltensperger, U., Barmet, P., Benduhn, F., Bianchi, F., Breitenlechner, M., Clarke, A., Curtius, J., Dommen, J., Donahue, N.M., Ehrhart, S., Flagan, R.C., Franchin, A., Guida, R., Hakala, J., Hansel, A., Heinritzi, M., Jokinen, T., Kangasluoma, J., Kirkby, J., Kulmala, M., Kupc, A., Lawler, M.J., Lehtipalo, K., Makhmutov, V., Mann, G., Mathot, S., Merikanto, J., Miettinen, P., Nenes, A., Onnela, A., Rap, A., Reddington, C.L.S., Riccobono, F., Richards, N.A.D., Rissanen, M.P., Rondo, L., Sarnela, N., Schobesberger, S., Sengupta, K., Simon, M., Sipilä, M., Smith, J.N., Stozkhov, Y., Tomé, A., Tröstl, J., Wagner, P.E., Wimmer, D., Winkler, P.M., Worsnop, D.R., Carslaw, K.S., 2016. Global atmospheric particle formation from CERN CLOUD measurements. Science 354, 1119–1124. https://doi.org/10.1126/science.aaf2649.

Duplissy, J., Enghoff, M.B., Aplin, K.L., Arnold, F., Aufmhoff, H., Avngaard, M., Baltensperger, U., Bondo, T., Bingham, R., Carslaw, K., Curtius, J., David, A., Fastrup, B., Gagné, S., Hahn, F., Harrison, R.G., Kellett, B., Kirkby, J., Kulmala, M., Laakso, L., Laaksonen, A., Lillestol, E., Lockwood, M., Mäkelä, J., Makhmutov, V., Marsh, N.D., Nieminen, T., Onnela, A., Pedersen, E., Pedersen, J.O.P., Polny, J., Reichl, U., Seinfeld, J.H., Sipilä, M., Stozhkov, Y., Stratmann, F., Svensmark, H., Svensmark, J., Veenhof, R., Verheggen, B., Viisanen, Y., Wagner, P.E., Wehrle, G., Weingartner, E., Wex, H., Wilhelmsson, M., Winkler, P.M., 2010. Results from the CERN pilot CLOUD experiment. Atmos. Chem. Phys. 10, 1635–1647. https://doi.org/10.5194/acp-10-1635-2010.

Emanuel, K.A., 1991. A scheme for representing cumulus convection in large-scale models. J. Atmos. Sci. 48, 2313–2335.

Emanuel, K.A., Zivkovic-Rothman, M., 1999. Development and evaluation of a convection scheme for use in climate models. J. Atmos. Sci. 56, 1766–1782.

Garfinkel, C.I., Son, S.-W., Song, K., Aquila, V., Oman, L.D., 2017. Stratospheric variability contributed to and sustained the recent hiatus in Eurasian winter warming. Geophys. Res. Lett. 44, 374–382. https://doi.org/10.1002/2016GL072035.

Gauss, M., Myhre, G., Isaksen, I.S.A., Grewe, V., Pitari, G., Wild, O., Collins, W.J., Dentener, F.J., Ellingsen, K., Gohar, L.K., Hauglustaine, D.A., Iachetti, D., Lamarque, F., Mancini, E., Mickley, L.J., Prather, M.J., Pyle, J.A., Sanderson, M.G., Shine, K.P., Stevenson, D.S., Sudo, K., Szopa, S., Zeng, G., 2006. Radiative forcing since preindustrial times due to ozone change in the troposphere and the lower stratosphere. Atmos. Chem. Phys. 6, 575–599. https://doi.org/10.5194/acp-6-575-2006.

Giorgi, F., Coppola, E., Solmon, F., Mariotti, L., Sylla, M.B., Bi, X., Elguindi, N., Diro, G.T., Nair, V., Giuliani, G., Turuncoglu, U.U., Cozzini, S., Güttler, I., O'Brien, T.A., Tawfik, A.B., Shalaby, A., Zakey, A.S., Steiner, A.L., Stordal, F., Sloan, L.C., Brankovic, C., 2012. RegCM4: model description and preliminary tests over multiple CORDEX domains. Clim. Res. 52, 7–29. https://doi.org/10.3354/cr01018.

Hansen, J., Sato, M., Ruedy, R., 1997. Radiative forcing and climate response. J. Geophys. Res. Atmos. 102, 6831–6864. https://doi.org/10.1029/96JD03436.

Hood, L.L., Soukharev, B.E., 2012. The lower-stratospheric response to 11-yr solar forcing: coupling to the troposphere–ocean response. J. Atmos. Sci. 69, 1841–1864. https://doi.org/10.1175/JAS-D-11-086.1.

Inamdar, A.K., Ramanathan, V., Loeb, N.G., 2004. Satellite observations of the water vapor greenhouse effect and column longwave cooling rates: relative roles of the continuum and vibration-rotation to pure rotation bands. J. Geophys. Res. Atmos. 109. https://doi.org/10.1029/2003JD003980.

Jokipii, J.R., 1989. The physics of cosmic-ray modulation. Adv. Space Res. 9, 105–119. https://doi.org/10.1016/0273-1177(89)90317-7.

Kenny, D.A., 1979. Correlation and Causality. John Wiley & Sons Inc., New York.

Kilifarska, N.A., 2011a. Long-term variations in the stratospheric winter time ozone variability—22 year cycle. C. R. Acad. Bulg. Sci. 64 (6), 867–874.

Kilifarska, N.A., 2011b. Nonlinear re-assessment of the long-term ozone variability during 20-th century. C. R. Acad. Bulg. Sci. 64 (10), 1479–1488.

Kilifarska, N.A., 2012. Mechanism of lower stratospheric ozone influence on climate. Int. Rev. Phys. 6 (3), 279–290.

Kilifarska, N.A., 2013. An autocatalytic cycle for ozone production in the lower stratosphere initiated by galactic cosmic rays. C. R. Acad. Bulg. Sci. 66 (2), 243–252.

Kilifarska, N.A., 2017. Hemispherical asymmetry of the lower stratospheric O_3 response to galactic cosmic rays forcing. ACS Earth Space Chem. 1. https://doi.org/10.1021/acsearthspacechem.6b00009.

Kirkby, J., Curtius, J., Almeida, J., Dunne, E., Duplissy, J., Ehrhart, S., Franchin, A., Gagné, S., Ickes, L., Kürten, A., Kupc, A., Metzger, A., Riccobono, F., Rondo, L., Schobesberger, S., Tsagkogeorgas, G., Wimmer, D., Amorim, A., Bianchi, F., Breitenlechner, M., David, A., Dommen, J., Downard, A., Ehn, M., Flagan, R.C., Haider, S., Hansel, A., Hauser, D., Jud, W., Junninen, H., Kreissl, F., Kvashin, A., Laaksonen, A., Lehtipalo, K., Lima, J., Lovejoy, E.R., Makhmutov, V., Mathot, S., Mikkilä, J., Minginette, P., Mogo, S., Nieminen, T., Onnela, A., Pereira, P., Petäjä, T., Schnitzhofer, R., Seinfeld, J.H., Sipilä, M., Stozhkov, Y., Stratmann, F., Tomé, A., Vanhanen, J., Viisanen, Y., Vrtala, A., Wagner, P.E., Walther, H., Weingartner, E., Wex, H., Winkler, P.M., Carslaw, K.S., Worsnop, D.R., Baltensperger, U., Kulmala, M., 2011. Role of sulphuric acid, ammonia and galactic cosmic rays in atmospheric aerosol nucleation. Nature 476, 429–433. https://doi.org/10.1038/nature10343.

Kobashi, T., Kawamura, K., Severinghaus, J.P., Barnola, J.-M., Nakaegawa, T., Vinther, B.M., Johnsen, S.J., Box, J.E., 2011. High variability of Greenland surface temperature over the past 4000 years estimated from trapped air in an ice core. Geophys. Res. Lett. 38. https://doi.org/10.1029/2011GL049444.

Köhler, P., Muscheler, R., Richter, K.-U., Snowball, I., Wolf-Gladrow, D.A., 2009. Comment on 'Magnetic effect on CO2 solubility in seawater: a possible link between geomagnetic field variations and climate' by Alexander Pazur and Michael Winklhofer. Geophys. Res. Lett. 36. https://doi.org/10.1029/2008GL036133.

Kristjánsson, J.E., Stjern, C.W., Stordal, F., Fjæraa, A.M., Myhre, G., Jónasson, K., 2008. Cosmic rays, cloud condensation nuclei and clouds—a reassessment using MODIS data. Atmos. Chem. Phys. 8, 7373–7387. https://doi.org/10.5194/acp-8-7373-2008.

Kug, J.-S., Jeong, J.-H., Jang, Y.-S., Kim, B.-M., Folland, C.K., Min, S.-K., Son, S.-W., 2015. Two distinct influences of Arctic warming on cold winters over North America and East Asia. Nat. Geosci. 8, 759–762. https://doi.org/10.1038/ngeo2517.

Kulmala, M., Riipinen, I., Nieminen, T., Hulkkonen, M., Sogacheva, L., Manninen, H.E., Paasonen, P., Petäjä, T., Maso, M.D., Aalto, P.P., Viljanen, A., Usoskin, I., Vainio, R., Mirme, S., Mirme, A., Minikin, A., Petzold, A., Hõrrak, U., Plaß-Dülmer, C., Birmili, W., Kerminen, V.-M., 2010. Atmospheric data over a solar cycle: no connection between galactic cosmic rays and new particle formation. Atmos. Chem. Phys. 10, 1885–1898. https://doi.org/10.5194/acp-10-1885-2010.

Li, C., Stevens, B., Marotzke, J., 2015. Eurasian winter cooling in the warming hiatus of 1998–2012. Geophys. Res. Lett. 42, 8131–8139. https://doi.org/10.1002/2015GL065327.

Luo, Y., Zheng, X., Zhao, T., Chen, J., 2014. A climatology of aerosol optical depth over China from recent 10 years of MODIS remote sensing data. Int. J. Climatol. 34, 863–870. https://doi.org/10.1002/joc.3728.

Manabe, S., Strickler, R.F., 1964. Thermal equilibrium of the atmosphere with a convective adjustment. J. Atmos. Sci. 21, 361–385. https://doi.org/10.1175/1520-0469(1964)021<0361:TEOTAW>2.0.CO;2.

Manabe, S., Wetherald, R.T., 1967. Thermal equilibrium of the atmosphere with a given distribution of relative humidity. J. Atmos. Sci. 24, 241–259. https://doi.org/10.1175/1520-0469(1967)024<0241:TEOTAW>2.0.CO;2.

North, G.R., Eruhimova, T.L., 2009. Atmospheric Thermodynamics: Elementary Physics and Chemistry. Cambridge University Press.

Pazur, A., Winklhofer, M., 2008. Magnetic effect on CO2 solubility in seawater: a possible link between geomagnetic field variations and climate. Geophys. Res. Lett. 35. https://doi.org/10.1029/2008GL034288.

Pudovkin, M.I., Babushkina, S.V., 1991. The influence of electromagnetic and corpuscular radiation of solar flares on the intensity of zonal atmosphere circulation. Geomagn. Aeron. 31 (3), 493.

Pudovkin, M.I., Raspopov, O.M., 1992. Physical mechanism of solar activity influence on the lower atmosphere and meteoparameters. Geomagn. Aeron. 32 (5), 3.

Pudovkin, M.I., Veretenenko, S.V., 1992. Variations of the meridional profile of atmospheric pressure during a geomagnetic disturbance. Geomagn. Aeron. 32, 118–122.

Ramanathan, V., Callis, L.B., Boughner, R.E., 1976. Sensitivity of surface temperature and atmospheric temperature to perturbations in the stratospheric concentration of ozone and nitrogen dioxide. J. Atmos. Sci. 33, 1092–1112. https://doi.org/10.1175/1520-0469(1976)033<1092:SOSTAA>2.0.CO;2.

Randel, W.J., Wu, F., Oltmans, S.J., Rosenlof, K., Nedoluha, G.E., 2004. Interannual changes of stratospheric water vapor and correlations with tropical tropopause temperatures. J. Atmos. Sci. 61, 2133–2148. https://doi.org/10.1175/1520-0469(2004)061<2133:ICOSWV>2.0.CO;2.

Randel, W.J., Wu, F., Forster, P., 2007. The extratropical tropopause inversion layer: global observations with GPS data, and a radiative forcing mechanism. J. Atmos. Sci. 64, 4489–4496.

Randel, W.J., Shine, K.P., Austin, J., Barnett, J., Claud, C., Gillett, N.P., Keckhut, P., Langematz, U., Lin, R., Long, C., Mears, C., Miller, A., Nash, J., Seidel, D.J., Thompson, D.W.J., Wu, F., Yoden, S., 2009. An update of observed stratospheric temperature trends. J. Geophys. Res. Atmos. 114. https://doi.org/10.1029/2008JD010421.

Roscoe, H.K., Rosenlof, K.H., 2011. Revisiting the lower stratospheric water vapour trend from the 1950s to 1970s. Atmos. Sci. Lett. 12, 321–324. https://doi.org/10.1002/asl.339.

Schaeler, B., Offermann, D., Kuell, V., Jarisch, M., 2009. Global water vapour distribution in the upper troposphere and lower stratosphere during CRISTA 2. Adv. Space Res. 43, 65–73. https://doi.org/10.1016/j.asr.2008.06.019.

Schmidt, G.A., Ruedy, R.A., Miller, R.L., Lacis, A.A., 2010. Attribution of the present-day total greenhouse effect. J. Geophys. Res. Atmos. 115. https://doi.org/10.1029/2010JD014287.

Schneider, D.P., Steig, E.J., Comiso, J.C., 2004. Recent climate variability in Antarctica from satellite-derived temperature data. J. Climate 17, 1569–1583. https://doi.org/10.1175/1520-0442(2004)017<1569:RCVIAF>2.0.CO;2.

Seidel, D.J., Randel, W.J., 2006. Variability and trends in the global tropopause estimated from radiosonde data. J. Geophys. Res. Atmos. 111. https://doi.org/10.1029/2006JD007363.

Shindell, D.T., Schmidt, G.A., Miller, R.L., Mann, M.E., 2003. Volcanic and solar forcing of climate change during the preindustrial era. J. Climate 16, 4094–4107. https://doi.org/10.1175/1520-0442(2003)016<4094:VASFOC>2.0.CO;2.

Sinha, A., Allen, M.R., 1994. Climate sensitivity and tropical moisture distribution. J. Geophys. Res. Atmos. 99, 3707–3716. https://doi.org/10.1029/93JD03195.

Sinha, A., Harries, J.E., 1995. Water vapour and greenhouse trapping: the role of far infrared absorption. Geophys. Res. Lett. 22, 2147–2150. https://doi.org/10.1029/95GL01891.

Sloan, T., Wolfendale, A.W., 2008. Testing the proposed causal link between cosmic rays and cloud cover. Environ. Res. Lett. 3, 024001. https://doi.org/10.1088/1748-9326/3/2/024001.

Soden, B.J., Jackson, D.L., Ramaswamy, V., Schwarzkopf, M.D., Huang, X., 2005. The radiative signature of upper tropospheric moistening. Science 310, 841–844. https://doi.org/10.1126/science.1115602.

Solomon, S., Rosenlof, K.H., Portmann, R.W., Daniel, J.S., Davis, S.M., Sanford, T.J., Plattner, G.-K., 2010. Contributions of stratospheric water vapor to decadal changes in the rate of global warming. Science 327, 1219–1223. https://doi.org/10.1126/science.1182488.

Spencer, R.W., Braswell, W.D., 1997. How dry is the tropical free troposphere? Implications for global warming theory. Bull. Am. Meteorol. Soc. 78, 1097–1106. https://doi.org/10.1175/1520-0477(1997)078<1097:HDITTF>2.0.CO;2.

Sprenger, M., Wernli, H., 2003. A northern hemispheric climatology of cross-tropopause exchange for the ERA15 time period (1979–1993). J. Geophys. Res. Atmos. 108. https://doi.org/10.1029/2002JD002636.

Stuber, N., Ponater, M., Sausen, R., 2001. Is the climate sensitivity to ozone perturbations enhanced by stratospheric water vapor feedback? Geophys. Res. Lett. 28, 2887–2890. https://doi.org/10.1029/2001GL013000.

Sun, X., Ren, G., Ren, Y., Fang, Y., Liu, Y., Xue, X., Zhang, P., 2018. A remarkable climate warming hiatus over Northeast China since 1998. Theor. Appl. Climatol. 133, 579–594. https://doi.org/10.1007/s00704-017-2205-7.

Svensmark, H., 1998. Influence of cosmic rays on Earth's climate. Phys. Rev. Lett. 81, 5027–5030. https://doi.org/10.1103/PhysRevLett.81.5027.

Svensmark, H., Friis-Christensen, E., 1997. Variation of cosmic ray flux and global cloud coverage—a missing link in solar-climate relationships. J. Atmos. Sol. Terr. Phys. 59, 1225–1232. https://doi.org/10.1016/S1364-6826(97)00001-1.

Tinslev, B.A., 1996. Solar wind modulation of the global electric circuit and apparent effects on cloud microphysics, latent heat release, and tropospheric dynamics. J. Geomag. Geoelec. 48, 165–175. https://doi.org/10.5636/jgg.48.165.

Tomasi, C., Cacciari, A., Vitale, V., Lupi, A., Lanconelli, C., Pellegrini, A., Grigioni, P., 2004. Mean vertical profiles of temperature and absolute humidity from a 12-year radiosounding data set at Terra Nova Bay (Antarctica). Atmos. Res. 71, 139–169.

Usoskin, I.G., Gladysheva, O.G., Kovaltsov, G.A., 2004. Cosmic ray-induced ionization in the atmosphere: spatial and temporal changes. J. Atmos. Sol. Terr. Phys. 66, 1791–1796.

Velinov, P.I.Y., Mateev, L., Kilifarska, N., 2005. 3-D model for cosmic ray planetary ionisation in the middle atmosphere. Ann. Geophys. 23, 3043–3046.

Wirth, V., 1993. Quasi-stationary planetary waves in total ozone and their correlation with lower stratospheric temperature. J. Geophys. Res. 98, 8873–8882.

Young, J.A., 2003. Static stability. In: Holton, J.R., Curry, J.A., Pyle, J.A. (Eds.), Encyclopaedia of Atmospheric Sciences. vol. 5. Academic Press, pp. 2114–2120.

CHAPTER

8

Geomagnetic field and internal climate modes

OUTLINE

Continuous interactions between different parts of the climate system (atmosphere, ocean, cryosphere, biosphere, etc.), influencing each other on different timescales, result in internal fluctuations; these are usually called natural or internal climate modes.

Observations from the last several decades have revealed some regional patterns of decadal-multidecadal climate variations, e.g. North Atlantic Oscillation (NAO), Atlantic Multidecadal Oscillation (AMO), El Niño-Southern Oscillation (ENSO), Pacific Decadal Oscillation (PDO), etc. Recent evidence suggests that some of these regional patterns may be components of global-scale, decadal-multidecadal climate variations.

Most climatic modes are defined by statistical classifications of the observed variability of surface temperature, sea-level pressure, precipitations, etc. They could be related to physical laws or to the spatial distribution of land and oceans, of mountains, etc. However, these statistical patterns may also be artefacts of nature, whereby they are not stable over long periods of time, or they may be statistical artefacts.

Due to the fact that the amplitude of decadal-centennial variability of incoming solar irradiance is very small (which undoubtedly is the main source of energy for Earth's climate system), climatic modes are usually attributed to the interaction of a vastly varying atmosphere with slowly varying components of the system (i.e. the ocean and the cryosphere).

https://doi.org/10.1016/B978-0-12-819346-4.00008-5
© 2020 Elsevier Inc. All rights reserved.

However, the synchronization between geomagnetic and climate variations, detected on different timescales (presented in this book), suggests that heterogeneous distribution of geomagnetic field intensity globally (slowly varying with time) could stand behind the regional patterns of decadal-centennial climate variability. Some evidence supporting this idea is provided in this last chapter of the book.

8.1 Relations between lower stratospheric ozone and the North Atlantic Oscillation

The North Atlantic Oscillation (NAO) is a measure of the sea level pressure variations between subpolar and subtropical latitudes in the North Atlantic region. It describes the mutual strengthening and weakening of the Azores High and the Icelandic Low. Under the positive mode of the NAO (corresponding to a deep Icelandic Low), positive temperature anomalies expands over the Eurasian Arctic, while negative anomalies occupy north-eastern Canada and the polar regions of the North Atlantic. This situation favours the expansion of North Atlantic cyclone tracks deeper into the Arctic Ocean, supplying more heat and moisture to Iceland and Scandinavia. At the same time, central-southern Europe and the Mediterranean experience drier conditions. Broadly opposing anomalies are associated with negative NAO states. In the North Atlantic Ocean, the NAO-associated wind and precipitation anomalies influence temperature, rate of oceanic convection, deep-water formation, and biological activity.

Temporal variability of NAO index (describing the fluctuations in the difference between sea surface pressure in Azores and Iceland) is more or less stochastic. However, the elimination of its interannual variations reveals that for most of the 20th century, the NAO has been in its negative phase. Only between 1900 and 1930 and from the mid-1980s to the 2000s has it been in a positive phase (see Fig. 8.1).

Bearing in mind the possibility for near tropopause ozone influence on the surface temperature and pressure (see Chapter 7), the temporal variations of NAO and ozone have been compared at the two centres of NAO mode (i.e. Iceland and Azores islands). The time series are shown in Fig. 8.2, showing fairly well the in phase covariance between lower stratospheric ozone and NAO over Iceland, and their antiphase centennial evolution over the Azores, during the 20th century and the first decade of the 21st century.

A quantitative estimation of the relation between NAO and O_3 density has been performed by the use of a lagged cross-correlation analysis. Two hypotheses have been examined: (i) the multidecadal variations of NAO drive the correspondent changes in O_3 spatial distribution; and (ii) the centennial variations of the lower stratospheric O_3 density influence the surface temperature and pressure, which in turn projects over different phases of the NAO index. The correlation coefficients have been calculated in a grid with 10-degree increments in latitude and longitude. In addition, the lagged correlation coefficients have been weighted by the autocorrelation functions of the NAO index, or of the ozone (calculated in each grid point). This procedure reduces the weight of the heavily lagged cross-correlation coefficients (Kenny, 1979). Furthermore, the correlations maps have been drawn from the statistically significant (at 2σ level) coefficients. In order to estimate the causality of NAO-ozone relations, the correlation coefficients with zero time lags have been ignored. The resulting maps are shown in

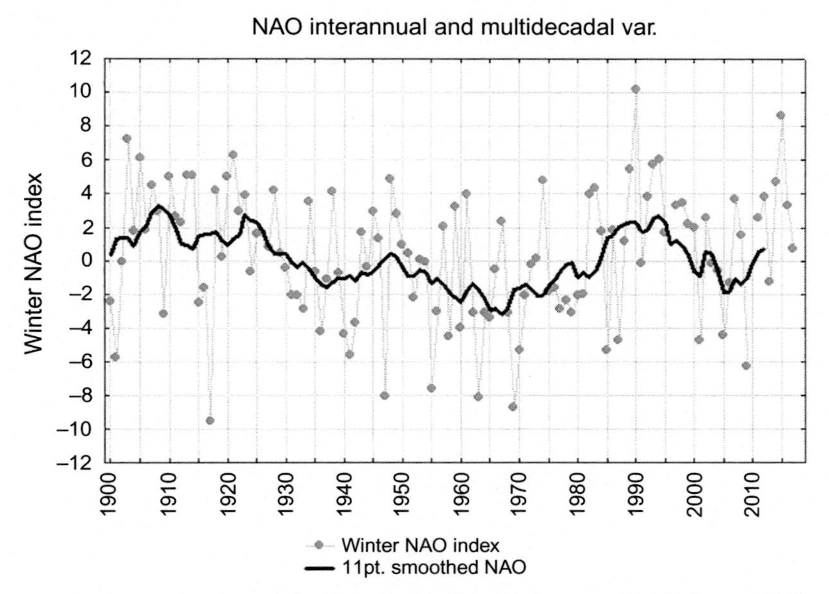

FIG. 8.1 Time series of normalized winter NAO index (i.e. (NAO(yr) − mean NAO)/mean NAO); *grey dotted line with full dots* illustrates the interannual variability, while the *thick black curve* represents the smoothed by 11-point moving window NAO index. *From Velichkova, T., Kilifarska, N., 2019. Lower stratospheric ozone's influence on the NAO climatic mode. C. R. Acad. Bulg. Sci. 72 (2), 219–225.*

Fig.8.3A, illustrating the strength of the correlation between NAO and ozone (i.e. the NAO index is an independent variable), and Fig. 8.3B, showing the opposite case, in which the forcing (independent) variable is the O_3 density.

Fig. 8.3B clearly shows the existence of two centres of ozone influence on the surface temperature and pressure. The centre of the maximal positive correlation is placed in the North Atlantic Ocean, between Greenland and Iceland, while that of the negative one is near the Azores islands. This means that ozone enhancement over Greenland and/or its depletion over the Azores is accompanied by a positive NAO index, a situation observed at the beginning and end of the 20th century (Fig. 8.1). In contrast, the negative NAO phase, which is observed in the middle of the past century, should be attributed to the O_3 reduction at polar latitudes, and particularly over Greenland and Iceland (Velichkova and Kilifarska, 2019).

Consequently, the NAO mode could be influenced by the variations of lower stratospheric ozone density in each centre of action (Azores or Iceland), or simultaneously in both of them. Analysis of the time delay of NAO response to ozone changes shows that surface temperature near the Icelandic Low respond with a delay of 1–2 years. In the subtropical centre of action, however, the atmospheric response is delayed approximately by a decade (see Fig. 8.3D).

The NAO mode influence on the spatial–temporal variability of the lower stratospheric ozone during the 20th century is much weaker (see Fig. 8.3A). The centre of their in phase covariance is detected over north-eastern Canada and south-western Greenland. The centre of negative correlation is found in south-eastern United States (see Fig. 8.3A).

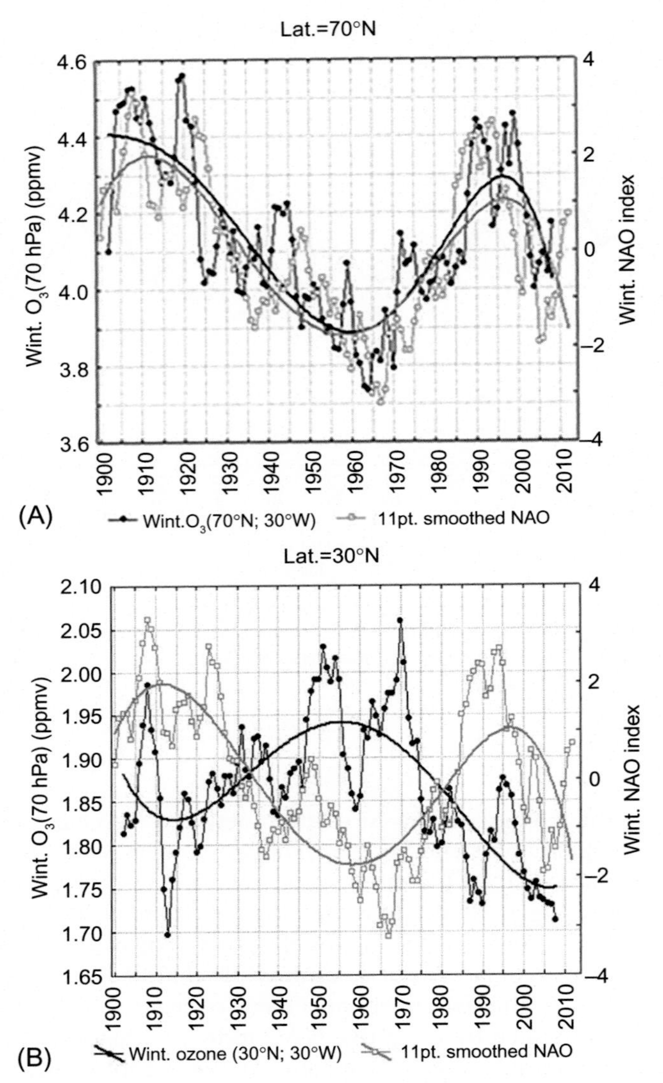

FIG. 8.2 Time series of winter NAO *(grey curves with open squares)*; smoothed (by 11-point moving window), and ozone mixing ratio at 70 hPa *(black curves with dots)* are compared for Iceland (A) and the Azores (B). The polynomial fits to each time series *(thicker grey and black curves)* are shown as well. *From Velichkova, T., Kilifarska, N., 2019. Lower stratospheric ozone's influence on the NAO climatic mode. C. R. Acad. Bulg. Sci. 72 (2), 219–225.*

In the context of the mechanism of ozone influence on the surface temperature (described in Chapter 7), the O_3 abundance over Iceland should be followed by a reduced surface temperature and pressure (in accordance with the ideal gas law: $\boldsymbol{p} = \boldsymbol{\rho} \cdot \boldsymbol{R} \cdot \boldsymbol{T}$, where $\boldsymbol{\rho}$ and $\boldsymbol{R} = 8.3144598$ (J mol^{-1} K^{-1}) are the density of the air mass and the atmospheric gas constant,

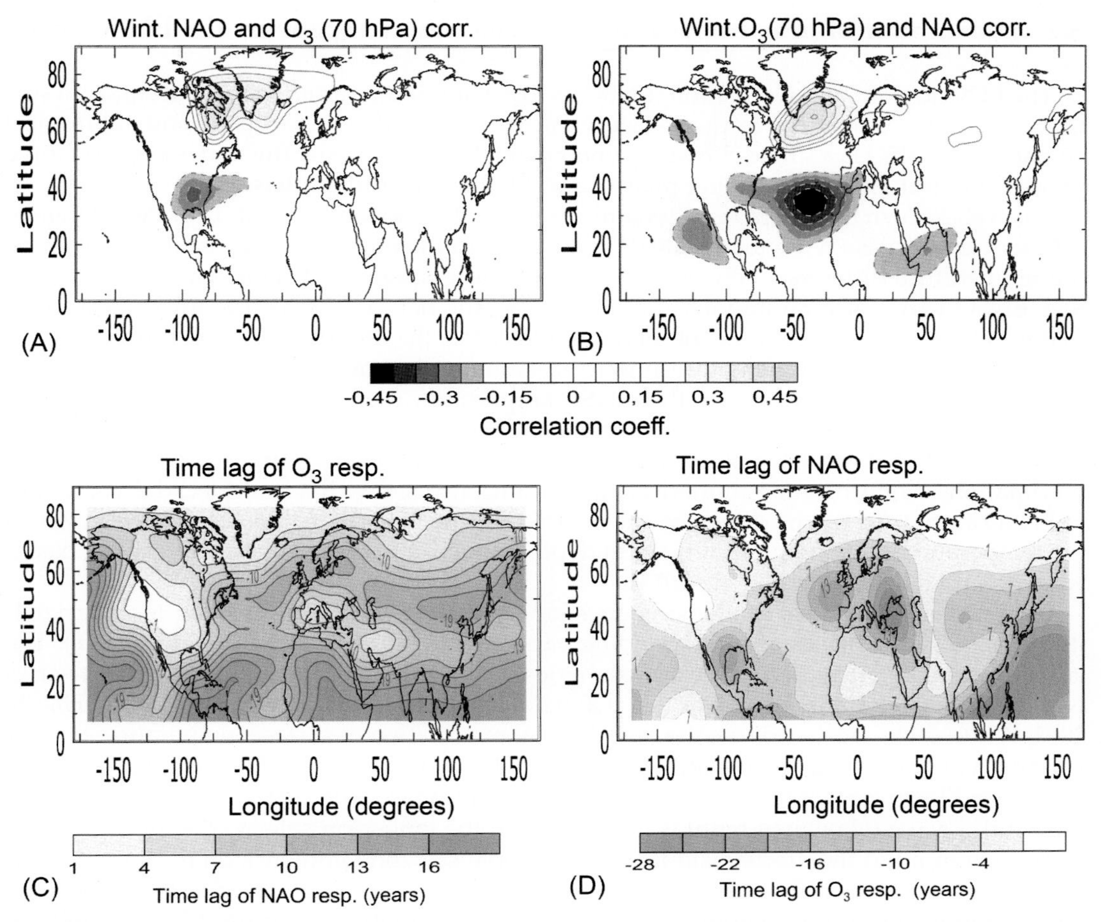

FIG. 8.3 Correlation maps of the winter ozone mixing ratio at 70 hPa and NAO index, calculated for the period 1900–2010. (A) The NAO index is a leading variable; (B) O_3 is the independent (leading) variable. (C and D) present the time lag of dependent variables response to the forcing factor. *After Velichkova, T., Kilifarska, N., 2019. Lower stratospheric ozone's influence on the NAO climatic mode. C. R. Acad. Bulg. Sci. 72 (2), 219–225.*

respectively, T is the temperature, and p is the atmospheric pressure). Consequently, in line with Jones et al.'s (1997) definition of NAO, this situation corresponds to a positive NAO phase. In contrast, the depletion of polar O_3 density stimulates regional surface warming and pressure rise, followed by a negative NAO phase.

Similarly, the O_3 depletion above the Azores should be accompanied by a surface warming and pressure rise, which corresponds to the positive NAO phase. In contrast, the ozone's enhancement in the extra-tropics will cool the surface, preparing in such a way the appearance of the negative NAO phase.

8.2 The imprint of near tropopause ozone and water vapour on the ENSO mode

The El Niño-Southern Oscillation (ENSO) is a quasiperiodic variation in winds and sea surface temperatures over the tropical eastern Pacific Ocean, affecting tropical and subtropical climate. The ENSO is a single climate phenomenon that periodically fluctuates between three phases: Neutral, La Niña (cooling phase), and El Niño (warming phase).

It is well-known that ENSO is a dynamic process which exhibits variability over a range of different timescales (Wittenberg, 2009). In addition to the interannual variability, driven by the temperature difference between the warmer west equatorial Pacific and cooler eastern one, ENSO varies also on interdecadal timescales, the nature of which is still unclear. Due to the lack of long enough direct observational records, the interdecadal variability of ENSO is studied mainly by coupled atmospheric-ocean numerical models.

Modelling studies have shown that ENSO amplitude varies over multidecadal timescales even in the absence of greenhouse warming (Rodgers et al., 2004; Wittenberg, 2009; Deser et al., 2011). These results suggest that the long-term ENSO variability is driven by natural modulations. Long-running numerical simulations also show that multidecadal variability of ENSO amplitude can be driven by the oceanic thermocline (i.e. a thin ocean layer in which the temperature is changing rapidly with depth) response to the changes of overlying basin-wide zonal winds (Borlace et al., 2013).

Consequently, the most challenging question for understanding ENSO long-term variability is to clarify the drives of corresponding variations in sea surface temperature (SST). The latter are usually attributed to the atmospheric-ocean coupling, the strength of which varies with time. Some authors, however (e.g. Labitzke and Loon, 1995; Tourre et al., 2001; Haigh, 2003; White, 2006, etc.), attribute the interdecadal modulation of surface temperature to solar variability. For example, Kirov and Georgieva (2002) have shown that decadal variations of the Indian Ocean surface temperature covariates in antiphase with solar activity, while that of the South Pacific (Hawaiian high) is in phase with the long-term solar variability.

Bearing in mind the mechanism for near tropopause ozone influence on the surface temperature (described in Chapter 7), we have examined the temporal covariance between the Nino 3.4 index (as a representative of the tropical SST variability in central Pacific) and the ozone mixing ratio at 70 hPa. Both data types have been smoothed by a 5-point moving window. The degree of relation between both variables has been estimated by the use of the classical lagged cross-correlation analysis, allowing identification of any delay in the dependent variable response. The spatial distribution of the degree of similarity between variables' temporal variability has been assessed by the correlation maps, drawn from the statistically significant correlations coefficients at 2σ level (i.e. with 95% confidence). The correlation coefficients have been calculated in each grid point with 10-degree steps in latitude and longitude. The lagged correlation coefficients are additionally weighted by the autocorrelation function of ENSO and ozone correspondingly, for a given time delay. Although this procedure reduces the strength of the relation, it allows us to compare the importance of correlations with different time lags (Kenny, 1979), based on the suggestion that reliability of covariance weakens with the rise of the dependent variable time lag. The spatial distribution and the strength of correlation between the Nino 3.4 index (i.e. the temporal variability of the tropical temperature in the central Pacific Basin) and lower stratospheric ozone for boreal winter (Dec–Apr) and summer (May–Sep) are shown in Fig. 8.4 and Fig. 8.5.

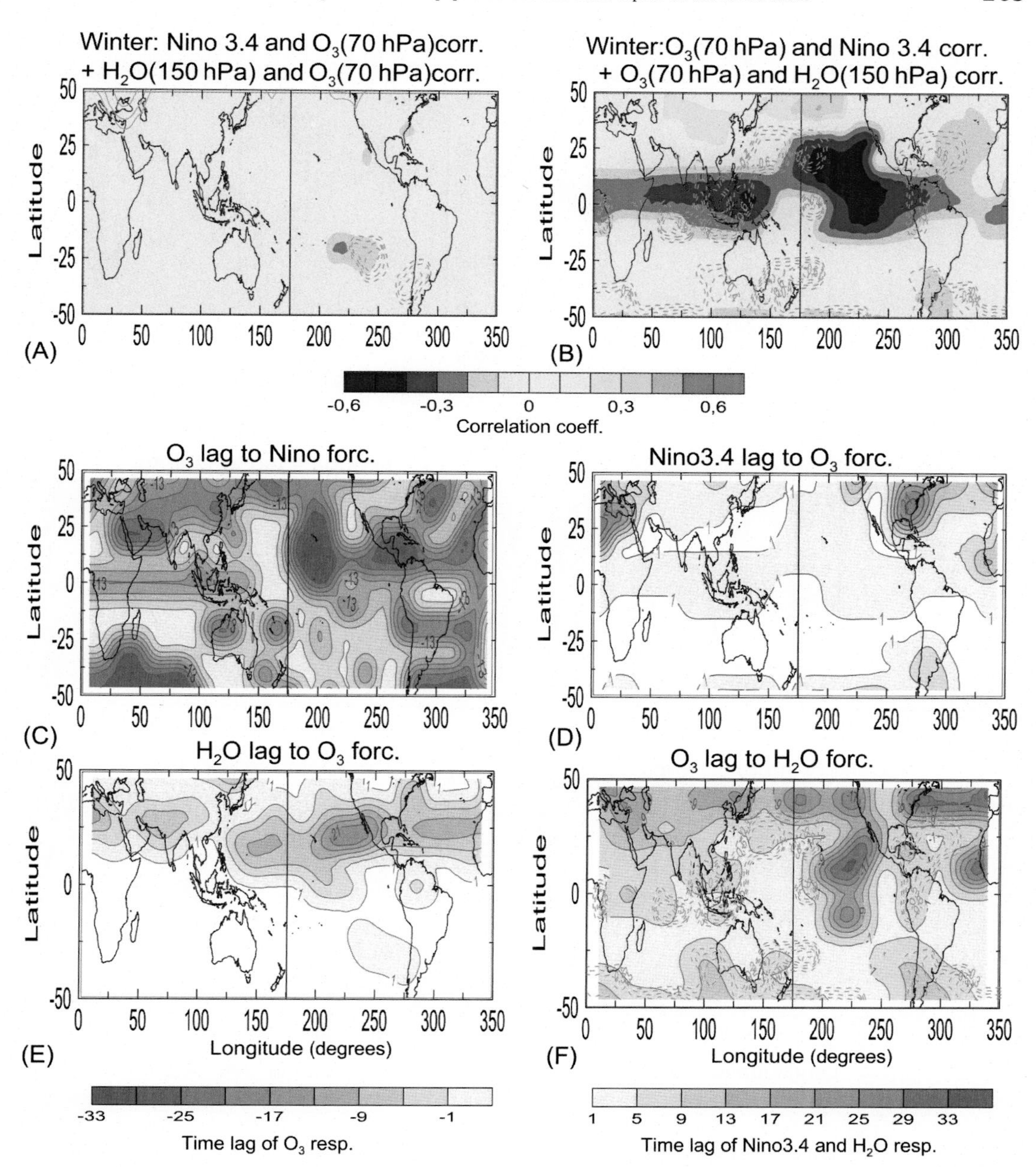

FIG. 8.4 Centennial correlations of boreal winter (Dec–Apr) ozone at 70 hPa with the Nino3.4 index *(shading)*, and water vapour at 150 hPa (contours). The O_3–H_2O correlation is calculated with O_3 being the leading factor. Panel (A) illustrates the impact of Nino3.4 on ozone variability, while panel (B) shows the opposite (ozone influence on Nino3.4). The *middle and lower panels* illustrate the time lag of dependent variables to the applied forcing: (C and E) time delay of ozone response to Nino3.4 and water vapour forcing; (D and F) Nino3.4 and H_2O (150 hPa) response to ozone changes at 70 hPa. The ozone–water vapour correlation *(dashed contours)* is overdrawn on the panel (F) to enable easier comparison between the strength of correlation and time delay of H_2O response. *Data sources: ERA 20th century reanalysis for ozone and specific humidity; data for Nino3.4 index are from https://www.esrl.noaa.gov/psd/gcos_wgsp/Timeseries/Data/nino34.long.data.*

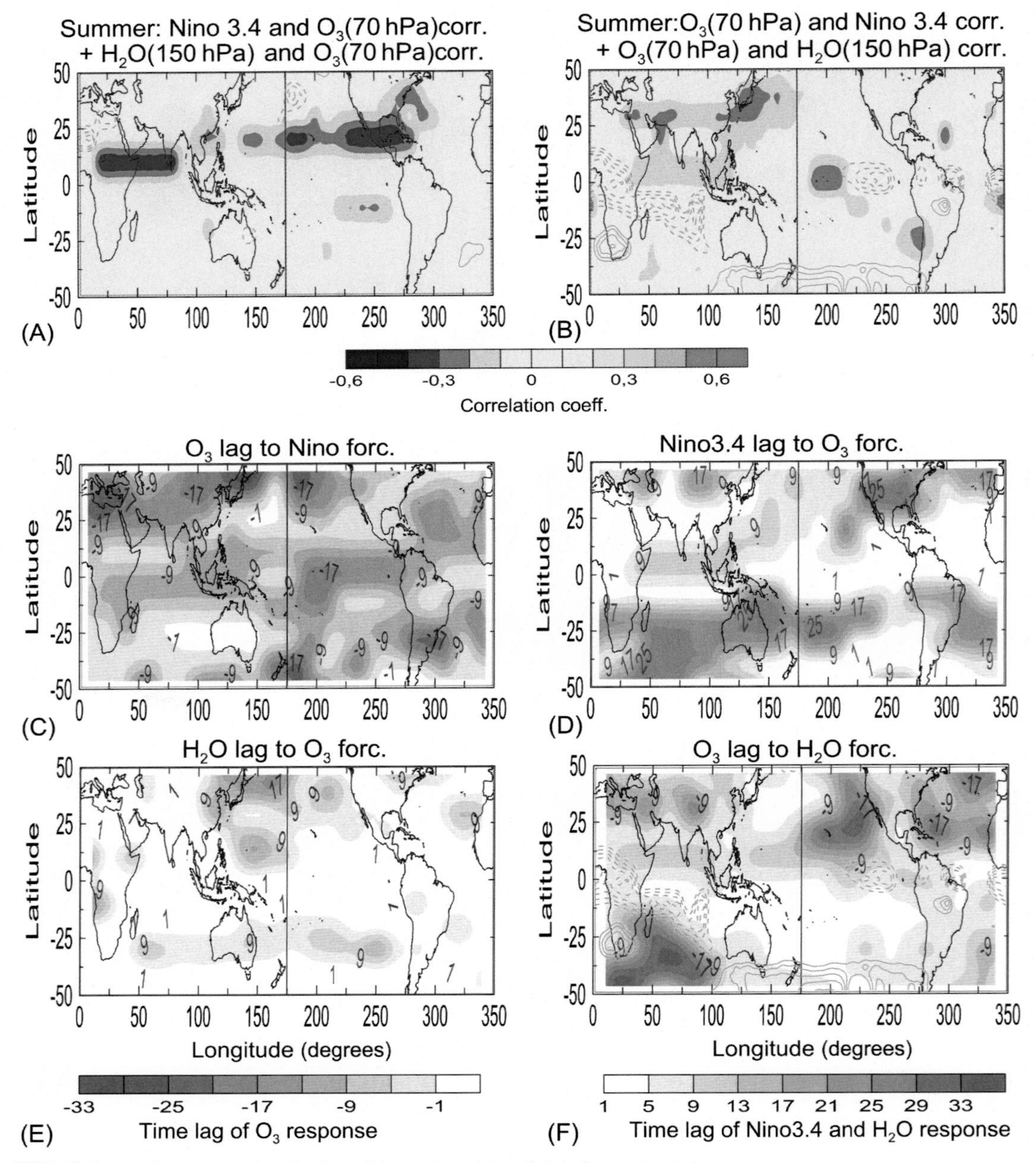

FIG. 8.5 As for Fig. 8.4, but for boreal summer season (May–September).

It is obvious that during boreal winter (Dec–Apr), the direction of influence is from ozone to the surface temperature (i.e. towards changes in the Nino 3.4 index). Fig. 8.4B reveals a strong relation between them over the maritime continent and the tropical–subtropical central Pacific Ocean, persisting for the entire 20th century. The delay of surface temperature response to ozone changes is ~1 year (Fig. 8.4D), which corresponds surprisingly well to the period of weakening of easterly trade winds preceding the appearance of the El Nino events (Niedzielski, 2014). In contrast, the ozone response to fluctuations in the Nino3.4 index is severely delayed, which reflects its negligible influence on the near tropopause ozone (Fig. 8.4C).

During boreal summer (May–Sep), the strength of the relation between ozone and the Nino3.4 index is substantially reduced (Fig. 8.5A and B), and displaced northwards. The response of both variables is significantly delayed from the time of applied forcing (Fig. 8.5C and D).

The mechanism of ozone influence on the surface temperature, described in Chapter 7, includes the modulation of near tropopause water vapour, due to the changes of the atmospheric moist adiabatic lapse rate. In order to examine the validity of this mechanism, the correlation maps of both variables (ozone at 70 hPa and water vapour at 150 hPa) have been overdrawn over the ozone-Nino3.4 correlations in Figs 8.4 and 8.5. It is obvious that regions of strong ozone-vapour coupling in Dec–Apr correspond fairly well to the regions with strong ozone influence on the tropical surface temperature, i.e. over Nino3.4 variability (see Fig. 8.4B). Such a coincidence is not revealed in May–Sep, which could be a reasonable explanation of the weaker ozone influence on the surface temperature.

The minimal delay of water vapour response to ozone changes over the Indian Ocean, Indonesia, and Central America (see Fig. 8.4F) suggests that these regions determine the interdecadal variability of the ENSO mode. On the other side, the synchronized variations of O_3 and H_2O vapour over the Eastern Pacific is without time delay, so the problem of causality could not be resolved in this region. The ozone-water coupling is well tracked also at subtropics, which could explain the distant influences (teleconnections) of ENSO variability.

These results suggest that centennial variations of the winter near tropopause ozone could be imprinted on the surface temperature, initiating its interdecadal changes (currently being attributed to the natural variability of the climate system). Spatial–temporal variations in the ozone density, itself, are initiated by irregularly distributed ionization in the Regener–Pfotzer maximum, due to geomagnetic lensing of cosmic rays in some regions over the globe (for details see Chapter 6). Thus, the spatial heterogeneity of the geomagnetic field, continuously varying with time, is projected down to the Earth's surface.

8.3 Concluding remarks

Unlike the well-understood influence of palaeoclimate on palaeomagnetic data, the opposite direction of the relation, i.e. geomagnetic influence on climate, still sounds heretical. This book does not simply offer one more hypothesis explaining such a relation. It proposes a new conceptual framework, based on some new discoveries, elucidating the missing links in the causal relation between the geomagnetic field and climate.

One of the main obstacles for understanding this relation is that the suggested mediator of geomagnetic influence, galactic cosmic rays (GCRs), inserts minuscule energy in the atmospheric system, which is insufficient to drive significant changes of the climate system. We have revealed, however, that lower energy electrons, created by GCRs near the tropopause (i.e. in the Regener–Pfotzer ionization layer) are able to activate various ion-molecular reactions, including ozone production or destruction. The access or deficiency of ozone near the tropopause is able to drive variations of tropopause temperature, and through modulation of the wet adiabatic lapse rate, to change the near tropopause humidity. Having the major contribution in the greenhouse warming of the planet, these changes are projected down to the earth surface (see Chapter 7). The mechanism for ozone-water vapour influence on climate has been validated by several experiments with the RegCM model, which reveals that changes of ozone density just above the tropopause lead to corresponding changes in the near tropopause water vapour and surface temperature (Kilifarska et al., 2018).

Another insurmountable problem is the existent evidence for geomagnetic-climate covariances, which seems mutually exclusive. This difficulty has been overcome by our discovery related to geomagnetic lensing of energetic particles propagating in the Earth's atmosphere, which ensures their heterogenously distributed impact in the chemical and thermodynamical balance of climate system (Kilifarska and Bojilova, 2019). This finding makes understandable the close response of the climate system to the drift of geomagnetic poles, geomagnetic excursions, and reversals reported by many scientists.

Moreover, the new conceptual framework is able to resolve some other problems like longitudinal variations and hemispherical asymmetry of the lower stratospheric ozone, its interrelation with the near tropopause H_2O vapour, the regional character of climate changes, and the interdecadal variability of climatic modes.

Juxtaposition of currently available knowledge reveals the necessity of its integration and synthesis into a single hypothesis, providing a mechanism for geomagnetic influence on the surface temperature. According to the newly created conceptual framework, the projection of geomagnetic influence on climate variability appears to be a result of the following processes:

- geomagnetic lensing of cosmic rays in certain regions over the world;
- temporal–spatial variation imposed on the near tropopause ozone by the low energy electrons in a Regener–Pfotzer layer of ionization, activating ozone production or destruction;
- regional changes of tropopause temperature (initiated by ozone variations), altering the wet adiabatic lapse rate, and as a consequence the near tropopause humidity;
- the decisive contribution of the upper tropospheric water vapour in the greenhouse effect modifies the variability of surface temperature, warming or cooling it.

Statistical analyses of the relations between all contributors to the above chain of links reveal many similarities in the spatial patterns of their temporal covariance. However, the last link—i.e. the relation between near tropopause ozone and water vapour—needs special attention, because of the various direction of the forcing impact. Both possibilities have been analysed: (i) ozone as a driver of H_2O variability; and (ii) ozone as a respondent to the water vapour variations.

The analyses show that in tropical regions, the coupling between ozone and H_2O vapour is strongest, with antiphase covariance and without time lag at decadal timescale, which does not allow determination of the direction of forcing. However, at timescales 6–11 years, the leading role of ozone has been detected over the Indian Ocean and Indonesia, corresponding fairly well to the strong connection between ozone and the Nino3.4 index (see Fig. 8.4). Another region of strong in phase coupling is found at extra-tropics, with dominating ozone forcing in the Northern Hemisphere, and water vapour in the Southern one. We attribute this hemispherical asymmetry in ozone-water vapour coupling to the asymmetrical geomagnetic field, controlling intensity and depth of penetration of cosmic rays in the atmosphere. As a result, various ion-molecular reactions are activated in the upper troposphere–lower stratosphere (see Chapter 6), creating or destroying the near tropopause ozone.

The extra-tropics are accessible to energetic particles trapped in Earth's radiation belts, which are subject to strong geomagnetic lensing. This could explain the existence of two belts around 55-degree latitudes with a positive trend of the lower stratospheric ozone, during the 20th century, in contradiction to the global depletion of atmospheric ozone. The strong coupling between ozone and water vapour corresponds well to these belts, and being projected down to the surface delineates fairly well the negative anomalies in the centennial climatology of the Northern Hemisphere surface temperature (see Fig. 6.19).

It is worth noting that statistical analysis of geomagnetic–climate relations, based on the period with direct instrumental measurements (i.e. 1900–2010), relies on the IGRF-12 model and ERA 20th century reanalysis because they provide gridded data for more than a century. Taking into account that the first half of the 20th century is less supplied by instrumental records, some deviations from the results presented in this book are possible if other data sets are used. Meanwhile, the same analyses have been initially elaborated by the use of merged ERA40 and ERA Interim reanalyses (providing data since 1957), and the general conclusions do not differ significantly. This increases our confidence that the chain of links between the geomagnetic field and climate is well-supported by the existing data records.

References

Borlace, S., Cai, W., Santoso, A., 2013. Multidecadal ENSO amplitude variability in a 1000-yr simulation of a coupled global climate model: implications for observed ENSO variability. J. Clim. 26, 9399–9407. https://doi.org/10.1175/JCLI-D-13-00281.1.

Deser, C., Phillips, A.S., Tomas, R.A., Okumura, Y.M., Alexander, M.A., Capotondi, A., Scott, J.D., Kwon, Y.-O., Ohba, M., 2011. ENSO and Pacific decadal variability in the community climate system model version 4. J. Clim. 25, 2622–2651. https://doi.org/10.1175/JCLI-D-11-00301.1.

Haigh, J.D., 2003. The effects of solar variability on the Earth's climate. Philos. Trans. R. Soc. Lond. Ser. A 361, 95–111. https://doi.org/10.1098/rsta.2002.1111.

Jones, P.D., Jonsson, T., Wheeler, D., 1997. Extension to the North Atlantic oscillation using early instrumental pressure observations from Gibraltar and south-west Iceland. Int. J. Climatol. 17, 1433–1450.

Kenny, D.A., 1979. Correlation and Causality. John Wiley & Sons Inc., New York

Kilifarska, N., Bojilova, R., 2019. Geomagnetic focusing of cosmic rays in the lower atmosphere—evidence and mechanism. C. R. Acad. Bulg. Sci. 72 (3), 365–374. https://doi.org/10.7546/CRABS.2019.03.11.

Kilifarska, N., Wang, T., Ganev, K., Xie, M., Zhuang, B., Li, S., 2018. Decadal cooling of East Asia—the role of aerosols and near tropopause ozone forcing. C. R. Acad. Bulg. Sci. 71 (6), 937–944.

Kirov, B., Georgieva, K., 2002. Long-term variations and interrelations of ENSO, NAO and solar activity. Phys. Chem. Earth A/B/C 27, 441–448. https://doi.org/10.1016/S1474-7065(02)00024-4.

Labitzke, K., Loon, H.V., 1995. Connection between the troposphere and stratosphere on a decadal scale. Tellus A 47, 275–286. https://doi.org/10.1034/j.1600-0870.1995.t01-1-00008.x.

Niedzielski, T., 2014. El Niño/Southern Oscillation and selected environmental consequences. In: Dmowska, R. (Ed.), Advances in Geophysics. Elsevier, pp. 77–122 (Chapter 2). https://doi.org/10.1016/bs.agph.2014.08.002.

Rodgers, K.B., Friederichs, P., Latif, M., 2004. Tropical Pacific decadal variability and its relation to decadal modulations of ENSO. J. Clim. 17, 3761–3774. https://doi.org/10.1175/1520-0442(2004)017<3761:TPDVAI>2.0.CO;2.
Tourre, Y.M., Rajagopalan, B., Kushnir, Y., Barlow, M., White, W.B., 2001. Patterns of coherent decadal and interdecadal climate signals in the Pacific Basin during the 20th century. Geophys. Res. Lett. 28, 2069–2072. https://doi.org/10.1029/2000GL012780.
Velichkova, T., Kilifarska, N., 2019. Lower stratospheric ozone's influence on the NAO climatic mode. C. R. Acad. Bulg. Sci. 72 (2), 219–225.
White, W.B., 2006. Response of tropical global ocean temperature to the Sun's quasi-decadal UV radiative forcing of the stratosphere. J. Geophys. Res. Oceans 111. https://doi.org/10.1029/2004JC002552.
Wittenberg, A.T., 2009. Are historical records sufficient to constrain ENSO simulations? Geophys. Res. Lett. 36. https://doi.org/10.1029/2009GL038710.

Index

Note: Page numbers followed by *f* indicate figures and *t* indicate tables.

Printed in the United States
By Bookmasters